FORDISM TRANSFORMED

General Editor
Professor Akiro Kudo, Faculty of Economics, University of Tokyo.

Series Adviser
Professor Mark Mason, Yale University.

FUJI CONFERENCE SERIES I

FORDISM TRANSFORMED

The Development of Production Methods in the Automobile Industry

Edited by

HARUHITO SHIOMI
and
KAZUO WADA

OXFORD UNIVERSITY PRESS
1995

Oxford University Press, Walton Street, Oxford OX2 6DP
Oxford New York
Athens Auckland Bangkok Bombay
Calcutta Cape Town Dar es Salaam Delhi
Florence Hong Kong Istanbul Karachi
Kuala Lumpur Madras Madrid Melbourne
Mexico City Nairobi Paris Singapore
Taipei Tokyo Toronto
and associated companies in
Berlin Ibadan

Oxford is a trade mark of Oxford University Press

Published in the United States
by Oxford University Press Inc., New York

British Library Cataloguing in Publication Data
Data available

Library of Congress Cataloging-in-Publication Data
Fordism transformed : the development of production methods in the
automobile industry / edited by Haruhito Shiomi and Kazuo Wada.
p. cm. — (Fuji conference series ; 1)
"Papers . . . presented at the 21st International Conference on
Business History which took place 5 to 8 January 1994 at the foot of
Mt. Fuji"—Acknowledgements.
Includes index.
1. Automobile industry and trade—Management—Congresses.
2. Automobile industry and trade—Japan—Management—Congresses.
3. Automobile industry and trade—Production control—Congresses.
I. Shiomi, Haruhito, 1943– . II. Wada, Kazuo, 1949– .
III. International Conference on Business History (21st : 1994)
IV. Series.
HD9710.A2F65 1995 95-18021
338.4'76292'0952—dc20
ISBN 0–19–828961–8

1 3 5 7 9 10 8 6 4 2

Typeset by Best-set Typesetter Ltd., Hong Kong
Printed in Great Britain
on acid-free paper by
Bookcraft (Bath) Ltd, Midsomer Norton, Avon

ACKNOWLEDGEMENTS

All the papers in this book, except for one, were presented at the 21st International Conference on Business History, which took place 5–8 January 1994 at the foot of Mt. Fuji. The paper by Mr Lee was not presented at the conference, although he attended the conference as an assistant. After the conference, the editors felt a study dealing with the Chinese automobile industry should be included in this book, and they asked Mr Lee to write a paper on the topic.

The International Conference on Business History has been held in Japan in January of every year since 1974 and has gained worldwide recognition under the name Fuji conferences. These conferences have been made possible by the gracious cooperation of business historians from all parts of the world. The continued generous support of the Taniguchi Foundation has made it possible to inaugurate a fifth series of these conferences, to run from January 1994 to January 1998. The Organizing Committee of the fifth series of the International Conference, chaired by Professor Akira Kudo, has decided to turn the spotlight in this fifth series onto the past tempo of enterprise growth after World War II in all industrial countries (including Japan) within a long-term perspective. Kazuo Wada and Haruhito Shiomi were entrusted with the task of organizing the first conference in the fifth series around a general topic relating to the production methods of the automobile industry. In line with the wishes of the Organizing Committee, Wada and Shiomi set the theme for the conference as 'The Development of Production Methods in the Automobile Industry: Comparative Perspectives on the Implementation and Modification of the Ford System'. The editors and the Organizing Committee then set about deciding which scholars should be invited to the conference. The selection was made extremely difficult by the fact that there are so many good scholars dealing with the chosen topic. In line with the stated policy of the Organizing Committee to invite as far as possible overseas scholars who have not been invited to the previous conferences, or at least scholars who have not participated in the last few conferences, and keeping in mind the Committee's preference for a small meeting in an atmosphere of friendliness, instead of a massive ceremonial convention, the editors and the Organizing Committee finally carried out their unenviable task, though not without regret that they could not invite all appropriate scholars to the conference.

The Taniguchi Foundation's financial support is also making it possible to publish the papers of the 21st International Conference. The editors, as well as the Organizing Committee, would like to express their deepest gratitude to the Taniguchi Foundation for its continuing sponsorship of the Conference. Professor Mark Mason kindly performed the role of adviser for the publication of the papers in this fifth series of the International Conference on Business History. His diligent support and expert advice greatly facilitated the publication of this book.

The papers in this book were revised in accordance with the discussions and suggestions brought up at the conference. This book is, in a sense, a joint product of both the papers' authors and those who attended the conference. Besides the authors of the papers, the following attended the conference: Akihiko Amemiya, Sumiaki Furukawa, Ken Hirano, Akira Kudo, Hidemasa Morikawa, Keiichiro Nakagawa, Kazuhide Nakamoto, Yuji Nishimuta, Tamotsu Nishizawa, Reiko Okayama, Koichi Shimokawa, Hiromi Shioji, Takamoto Sugisaki, Hisashi Watanabe, and Seiichiro Yonekura. The editors would like to thank them for their active participation and much-valued contribution.

To hold the International Conference requires much administrative and secretarial support. Such support was ably provided by the following staff: Mikyung Han, Chunli Lee, Hiroshi Nishikawa. Two interpreters, R. Hanna Brendon and Eiichiro Kazumori, helped ensure smooth communication among the participants. Edmund R. Skrzypczak of Nanzan University checked and polished the English of all papers in the book; without his diligent efforts, readers might have had to cope with numerous eccentric expressions, especially in the non-English speakers' papers. Last but not least, the editors would like to express their gratitude to many business historians, not mentioned above, who helped us in many ways.

K.W.

While this book was in the process of being printed, the editors learned of the sudden passing away of Mr Toyosaburo Taniguchi, the founder of the Taniguchi Foundation. On behalf of the Organizing Committee of the Fuji Conference and the Business History Society of Japan, we wish to convey to the family of Mr Toyosaburo Taniguchi our sincere condolences. His memory and his wonderful contributions to the development of business history and to mutual understanding among business historians will remain with us.

CONTENTS

LIST OF CONTRIBUTORS

WERNER ABELSHAUSER is Professor of Economic and Social History at the University of Bielefeld.

PATRICK FRIDENSON is Professor of History, Centre de Recherches Historiques, École des Hautes Études en Sciences Sociales, Paris.

TAKAHIRO FUJIMOTO is associate professor in the Faculty of Economics at Tokyo University.

DAVID A. HOUNSHELL is the Henry R. Luce Professor of Technology and Social Change at Carnegie Mellon University.

NILS KINCH is professor in the Department of Business Studies at Uppsala University.

CHUNLI LEE is doctoral student at the Graduate School of Tokyo University and an affiliated research fellow of the International Motor Vehicle Programme, MIT.

WAYNE A. LEWCHUK is professor at McMaster University.

STEFANO MUSSO is professor at the Regional Institute for Educational Research, Piemonte, Italy.

IZUMI NONAKA is associate professor in the Faculty of Economics at Josai University.

HARUHITO SHIOMI is professor in the Faculty of Economics at Nagoya City University.

KAZUO WADA is associate professor in the Faculty of Economics at Tokyo University.

TIMOTHY R. WHISLER is assistant professor in the Department of History and Political Science at Saint Francis College, USA.

Introduction

HARUHITO SHIOMI

The one-car-per-minute production method that was implemented by the Ford Motor Company at its Highland Park plant from April of 1914 had a profound effect, not only on the automobile industry, but on virtually every other industry as well. The success of the Ford innovation put strong pressure on existing competitors to adopt mass production methods, and provided a model for later entrants that could not be ignored. Makers in other countries also made a thorough study of the Ford method, often making adjustments in the light of their own country's historical and social conditions.

Accordingly, the aim of the current work is to examine the global development and long-term effects of the Ford system of automobile production through to the present day, and to give an overall picture of the various ways the Ford system was received and modified.

In addition to the USA, the birthplace of the Ford system, seven other major auto-producing nations are looked at: England, France, both Germanies (counted as one), Italy, Sweden, China, and Japan. Case studies of representative manufacturers are taken up in a multidimensional approach that highlights the various Ford systems that developed in those countries.

The Original Ford System

In examining the far-reaching effects of the Ford system and the changes that it underwent, it is probably best to begin with a consideration of the original prototype production method that served as a starting-point.

Shortly after the implementation of the moving assembly line at Ford's Highland Park plant, *The Engineering Magazine* began a sixteen-part report on the plant; running from April 1914 to July 1915 inclusive, it was written by Horace L. Arnold and Fay L. Faurote and entitled 'Ford Methods and the Ford Shops'. A few years later, *Industrial Management* ran a thirteen-part report from September 1922 to September 1923 inclusive on the River Rouge plant, written by John H. Van Deventer and entitled 'Ford Principles and Practice at River Rouge'. These important historical materials provide us with a detailed look at the structure of the company, including virtually every shop, division, and office, from an industrial engineering standpoint.[1]

According to these sources, the production of main parts such as cylinder blocks, brake drums, and crankcases was conducted on separate conveyor-belt equipped casting and machining lines for each type of part, and these separate sets of production lines were synchronized with the 70-second cycle time of the main assembly line. The total length of the belt conveyors in the Highland Park plant in April 1915 was an impressive 50 miles, [2] and an observer of the River Rouge complex likened it to a 'great water supply system 'with mains and many feeder-pipes.[3] According to the recollections of Charles E. Sorensen, the well-known engineer who worked with Henry Ford from 1905 to 1944, the time required to process basic materials into a completed car turned over to the dealer was originally 21 days at Highland Park, subsequently reduced to 14 days. However, at River Rouge, which had its own steel mill, this was brought down to only 4 days.[4]

It should be noted that it was not long before André-Gustave Citroën and Louis Renault of France, Giovanni Agnelli (Fiat) of Italy, Herbert Austin and William Morris of England, and other managers and engineers from a variety of countries were visiting the Highland Park and River Rouge plants, and also not long before workers and engineers with experience at Ford were taking their knowledge and experience back to their home country, as happened with Swedish personnel. Kiichiro Toyoda, the founder of Toyota, was a relatively late visitor, not going to River Rouge until 1929, but after the war his successor, Eiji Toyoda, and the director of manufacturing, Shoichi Saito, led two exhaustive observation tours of the plant for a total of seven months from July 1950 before returning to Japan to begin full-scale production of passenger cars. In this way, it was through personal visits and observation that the Ford system was transmitted directly to countries that have since become significant auto producers.[5]

At the time of the Ford Motor Company's fiftieth anniversary, the company proudly claimed that the Ford system was a product of America's industrial history, and was made up of seven elements: power, accuracy, economy, system, continuity, speed, and repetition.[6] Many questions can be asked, however, about what the many observers of Ford factories learned from the experience, what elements they adopted, what elements they modified, what features they added, and, finally, what kinds of production systems they ended up creating.

Various Ford Systems

In 1916, less than two years after the appearance of the Ford system, annual US automobile production reached 1-million units, a level not destined to be reached by any other nation until England achieved that figure in 1954, well after World War II. West Germany hit the 1-million unit mark in 1956, followed by France in 1958, and by Italy and Japan in 1963. Thus was formed the club of the six major automobile-producing countries.

However, with respect to the element of speed that was part of the Ford system, the one-car-per-minute pace established by Ford in 1914, it would not be until the late 1960s that leading makers in other countries arrived at this benchmark. At Toyota, for example, its first facility dedicated to passenger car assembly, the Motomachi plant, began with a monthly production run of 5,000 units on its one assembly line, which was operational only one shift a day. With the expansion of its second assembly plant at Takaoka in 1970 Toyota became the first Japanese car-maker to realize the one-car-per-minute cycle time. Each line was running two shifts and was producing 20,000 units per month.[7]

In order to achieve the level of mass production set up as a target by the Ford cycle time, auto makers around the world had to overcome a host of problems in the areas of parts and materials purchasing, sub-assembly inventory, quality control, equipment maintenance, and process management, in addition to sorting out various employment-related issues such as hiring, training, and discipline. The kind of systematic effort needed to implement and modify the Ford system was, it would seem fair to say, an ongoing process that went on in the various countries well into the 1960s.

The first section of the current work takes up this implementation and modification process leading up to the late 1960s in terms of three trends. First is the trend towards mass production of different models with a variety of specifications (multiple versions). Ford's Highland Park and River Rouge plants were specially designed to produce only one model, the Model T, and required a complete shutdown of about one year for refitting when the changeover to the 1928 Model A was undertaken. GM's full-line strategy required five separate specialized assembly plants, one for each model produced. The subsequent challenge was to maintain the speed of the dedicated assembly line while introducing into it the flexibility needed for producing a number of vehicles with a variety of specifications. Concerning this aspect, the present volume will focus on the case of the Japanese auto industry, examining Toyota's efforts to establish a mixed assembly line. These efforts will be divided into two periods, pre-1955 and post-1955, and will be traced from before the war through to the present day.

The second trend visible in the process of the international implementation and modification of the Ford system leading up to the late 1960s is the development and application of transfer automation. The Highland Park and River Rouge facilities employed 15,000 and 35,000 workers respectively, reaching very high levels of integration, but the majority of these were semi-skilled machine tenders and assemblers who performed simple, repetitive work. The challenge of eliminating these semi-skilled workers was initially taken up in machine and press shops following World War II, and the present volume shows the leading role played by the Ford Motor Company itself in this area of change. This stage is a precursor to today's robotized factories, representing the imposition of rather inflexible automation directly on traditional dedicated production lines.

The third trend is the application of the Ford system to small lot assembly, addressed in the present work with reference to sports car production by BLMC of England and Nissan of Japan, and luxury car production by Volvo of Sweden. Today, Nissan is able to produce niche cars in lots less than ten, but this is completely different from the craft production techniques that were used before the adoption of the Ford system; partial, or skilfully modified use of some of the elements making up the Ford system can be seen.

The above three trends have previously been treated separately as lean production, mass production, and craft production respectively, but the present work treats them as aspects of the implementation and modification of the Ford system, providing the historical background to their development.[8]

The Ford System and Production Management

For our purposes here, the most important of these three trends is the modification of the original Ford system to accommodate simultaneous production of different models with a variety of specifications. This is because modification in this area, particularly with regard to production management and control, led to a number of changes in the methods traditionally used. Of these, we will focus here on two specific important aspects: quality control activities, and parts production.

Previous research has not given sufficient consideration to the production management and control side of the original Ford system, possibly because of the convenience of the 'great water supply system' metaphor and subscription to the myth that Henry Ford disliked organization. However, at the entrance to the Highland Park plant there was a large four-storey headquarters building equipped with the latest office equipment, with a total of 864 salaried employees in 1915, rising to 1,398 in 1916. The employment department alone had 526 people.[9] The Ford Motor Company made the monthly production plan to cope with the fluctuation of demands. The production office had to allocate this monthly car-production number to Detroit and twenty-two other domestic, and two foreign plants, and moreover coordinate the daily flows of 6-million materials and parts (5,000 per unit, 1,200 units per day) on one-car-per-minute cycle time in the Highland Park plant.[10] With regard to the quality control and parts production aspects examined in the current work, there were 600 inspectors in the Highland Park quality control department to check on all in-house produced and externally supplied parts, and many outside suppliers supplied bodies, wheels, tires, coil-box units, carburettors, lamps, 90 per cent of the car painting, all drop forgings, all roller bearings, grease caps, spark plugs, electric conductors, gaskets, hose connections and hose clips, horns, fan belts, muffler pipes, and a considerable part of the bolts and nuts.[11]

What becomes clear from the studies found in this volume is that the shift to mass production of different models to a variety of specifications was caused not by technological progress and innovation, but rather by progress and innovation in management techniques in the specific areas of quality control and parts production.

The second section of the current work identifies two types of quality control that developed alongside the Ford system. The first was SQC, examined here as it paralleled the introduction of the Ford system in France. The second was the development of TQC in Japan, as seen in the cases of Toyota and Nissan.

By way of a follow-on from the above, new Japanese parts supply systems are compared with traditional American methods, through a consideration of the hierarchical structure of parts suppliers in Japan and the process of development of the approved drawings system.

The Ford System and Socio-Economic Systems

While the first two sections examine various developments in the production system itself, the third section offers a consideration of the production system as it existed within broader socio-economic systems.

During the period between the world wars, the Ford Motor Company set up transplant manufacturing facilities in Copenhagen (1919), São Paulo (1921), Trieste, Italy (1922), Yokohama and Buenos Aires (1924), a suburb of Paris and the German city of Cologne (1925), and the London suburb of Dagenham (1928). Meanwhile, the Canadian Ford Company was establishing plants in Australia, New Zealand, South Africa, and other countries. Of these, the Dagenham and Cologne plants were essentially miniature versions of River Rouge with integrated production systems.[12] Built in order to facilitate significant expansion in the Far East, the Yokohama plant was equipped with modern conveyor systems for chassis and body assembly, and it was capable of producing a maximum of 200 cars a day, or 20,000 cars per year.[13]

Here, I wish to mention two points concerning the interwar Ford system that require confirmation. First, it would seem that the Ford system had by this time already established itself quite successfully in a variety of historical and social environments around the world. Perhaps because it was a new and supplementary industry, and perhaps also because of the special circumstances of wartime economies, there was unexpectedly little labour conflict in its spread around the world.

Second, one gets the impression that, as seen in the daily production record of the Yokohama plant, the conveyors ran at a comparatively slow pace and did not take full advantage of the system's possibilities. It can also easily be supposed that the seven elements in the original Ford system were only partially implemented.

Thus, the question arises as to what constraints were placed on the interwar Ford system by the socio-economic systems then in place. Alternatively, it can be asked what influence the Ford system had on socio-economic systems. The system's mutual influence is an important area for consideration, and the current work treats this issue from the standpoint of the historical development of the male domain, labour-union activities, and the general macroeconomic system.

The present work, then, utilizes the above-stated multifaceted framework to throw light on the varying ways in which Fordism was transformed in the long process of implementing and modifying the Ford system, right through the 1960s, and provides an international comparison of the business history of the automobile industry.

NOTES

1. Arnold and Faurote's report was immediately published in book form: *Ford Methods and the Ford Shops* (New York, 1915).
2. *Ford Times*, 8/7 (Apr. 1915), 311.
3. J. H. Van Deventer, 'Ford Principles and Practice at River Rouge', *Industrial Management*, 64/4 (Oct. 1922), 196.
4. C. E. Sorensen, *My Forty Years with Ford* (New York, 1956), 166.
5. Toyota Motor Corporation, *Toyota Jidosha 30nen shi* [A 30-year history of Toyota Motor] (1967), 327–8, and *Sozo kagiri naku* [Creation without limit] (1987), 51.

6. Ford Motor Company, *The Evolution of Mass Production* (Dearborn, 1956), 14.

7. *Toyota Jidosha 30nen shi*, 468, supplemented with interviews.

8. For more on previous treatment of this subject, see D. Roos, J. P. Womack, and D. Jones, *The Machine That Changed the World* (New York, 1990); also, M. J. Piore and C. F. Sabel, *The Second Industrial Divide* (New York, 1984).

9. See *Ford Times*, No. 3 (July 1910), 424. Also, A. Nevins and F. E. Hill, *Ford: Expansion and Challenge 1915–33* (New York, 1957), 687, and Arnold and Faurote, *Ford Methods and the Ford Shops*, 47.

10. Ibid. 33–4.

11. Ibid. 97–9, 28–9.

12. Nevins and Hill, *Ford*, ch. 14.

13. Nissan Motor Corporation, *Nissan Jidosha 30nen shi* [A 30-year history of Nissan Motor] (1965), 14.

PART I

THE VARIOUS FORD SYSTEMS

1

The Emergence of the 'Flow Production' Method in Japan

KAZUO WADA

1.1. Introduction

In 1955, Toyota marketed the Crown, which it advertised as the first 'purely domestic-made passenger car' in Japan.[1] In the same year Nissan and other assemblers also marketed their own new models of cars, made in technological cooperation with foreign assemblers. Indeed, this showed that Japanese automobile assemblers had managed to acquire the technological capability to produce passenger cars by that time.

But car production in Japan was far below that in the USA. In 1955 the USA produced about 9.2 million vehicles; Japan, just about 69,000 vehicles. After the early 1960s Japanese automobile assemblers constructed new dedicated plants for passenger cars. In 1980 Japan produced over 11 million vehicles and became the largest car producing country in the world, passing the USA. In the same year Toyota and Nissan made 3.2 million and 2.6 million vehicles respectively.[2]

Many researchers have concentrated their analyses mainly on the period after 1960, in order to explain the reasons for the spectacular growth. (The following few contributions to this book will, in fact, also analyse the development of the Japanese automobile industry after around 1960, focusing mainly on Toyota.)[3] As a result, the period before 1955 has not been investigated fully, or even in some detail, by many researchers. However, we should pay more attention to the development of production methods in those earlier years. As Cusumano insisted: 'The Japanese automobile industry did not develop so quickly.'[4] It took decades to acquire the technological capability to produce automobiles. The process of acquiring it had a galvanizing effect on the development of production methods at Japanese automobile manufacturers. This paper aims to clarify that process by focusing on Toyota.

Toyota and Nissan took different approaches to establish themselves as automobile manufacturers. Cusumano sharply differentiates both com-

panies' initial attempts: Nissan's approach was 'direct technology transfer' while Toyota's was 'indirect technology transfer'.[5] Nissan purchased
many machines from the Graham Paige Company and directly trans-
planted the American way of manufacturing motor vehicles. It manufactured trucks that were designed by Graham Paige Co. It imported both
manufacturing technology and product design directly from the USA as
one package. However, even Nissan's approach did not mean a smooth
adoption of the mass production method. Lack of supplementary know-
how to run machine tools and so on, as well as a shortage of skilled
workers, caused trouble in the implementation of mass production. In
contrast to Nissan, Toyota purchased many foreign machine tools, but it
did not transplant other countries' technology and products as a part of
one package. Instead, it tried to adapt the imported technology to existing
conditions. The case of Toyota would better serve to clarify the difficulties
of establishing the automobile manufacturing business than would the
case of Nissan. For this reason, this paper will focus on Toyota.

The next section, however, will deal with the development of production methods in other manufacturing industries, especially the aircraft
industry. As a late-developing country, Japan faced many obstacles to
establishing the automobile industry in its economy. Japanese engineers
faced an uphill struggle particularly in implementing a mass production
method, not only in the automobile industry but also in the manufacturing industries in general. In its nascent years, the Japanese automobile
industry lagged far behind its American counterpart. Even before World
War II, the idea of the 'Ford Production System' was spread among
Japanese engineers. It had a galvanizing effect on the formation and
development of production methods in Japan. Its influences were not
confined to the sphere of auto manufacturing. Engineers working in many
industries looked carefully into the idea of the 'Ford Production System'
and adapted it to conditions in their own areas. In many places they also
experimented with, and implemented, their own ideas before and during
World War II, although their attempts did not prove successful. Their
efforts increasingly focused on ways to implement the 'flow production'
method as an intermediate step to mass production. Such experiences had
a significant effect on the postwar development of the Japanese automobile industry. This paper deals, therefore, with the development of the
'flow production' method not just in the Japanese automobile industry,
but in Japan's manufacturing industry as a whole.

1.2. 'Flow Production' as a Substitute for 'Mass Production'

War with China broke out in 1937. This war changed the Japanese
economy significantly. The automobile industry, like everything else, was

under the tight control of the government. In 1938 the Ministry of Commerce and Industry forbade the production of passenger cars other than for military use, and consequently vehicle production concentrated on trucks.[6] Trucks were required to move troops and military goods within China. Automobiles were included among the fifteen major items that the government targeted for increased production under a four-year plan announced in 1938. However, the motor vehicle's strategic-item status was questioned as the war front widened, especially after confrontation with Great Britain and the USA began in 1941. Aircraft took over as an important strategic item for doing battle, especially against the United States. The government endeavoured to increase aircraft production as much as possible, giving planes priorities over trucks. As a result of this, engineers endeavoured to find a way to increase aircraft production. They focused their attention on the question of production methods in order to find a way to increase output.

Japanese engineers had carried out research on time and motion studies even before the war. They also had studied Taylor's work on scientific management in detail.[7] The achievements of Ford were well known among them. In the face of a depression, an Industrial Rationalization Movement was inaugurated in the 1920s. In order to achieve the aims of the movement, the Ministry of Commerce and Industry set up a Temporary Industry Rationalization Bureau as a cabinet-level council in early 1930.[8] Pursuit of the aims of the movement, as well as research on time and motion studies, did not result in any extensive adoption of a 'mass production' method in Japan.

Some engineers fully understood the importance of mass production even before the war. Ichiro Sakuma, an engineer for a major aircraft company, toured Europe and the United States before the war. He was strongly impressed by the fact that in these countries the aircraft industry as well as the automobile industry produced a large volume of units through the mass production method. He concluded that Japanese factories produced smaller unit volumes, not because of a shortage of men and equipment, but because the men and equipment were not fully utilized. In his opinion, therefore, just putting more resources into plant would not ensure any increase in production at all. Changing the production method seemed to him to be the key to increasing output.

As demand for aircraft increased, engineers had to figure out how to increase output. Their efforts can be traced in engineering journals from that time, even though they could not describe everything in detail because of the need to keep production methods secret during the war. As the war proceeded, engineers increasingly used the term 'taryo seisan' (high-volume production), rarely using the term 'tairyo seisan' (mass production). Nobuo Noda, one of the pioneers who introduced 'scientific management methods' in Japan, pointed out the difference between the two terms:[9]

1. 'high-volume production' requires the absolute interchangeability of parts. Therefore, there is no fitter in 'high-volume production'.

2. However, 'high-volume production' would not be equal to genuine 'mass production' with an extensive use of special-purpose machine tools. The present condition of Japan would not allow engineers to install a large number of machines.

3. The essence of 'mass production' would be a smooth flow of production. A smooth flow of production could be achieved without any mechanization.

On the basis of these ideas, engineers endeavoured to find a way of implementing the smooth flow of production at plants in order to increase output.

In 1941, the Japan Society for the Promotion of Science carried out a survey on the extent to which major Japanese companies had adopted the 'flow production' method and how they implemented the system in their own plants. This survey revealed that the 'flow production' method was not in general use even among major companies. In 1943, an engineer also lamented that there were no major companies implementing the genuine 'flow production' method in their machine shops.[10]

Engineers struggled to have a smooth flow of production implemented at plants, especially at aircraft plants, during the war. Actually, they did manage to have a smooth flow of production realized to some extent, although the speed of the flow was slow and the extent of its application was limited.

Engineers in the engine departments in the aircraft industry were the first to achieve a smooth production flow. Engineers modified a 'tact system' that had been adopted at major German aircraft companies. This modified 'tact system' was called the 'zenshin-shiki' (pushing-line) method, and it was highly praised as an important step for increasing the output of aircraft. Mention of the achievements of this method appeared repeatedly in books and journals. The body of an aircraft was intermittently moved from one operation-process to another. Japanese engineers reported that the intermittent time was about one hour in Germany, whereas we know from recent research that in Japan it was 4.5 to 10 hours.[11] Even though flow in production had been created, the flow was extremely slow in comparison with Germany's. Yet the method was highly praised as being an innovative way to increase output.[12] This clearly shows how backward Japan was and how slow to achieve the smooth flow of production at plants.

According to engineering journals and other testimony from engineers, the first plant to implement this method at its body shop was the Mitsubishi Nagoya plant. The experimental implementation of the tact system was successful in the September of 1941. It took almost two years to implement the method in the body assembly shop.[13] In 1943 the Army officially commended Wataru Sasaki, Chief Engineer, for his success in implementing the tact system. This shows how much the method became appreciated as a solution to the problem of increasing output. The official commendation was intended as a way of publicizing the method to other companies, although the wartime shortage of materials and other un-favourable conditions prevented the quick spread of the method. A report published after the war revealed the fact that Kawasaki Aircraft Co. intro-duced the tact system in 1942 and several companies tried to establish a smooth flow of production during the war.[14] In the Japanese aircraft industry, the efforts of engineers to implement a smooth flow of pro-duction seem to have been moderately successful.

Once the tact system was implemented in body-assembly departments, and the intermittent time was tightly controlled, a smooth production flow was assured in the body-assembly shop. However, the situation in the parts-making shops at body plants was still chaotic. Parts were not being supplied smoothly to the body-assembly department. Engineers therefore turned their attention to the process of parts-making.

A large number of parts were manufactured under a traditional ma-chine layout, in which the same kinds of machines were clustered to-gether. Parts were passed along from one kind of machine to another kind of machine, depending on the machining processes required. Thus, parts were often moved here and there within a plant. Further, a wide variety of parts were made in the same plant. As a result, different kinds of parts were going this way and that, forward and backward, within the plant. The paths of parts criss-crossed within the plant, and sometimes they became tangled. As output increased, the situation went from bad to worse. Foremen often forgot exactly where certain parts were within the machining process. Written plans for parts-making often did not corre-spond to the actual process of parts-making. Therefore, if amendments were made to some parts, foremen often had to go and hunt for the parts within the plant with unreliable plans. As amendments to some parts were made often, foremen had to spend a lot of time running after those parts. This made the situation in a parts-making plant chaotic.

In order to improve the situation, engineers attempted various sol-utions in different divisions of the industry. One was the placing of a team of administrative staff on the job site; another was the setting up of warehouses on the job site; and a third was the introduction of a quasi-sequential layout of machine tools. These were tried independently or jointly at different firms.

Engineers placed a team of adminisrtative staff who were responsible for expediting the progression of parts-making on the job site.[15] This team normally consisted of three people: a leader, a person to control the progression, and a person to inspect the quality of in-process parts.

When many kinds of parts were made in the same plant, it was extremely difficult to synchronize the processing time for every kind of part. As a result, unfinished parts were piled up, awaiting the next machining process. Engineers proposed that the piles of unfinished parts be stored in warehouses. Thus, when some parts finished going through the drilling-machine process and were waiting for the next step in the process, the grinding machine, they were to be stored in a warehouse after being inspected. When it was time for them to undergo the next machining operations, the team of administrative staff would bring them to the job site from the warehouse. Storing unfinished parts in warehouses, rather than just piling them on the floor in disorder, seemed to make it easier to trace exactly where all unfinished parts were. Parts-making could proceed in an orderly manner. Storing unfinished parts in warehouses also revealed exactly how many unfinished parts existed. These unfinished parts could be eliminated if the differences in processing times for different kinds of parts were reduced. Thus, the amount of unfinished parts in warehouses could show the extent to which differences in processing time had been eliminated.

The establishment of warehouses and placing a team of administrative staff on the job site apparently had a significant effect on shop-floor management. Inspection of unfinished parts was made before they were stored in the warehouses. This meant that inspection was carried out after every machining process was finished, not after the end of the whole manufacturing process. Until the warehouses were set up, people in charge of inspection were based in the central office of the plant. With implementation of this new approach, the function of inspection was shifted to the job site.

Engineers were well aware of the fact that machine layout would have a significant effect on productivity. In most cases of parts-making, however, the scale of operations would not permit rearranging machines into line-organization in order to exploit the merits of machine specialization. Ichiro Sakuma at Nakajima Aircraft Co. tried to implement a smooth production flow in the parts-making division of an engine manufacturing facility.[16] He proposed to rearrange the layout of machine tools even at the lower production level, which normally would not permit a sequential layout. He suggested that a wide variety of parts be classified into several groups in accordance with the manufacturing processes: parts with the same or a similar machining process were put into the same group. Each group of parts was allocated to a separate workshop, if possible. As a result, within any given workshop, similar manufacturing processes would be carried out. Each workshop was divided into several sections, in which the same or similar operations were carried out. The result was that the same kinds of machines were brought together in each section. The arrangement of operation sections roughly followed the flow of production. The number of machines in each operation section was adjusted to equalize the duration of operations in each of the sections.

In addition, engineers tried to manufacture in small lot-sizes, so that they could easily trace the exact whereabouts of unfinished parts. They also allocated a serial number to every production-lot, and these numbers were used to keep track of everything going into an aircraft, right down to the parts from which they were assembled. The progress of production was monitored by using the number assigned to the parts in any process at any time.

By the end of the war, Japanese engineers had closely analysed the process of operations at aircraft plants and sometimes managed to implement a smooth production flow despite the lack of resources. These engineers employed in aircraft firms suddenly lost their jobs at the end of the war; many found jobs at automobile companies,[17] while some worked as consultants affiliated with engineering associations such as Noritsu Kyokai (The Association of Efficiency).

1.3. The Emergence of 'Flow Production' at Toyota

1.3.1. Difficulties in Establishing the Automobile-Manufacturing Business

Aircraft engineers endeavoured to find a way to increase the number of aircraft produced. They increasingly focused on how to achieve a smooth flow of production at their plants, working on the basis of a close examination of the production process. In the case of automobiles, however, engineers faced a more fundamental problem: how to produce a prototype automobile. As early as 1902, the first attempt was made to assemble a motor vehicle in Japan: Shintaro Yoshida, a bicycle trader, and Komanosuke Uchiyama, an engineer, produced passenger cars and buses using foreign-made engines. Several other attempts also followed. Despite these early attempts, as late as the 1920s Japanese engineers could not produce automobiles on a solid commercial basis. The use of automobiles also remained limited in Japan. This significantly changed in 1923 after a major earthquake devastated the Tokyo area and destroyed railway and other transportation networks. The City of Tokyo imported 800 Ford Model T trucks, which served as buses. This order led Ford to establish a fully-owned subsidiary in Japan. In 1925 Ford of Japan began to assemble knock-down parts in Yokohama. General Motors also set up a subsidiary in 1927 and established an assembly plant in Osaka. By 1930 the demand for motor vehicles had increased, and the existing Japanese companies suffered from the domination of American subsidiaries in the market. But the growing demand for automobiles stimulated some entrepreneurs to consider entering the business of automobile manufacturing. One of them was Kiichiro Toyoda, the founder of Toyota Automobile Company.

It is not known exactly when Kiichiro conceived the idea of manufacturing automobiles. By the end of the 1920s, however, he seemed to have made up his mind. In 1929, when he visited England to conclude patent negotiations with Platt Brothers, he was impressed by the widespread use of automobiles. Platt Brothers paid £100,000 to Toyoda Automatic Loom Works, Ltd. for the patent rights to manufacture and sell the automatic loom invented by Sakichi Toyoda, father of Kiichiro.[18] This fee gave Kiichiro the financial basis he needed to begin manufacturing automobiles. Even before obtaining the fee, Kiichiro had purchased many precision machine tools that were not required for manufacturing textile machinery. After obtaining the fee, he speeded up the rate of accumulating the technological capabilities needed for manufacturing automobiles. In 1930 he set about manufacturing gasoline motors on an experimental basis. He and his staff learned by taking apart the cylinders of motorcycles.[19] In the summer of 1933 they at last succeeded in manufacturing a four horsepower gasoline engine.[20] Later that year an emergency board of directors meeting of the Toyoda Automatic Loom Works, Ltd. formally endorsed the plan for manufacturing automobiles.[21] Immediately after the board's decision, Kiichiro sent Risaburo Oshima to Europe to purchase machine tools for automobiles. Kiichiro also invited Takatoshi Kan to the company.[22] Kan once participated in a pioneering automobile manufacturing project that was abandoned because it was not economically feasible. Kan was one of only a handful of engineers experienced in manufacturing automobiles in those days.

Kiichiro organized a group of engineers and stocked up on machine tools at the company. Nevertheless, he faced an uphill struggle. The first thing they had to do was to produce a prototype car as soon as possible. At the same time, they had to set up the plant for manufacturing automobiles on an economic basis. Kiichiro carried out two surveys of parts suppliers for automobiles, in 1931 and 1933. He found the quality of parts made in Japan was poor, the main cause being inferior steel materials. This prompted him to set up a steel manufacturing department within Toyoda Automatic Loom in 1934.[23]

Kiichiro's goal was the manufacture of passenger cars. In May 1935 the company produced a prototype passenger car. It was to produce just five prototype passenger cars in all, before it was forced to concentrate on the manufacture of trucks. The difficulty of manufacturing passenger cars might have prompted the change. It took, at least, two years to produce a prototype car after the company formally embarked on the business of manufacturing automobiles. Even this prototype car, a product of two years of intensive work, was not 'purely domestic-made' at all: it was constructed by using many Chevrolet parts. Kiichiro and Kan intentionally adopted this approach because the company could not, either financially or technically, produce every part. Their survey of partsmakers had revealed the poor quality of parts made in Japan. In order to shorten the time for making a prototype car and also to make it possible to use reliable genuine Chevrolet parts, therefore, they intentionally designed their prototype car with the same-sized parts as Chevrolet's. This design strategy smoothed the way for them to start their automobile business. Yet, parts made in Japan were used in the assembly of their cars to some extent and caused the breakdown of the process in its early days. As a result, Toyoda had to produce these parts at their own plant.[24]

1.3.2. *Koromo Plant as a Pilot Plant for 'Flow Production'*

When Kan moved to the Toyoda company in 1933, Kan and Kiichiro started to draw up plans for a pilot plant, later known as Kariya Plant. The task seemed to be insuperable: they had little detailed knowledge of the automobile production process, nor the processes or materials involved in parts-making. Engineers at aircraft companies knew how to manufacture products, but not how to produce products on a large scale. In the automobile industry, engineers have to acquire a completely difficult type of technological capability.

Kan went to the USA in order to purchase machines for Kariya Plant in January of 1934. Before he left, he drew up a skeleton outline of an operation sheet for auto manufacturing. In accordance with instructions from Kiichiro that he visit a number of factories during his five-month stay, Kan visited Ford, Chrysler, and Graham Paige factories. Kan observed their operations carefully, analysing the manufacturing process and the flow of production. When he visited machine-tool companies to purchase machines, he requested detailed explanations regarding materials and the ways of processing them. He made revisions to his outline of an operation sheet, as well as to the tool layout sheet he had drawn up. After he returned to Japan, he made further revisions. Finally, by the spring of 1936, the plans were completed.

While Kan was engaged in completing his operation sheet and tool layout sheet, the company managed, as I pointed out above, to produce five passenger cars on an experimental basis by 1935. In March of 1936 Kiichiro Toyoda directed Kan to design a new plant, Koromo Plant, which began operations in November of 1938 with the capacity to produce 2,000 cars per month. This factory, in a sense, encapsulated Kan's observations of the American manufacturers' production methods and Kiichiro's ideas about auto manufacturing.

At Koromo Plant Toyota did not install a large number of special-purpose machines. Instead, it installed 'adjustable' special-purpose machines in the machine shop; these were adjustable to design changes. As the production level at Toyota would be far lower than that at Ford, Toyota consciously chose different types of machines. Toyota also installed machines on floorboards in order to make it easier to change the layout of machines, which was not to be inflexible. The layout of machines was intended to follow the flow of work. By adjusting the layout of machines, Toyota tried to establish a smooth flow of production. The adoption of a conveyor system was, however, limited. Although Kan knew the importance of a conveyor system for handling materials, financial stringencies prevented the adoption of conveyor systems throughout the whole process, so the Koromo Plant was equipped with a conveyor system only at the final-assembly line. Thus, many steps were taken to implement 'flow production' at Koromo Plant. Toyota's official history states: 'Flow production was employed, whereby the progress of each phase of manufacture was carried out in synchronization with the final stage; this was aiming toward realization of the "Just-in-Time System".'[25] However, whether flow production was thoroughly employed or not was disputable. In another official history of Toyota, a worker recalls the conditions at Koromo Plant in those days as follows: 'Process flows were often disturbed, work-in-process inventories piled up, and there was a lack of balance in machine utilization.'[26] Toyota quite obviously did not achieve 'flow production' throughout the whole process at Koromo Plant.

Nevertheless, Koromo Plant certainly introduced a new production system. Many researchers have pointed out the fact that Kiichiro Toyoda tried to introduce the 'just-in-time' system (non-inventory production method) at Koromo Plant. But it was 'not possible to bring about completely the just-in-time System as conceived by Kiichiro'.[27] According to Toyota's official history, Toyota tackled 'the task of reforming the production method' by implementing a new production system at Koromo Plant. The official history describes this method thus:

Under the new production system [at Koromo plant], a series number was allocated to every 10 vehicles produced, and these numbers were used to keep track of everything [that belonged] to those vehicles, right down to the parts from which they were assembled. The progress of manufacturing could thus be monitored by checking the number assigned to the parts in each process at any given time.[28]

This 'new production system' does not suggest the implementation of flow production at all. This just meant that Toyota adopted the same method as the Japanese aircraft companies did during the war. As we have seen, when engineers at the aircraft companies had to increase output, they increasingly found it difficult to control the progression of production. They segmented the production processes into several parts, and tried to synchronize the processing time for every part of the production process. In addition to this, engineers at aircraft companies allocated a serial number to every production-lot, and these numbers were used to keep track of everything belonging to the aircraft, right down to the parts from which they were assembled. They apparently tried to try to control the progression of production without using tags or documents describing the processing schedule, by numbering each small lot of parts. The scarcity of resource materials for this period made it impossible at this stage to determine which method was adopted earlier and the extent to which one influenced the other. Nevertheless, it would be reasonable to assume that Toyota's attempts at implementing 'flow production' were in line with similar attempts by engineers at aircraft manufacturing companies.

1.3.3. Implementing a 'Flow Production' Method at the Machining Shop

After the war, Toyota faced up to the problem of implementing 'flow production' throughout the whole process. Crucial attempts were made in the machining shop by Taiichi Ohno, who is widely known as one of the earliest proponents of the 'just-in-time' method at Toyota. He recalled the situation after he took charge of the transmission and suspension shop in 1946:

> The shop floor in those days was controlled by foremen-craftsmen. Division managers and section managers could not control the shop floor, and they were always making excuses for production delay. So the first thing we did was to make manuals of standard operation procedures, and we posted them above the work stations so that supervisors could see at a glance if the workers were following the standard operations or not.[29]

When Ohno's efforts to standardize jobs were officially approved in 1948, they became an important test case for standardizing jobs at Toyota as a whole.

Shoichi Saito, Managing Director, thought the company would not survive if it could not reduce costs. In those days, however, the company did not have the costs data for each manufacturing process. In order to proceed with rationalizing the factory operations, Saito felt the company should have detailed data. So he decided to begin by analysing the costs at the transmission and suspension shop. In July 1948 he ordered the head of the transmission and suspension shop to 'rationalize the factories'. Responding to this, Ohno introduced a new reporting system at the transmission and suspension shop.[30] Under the new system, the numerous documents that were previously used in the shop were integrated into just two reports, a daily report on operations and a daily report on inspection.

In November 1948 the new reporting system was implemented, and Ohno and his staff analysed the data accumulated through the daily reports. They estimated the length of time it took to make each part and evaluated the efficiency of each working group on the basis of the time studies. By analysing the data thus collected, they improved the process step by step. This would be the first attempt at trying to analyse cost data systematically throughout the whole process at Toyota.

When an organizational change was made in 1949 to integrate the engine machining shop and the transmission and suspension shop into one shop, called the machining shop, Ohno was put in charge of it. He and his staff strove to standardize jobs and to improve the processes in the machining shop. Ohno's ultimate goal, however, was not just to standardize jobs, but to create a flow of production in Toyota. In order to implement this flow of production, he and his staff changed the layout of machines and proceeded with a system of multi-task operations by a single worker while standardizing. He wrote:

In 1949 and 1950, when I was in charge of the machining shop, I tried to create a 'flow of production' as the first step for introducing a 'Just-in-Time' system. In order to create the flow, I changed the layout of machines and tried out what we called 'multi-task operations', in which a single worker handled several [different kinds of] machines.[31]

On the basis of his interview with Ohno, Fujimoto wrote about changing the layout of machines in those days:

The engine machining shop had already adopted a product-focused machine layout (i.e., installing machine tools according to process sequence for a particular product group), but the transmission and suspension shop was still organized by types of machines (e.g., balling, lathe, milling, grinding). Consequently, the level of in-process inventories was high. It took Ohno and his staff two years to convert the layout to a product-focused layout.[32]

Ohno attempted to change the layout of machine tools, just as engineers in the aircraft industry had placed machines tools according to sequential operations on different parts in order to create a 'flow of production' in the war years.

Ohno found that multi-task operations by a single worker were essential for creating a flow of production. He also recognized that the 'layout by types of machines' would not create a flow of production. But even if the company successfully changed the layout of machines to make it 'product-focused', this would not result in increased productivity. He thought the 'product-focused' layout would be beneficial to the company only if the workers handled several different kinds of machines. In the 'product-focused' machine layout, machines were placed according to sequential operations, 'production flow'. Several different kinds of machines were clustered together. If a worker could not handle different kinds of machines, the 'product-focused' machine layout would force the company to employ more employees in order to handle all the machines. If, however, machines were arranged by types of machines, multi-machine handling by a single worker could decrease the number of workers needed to handle machines and consequently the company could increase productivity.

According to Toyota's official history, 'By 1949, the multi-machine handling system had developed considerably, and productivity improved as a result.'[33] Under the system, each employee was responsible for a group of machines of the same type. One problem, however, was that different kinds of machines did not perform evenly. In order to resolve this problem, a worker was put in charge of handling more than one process in the order of the work flow. The introduction of this multi-process handling would advance the production system at Toyota toward 'flow production' considerably.[34] However, it was not easy to introduce the practice of multi-process handling by a single worker. Ohno proposed the company should control the jigs. As long as workers controlled the jigs, the company could not increase the number or type of jobs that each worker would carry out. However, foremen-type workers prided themselves on grinding their own jigs. Therefore, when the company proposed that the company should have responsibility for controlling and grinding the jigs, workers stiffly resisted the proposal. They thought the company's proposal was denying them their inherent right as craftsmen. As a result, during the labour dispute of 1950 workers bitterly criticized the 'Ohno line', in which he tried out his 'flow of production' idea. The labour dispute was caused by the financial crisis (Toyota was on the verge of bankruptcy), but in a sense the dispute was one of craftsmen-type workers resisting the 'Ohno line', which seemed aimed at depriving workers of their privileges. The end of the dispute made it easier for Ohno and his staff to implement the changes within their shop.

Further, we should also point out that Toyota's 'five-year plan' accelerated the spread of Ohno's experimental efforts to wider areas. In 1950, a turbulent year for the company, Toyota sent two managing directors, Eiji Toyoda and Shoichi Saito, to Ford's River Rouge Plant. 'There they sudied the facilities needed for mass production, production technology, and methods of production control and plant management.'[35] On the basis of their reports, Toyota formulated a five-year plan for modernizing its production control as well as production facillities. It aimed at doubling production capacity to 3,000 units per month without increasing employees, and it required 'the full-scale introduction of a flow assembly system'.[36] This 'five-year plan' gave a fresh impetus to Ohno's attempts at implementing a 'flow of production'. In line with the plan, Toyota purchased many new machines and replaced older ones; it equipped the new machines with devices that would automatically stop the operations after the intended machining operation was finished. Machines equipped with this device made it easier for workers to handle several machines. According to Toyota's official history,[37] productivity at the machining shop greatly increased after the introduction of this device and the practice of multi-machine handling. As a result, the layout of machines was completely rearranged into one according to 'sequential operations on various machines rather than by types of machine'.[38] As the rearrangement of machines spread, so multi-task operations also spread. The experimental attempts to implement a 'flow production' method bore fruit, especially after the five-year plan lent support to these attempts by providing the material basis, as well as by authorizing the goal of Ohno and his team as a target for the company as a whole.

Although experiences in the machining shop had a galvanizing effect on the implementation of a 'flow production' at Toyota, the implementation spread unevenly among shops. From 1954 on, Toyota began to implement the so-called supermarket method, in which parts were 'supplied from a preceding process to a subsequent process in the same way that items were supplied by a supermarket to customers—at the right time and in the required amounts.'[39] When the company marketed Crown in 1955, Toyota was using the 'tact system' just in the body-assembly process.

As we pointed out in Section 2, Nobuo Noda distinguished between 'high-volume production' and 'mass production'. Faced with a shortage of machine tools during the war, Japanese engineers tried to realize 'high-volume production' through establishing a smooth flow of production at plants. Many years later, Toyota was equipped with a large number of machine tools under the five-year plan. The spread of the 'supermarket' method meant the establishment of the 'mass production' method which Noda had in mind, thanks to the large number of machine tools.

If we look at the Toyota Group as a whole, we can trace other direct influences of the method that emerged from the experiences of aircraft manufacturers in the wartime years. In 1954 the Toyota Auto Body Co. thoroughly rearranged its production line with the assistance of Noritsu Kyokai, an association of which many aircraft engineers were members. Further, Aichi Prefecture carried out an investigation into the relations between Toyota and its suppliers and gave managerial guidance to them in 1952 and 1953. This Keiretsu Diagnosis led to a reshaping of their relations. This diagnosis was carried out under the strong leadership of Mr Murai, an engineer who worked for an aircraft manufacturer during the war and who published many articles on ways to improve relations between a major company and its suppliers.[40]

1.4. Conclusion

The achievements of the Japanese automobile industry before World War II and a decade after do not seem very impressive, whether in terms of the volume of production or in terms of the quality of vehicles. The United States Strategic Bombing Survey shrewdly sized up the state of the Japanese automobile industry:

Use of motor vehicles in Japan has never been extensive. It is a mountainous country with a poor road network and is dependent on shipping and rail facilities for long-haul transportation. . . . Domestic production of motor vehicles in Japan did not become significant until after 1935, and the country was largely dependent on imports until 1938.

In view of the development of production methods at Toyota, however, we should pay more attention to what happened before the 1960s.

As we have seen, production methods at the automobile companies were strongly influenced by the wartime experiences at aircraft companies. A historian of technology has something similar to say about wartime experiences in the United States:

The composite picture of American aircraft production during the war is one of automotive mass production modified by circumstances and vision. Circumstances created an extensive system of subcontracting, of limited inventories, of tight co-ordination between need and production. Industry leaders tried to maintain a system of high levels of quality control, trained workmen, and technical versatility. All of these elements suggest movement toward a system of mass production similar to modern Japanese methods. But the co-operative mechanisms of the war era disappeared in the immediate aftermath of the war and few tried to reconstruct the system.[41]

This suggests that the production methods followed by aircraft companies might have provided the right insights on how best to renovate production methods at automobile companies in the United States, too.

NOTES

1. Toyota Motor Corporation, *Toyota: A History of the First 50 Years* (privately published, 1988), 134.
2. See M. A. Cusumano, *The Japanese Automobile Industry: Technology and Management at Nissan and Toyota* (Cambridge, Mass., 1985).
3. See also K. Wada, 'The Development of Tiered Inter-Firm Relationships in the Automobile Industry: A Case Study of Toyota Motor Corporation', *Japanese Yearbook on Business History*, 8 (1991).
4. Cusumano, *The Japanese Automobile Industry*, 1.

5. Ibid., ch. 1.
6. See M. Mason, *American Multinationals and Japan: The Political Economy of Japanese Capital Controls, 1899–1980* (Cambridge, Mass., 1992).
7. See K. Okuda, *Hito to keiei: Nihon keiei kanri-shi kenkyu* [People and management: studies in the history of Japanese management control] (Tokyo, 1985).
8. See Mason, *American Multinationals and Japan*.
9. See N. Noda, 'Zosan kessen to taryo seisan' [A decisive battle for increasing output and the high-volume production], *Nihon Noritsu* [Efficiency in Japan], 2/5 (1943).
10. S. Mashima, 'Seizo keikaku ni tsuite' [On the manufacturing plan] (mimeograph, privately circulated, 1943), 43. This document consists of notes of a lecture that Mashima, a researcher on production engineering, delivered at the Society of the Naval Shipbuilding Industry in June 1943.
11. K. Yamamoto, *Nihon ni okeru shokuba no gijutu rodoshi* [Technological and labour history of workshops in Japan] (University of Tokyo Press, 1994), ch. 4.
12. H. Aikawa, *Gijutsu oyobi gino kanri* [Technology and skill control] (Tokyo, 1944), 107.
13. Ibid. 193.
14. See Koku Kogyoshi Hensan Iinkai [Editorial committee for the aircraft industry] (ed.), *Minkan kokuki kogyoshi* [The history of the civil aircraft industry] (mimeograph, privately circulated, 1948).
15. In 1944 Kuninobu Niizaki publicized this idea in his article 'Kotei kanri no ichi hoshiki' [A way of controlling work-process], *Nihon Noritsu*, 3/5 (1944). This article seems to have been the first and only one to describe the idea in detail during the war. According to Niizaki, the idea was already being tried in several plants in order to improve productivity in parts-making.
16. Sakuma named his concept 'semi-flow production'. The description of his idea is mainly based on the article by I. Sakuma, 'Han nagare sagyo seisan hoshiki ni kansuru kenkyu' [Study on the semi-flow production method], *Nippon Noritsu*, 2/7 (1943).
17. See Cusumano, *The Japanese Automobile Industry*, 120.
18. The patent rights covered everywhere except Japan, China, and the USA. On this deal, see Y. Taniguchi, '1930 nen zengo no boshoku kikai kogyo ni okeru nichiei kankei no ichidanmen' [One aspect of the relationship between Japan and Britain in the textile-machine industry around the 1930s] in K. Oishi (ed.), *Senkanki Nihon no taigai keizai kankei* [Japan's economic relationship with foreign countries between the Wars] (Tokyo, 1992).
19. Toyota Motor Company History Editorial Committee, *Toyota Jidosha 20-nenshi* [The first 20 years of Toyota Motor Company] (privately published, 1958), 25.
20. Ibid.
21. In January 1934 the special meeting of shareholders formally approved the company's entry into the business of manufacturing automobiles.
22. Cusumano wrongly refers to Kan's name as 'Suda' throughout his book.
23. This department became an independent firm in 1940, the Aichi Steel Co.
24. This experience later led Toyoda to bring up suppliers under their strong influence. On the development of the relationship between Toyota and its suppliers, see Wada, 'The Development of Tiered Inter-Firm Relationships in the Automobile Industry'.

25. *Toyota: A History of the First 50 Years*, 72.
26. Toyota Motor Company History Editorial Committee, *Toyota no alumni* [The history of Toyota] (privately printed, 1978), 94.
27. Ibid.
28. *Toyota: A History of the First 50 Years*, 72.
29. Interview with Ohno by Koichi Shimokawa and Takahiro Fujimoto, 16 July 1984. This interview is quoted in T. Fujimoto and J. Tidd, 'Ford shisutemu no donyu to genchi tekio: nichiei jidosha sangyo no hikaku kenkyu' [The adoption and adaptation of the Ford system: A comparative study of the Japanese and British automobile industries], *Keizaigaku ronshu* [The Joúrnal of Economics, University of Tokyo], 59/3 (1993), 38–9.
30. Toyota Motor Company History Editorial Committee, *Toyota Jidosha 20 nens-hi* [Twenty years of Toyota], (privately published, 1958), 289.
31. Ohno Taiichi, *Toyota seisan hoshiki* [Toyota production method] (Tokyo, 1978), 58.
32. Ibid.
33. *Toyota: A History of the First 50 Years*, 142.
34. Ibid. 143.
35. Ibid. 113.
36. Ibid. 112.
37. Toyota Motor Company History Editorial Committee, *Sozo Kagirinaku: Toyota Jidosha 50 nen-shi* [Unlimited creativity: Fiftieth anniversary history of Toyota Motor Company], 255.
38. David Hounshell, *From American System to Mass Production, 1800–1932: The Development of Manufacturing Technology in the United States* (Baltimore, 1984), 221.
39. *Toyota: A History of the First 50 Years*, 143.
40. See Wada, *'The Development of Tiered Inter-Firm Relationships in the Automobile Industry'*.
41. G. C. Sinclair, 'Public and Private Agent of Technical Exchange', in Japan Science Foundation (ed.), *Kokusai eki na gijutsu koryu ni tsuite no chosa kenkyu* [Research on international exchanges in technology] (privately published, 1992), 141.

2

The Formation of Assembler Networks in the Automobile Industry: *The Case of Toyota Motor Company (1955–80)*

HARUHITO SHIOMI

2.1. Introduction

It was not until 1955 that Japan could lay claim to a completely domestically produced passenger automobile. In January of that year the Toyota Motor company (Toyota hereafter), which had elected not to ally itself with any foreign manufacturer for passenger auto production, released the Crown. This model was decisive in convincing many that, with domestic cars having come so far, foreign cars were no longer necessary. In fact, 1955 can be said to have been the Japanese automobile industry's version of America's 1908, when the epoch-making everyman's Model T Ford made its appearance.

As indicated in Figure 2.1, which compares the movement of production figures for Toyota and Ford, with these two years taken as respective starting points, it took Ford 15 years, until 1923, to reach annual production of two million units after the first Model T, while it took Toyota 17 years after the first Crown to reach the same level in 1972.[1] In terms of volume, then, Toyota followed essentially the same growth process as Ford in realizing mass production, but approximately a half century later. The two companies' responses to the respective domestic markets that had supported their growth, however, were quite different.

First, while Ford had reached the two-million-unit production level solely with the Model T, most of Toyota's growth during its first 29 years, from its founding in 1937 to 1965 inclusive, came from the production of trucks and buses. Initially, the Crown too was used primarily by taxi companies and for other commercial purposes.

Trans. Brendon R. Hanna, doctoral candidate, Kyoto University Faculty of Economics.

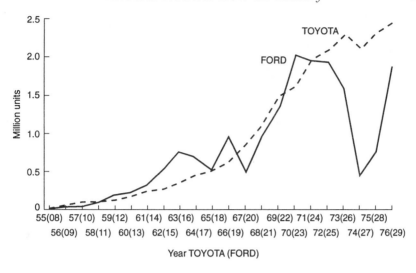

Fig. 2.1. *Growth to 2 million units per year*

Secondly, it was not until 1966, known as 'year one of the era of private car ownership' (in mock reference to Japan's system of imperial eras) that Toyota's volume of passenger auto production overtook that of other vehicles, reaching 54 per cent of the total. Thereafter, Toyota's rapid growth depended on explosive demand in the market for affordable private automobiles, and the company became the third car maker in the world after Ford and GM to reach annual production of two million units. It should be noted that Toyota's reaction to this boom in new demand was to adopt the full line strategy that GM had developed so effectively in the saturation/replacement market of the 1920s; by 1970 Toyota had developed a 'full line, wide selection' system with nine different passenger cars, each of which was produced in a number of different versions.

Although a number of demonstrative studies have been done concerning the parts supply system that allowed Toyota to reach the two-million-unit level of production, there has been virtually no consideration of the situation with regard to final assembly, an area that is closely related to marketing strategy.[2] However, because the introduction of the moving assembly line to the final assembly process is the most important index of acceptance of the Ford system, the structure of the final assembly portion is central to any analysis of the Japanese process of modification of the Ford system. The purpose of the present paper, then, is to outline an historical consideration of the set of final assembly plants that Toyota built up in its maturation toward the 'full line, wide selection' system of the early 1970s. Consideration of these long-term dynamics may also lead to better consolidation of existing analyses of the parts supply system.

2.2. Assembly Consignment and its Various Patterns—Preliminary Consideration

Toyota does not produce all Toyota automobiles itself; it was able to become a two-million-unit producer by building up a network of eight consignment assemblers. Before proceeding further, let us delineate this arrangement.

2.2.1. Share of Consignment Production

It is extremely difficult to verify from publicly available materials the number of Toyota vehicles that are produced on consignment. Table 2.1, then, was constructed through examination of the corporate histories of the eight assemblers, in some cases supplemented by interviews, and simply comparing the resulting consignment production figures with Toyota's total production.[3]

Two points should be noted here. First, it was in September of 1949, for its Crown model, that Toyota began to make its own car bodies in house; previously it had employed a system of contracting out its bodies. In fact, it still uses this system for its trucks and buses, with almost all of these types of vehicles being completed by consignment assemblers, but with Toyota chassis having been transported to the assemblers.[4] In these cases, the number of Toyota-built chassis units overlaps with the number of assembler-built body units, thus inflating the apparent share of consignment production. This is particularly true of the figures for 1960, which saw the peak of truck production for the US Army Procurement Agency in Japan.

Secondly, concerning passenger automobiles, all models from the Crown onward (with the exception of the luxury Century) were manufactured using unit body (monocoque) construction; final assembly became a matter of attaching parts to the finished body and this process was integrated with body production.[5] Consequently, except for a portion of the original Crowns and the comparatively small numbers of Centuries produced, there is little danger of double counting.

Having noted these points, we can safely say that Toyota has, from the beginning of small lot production through to the present day, developed a production system based on a comprehensive and consistent policy of consignment of final assembly for nearly half the total units produced. Thus was the assembler network formed.

TABLE 2.1. *Car Production of TOYOTA and its Assemblers*

	TOYOTA			OEM Supply from Assemblers							
	passenger car	bus & truck	total	TOYODA A.L.W.	TOYOTA A.B.	KANTO A.W.	WIND MOTORS	DAIHATSU M.	CENTRAL M.	ARACO	GIFU A.B.
1955	7,403	15,383	22,786	763	7,605	4,642				1,137	
1960	42,118	112,652	154,770	8,863	74,286	29,537			4,152	7,960	364
1965	236,151	241,492	477,643	1,068	104,361	77,245			11,402	14,694	4,166
1970	1,068,321	540,869	1,609,190	49,991	291,413	271,724	104,789	33,993	52,803	41,323	22,000
1975	1,714,836	621,217	2,336,053	98,254	356,949	314,562	164,380	98,462	61,484	64,673	35,545
1980	2,303,284	999,060	3,293,344	182,626	432,302	394,518	280,106	146,059	71,725	109,677	55,903
1985	2,569,284	1,096,338	3,665,622	176,259	459,019	422,753	351,105	149,652	85,462	135,222	60,136

2.2.2. *Various Patterns of Assembly Consignment*

As to which assembly was contracted out, Table 2.2 indicates the dynamics of the division of labour among Toyota and its eight assemblers, and shows that there are four patterns of assembly consignment.[6]

Pattern 1: contracting out of commercial versions of leading models. The various model lines typically contain vans, wagons, pickups, or other vehicles designed for commercial use. In the case of the Crown line, for example, commercial vehicles made up 28 per cent of sales in 1956, immediately following their release. But by the time Toyota sales hit the two-million mark in 1972, this figure had dropped to only 5 per cent. All of these units Toyota contracts out.

The commercial Crown models were consigned to Kanto Auto Works simultaneously with their introduction, and two years later to Central Motor as well. The Corona line, first marketed in 1957, was switched to this pattern in 1960, with Kanto initially taking on production of commercial models and Central joining in from 1964. This arrangement can also be found with the Publica, on sale from 1961, and with commercial models being consigned to Toyota Automatic Loom Works and Hino Motors from 1967; with the Corolla, introduced in 1966, and with commercial models going to Kanto the following year; and with the Mark II, released in 1968 with the same models taken on immediately by Central.

Pattern 2: contracting out of standard (sedan) versions of leading models. This arrangement provides for multiple site production and was adopted in response to rapidly expanding demand for the mainstay standard versions of popular model lines.

Except in the period immediately following its introduction, standard type Crowns were produced at two sites when part of the work was done by Kanto Auto Works from 1969. In gearing up for two million unit production, standard Coronas were consigned to Toyota Auto Body in 1965 and to Kanto in 1970, for a total of three assembly locations; assembly of standard Corollas was begun by Kanto in 1969, to provide two manufacturing bases for it; and Mark II sedans were contracted out to Toyota Auto Body in 1968 and, additionally, to Kanto in 1970, for a sum of three manufacturing bases.

Pattern 3: contracting out of low volume models. With the expansion of the full line system, models whose sales dropped off in a fluctuating market began to be contracted out.

The Publica, Toyota's first compact car, ceased to be assembled by Toyota in 1970, and the main centre of production of the Corona, which saw a declining trend in sales from the late 1970s, was also shifted outside Toyota's own plants.

Pattern 4: complete contracting out. Certain Toyota cars are not made at all by Toyota.

TABLE 2.2. *Allocation of Passenger Car Models*

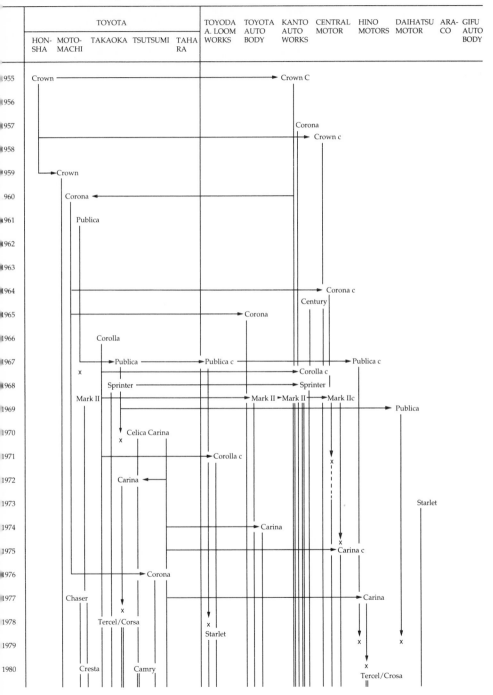

note, c: commercial type x: discontinued

The first-generation Corona, developed by Kanto Auto Works, was produced only by that company until the 1960 model change; the top-of-the-line luxury Century, developed by Toyota, was also built only by Kanto from the time of its original release in 1964; and the Starlet, developed jointly by Toyota and Daihatsu Motor, was assembled solely by Daihatsu from the model's introduction in 1973 until 1978, when Daihatsu was joined by Toyota Automatic Loom Works to provide for dual assembly.

2.3. Maturation of the Assembler Network

Given the above observations, let us now examine the development of Toyota-centred interfirm relations into a proper assembler network.

Taking into account the government's announcement in June of 1960 of the 'fundamental principles of the plan to liberalize trade and foreign exchange', Toyota proposed the following month a long-term capital investment schedule aiming for monthly production of 50,000 units after five years, in 1965, and with a preliminary target of 30,000 units within three years. This policy, known as the '30,000 Unit Plan', attempted to modify the optimum scale of car production (according to George Maxcy and Aubrey Silberston) to reflect the situation at Toyota, and 'with great difficulty a tentative solution was arrived at after about eight months'. Based on the results of this analysis, the optimum scale of production at each facility was slightly under the Maxcy/Silberston recommendation, with monthly production at assembly plants scheduled for between 8,000 and 15,000 units (leading to annual production estimates of between 80,000 and 150,000). Significantly, these figures were subsequently used by Toyota as criteria for its 'Plant Unit Capital Investment Plan', which was used in turn for the relentless pursuit of economies of scale. Further, 'the same line of thought was introduced to assembly consignment and purchasing policy with regard to related and "cooperating" companies'.[7]

As is generally known, the Maxcy/Silberston optimum scale was calculated assuming the postwar American auto industry's latest technology.[8] Consequently, the year 1960 can be considered the starting-point in Toyota's creation of an assembler network that would allow it to become cost competitive with America's Big Three.

2.3.1. *Formation of Essential Preconditions*

At the time the '30,000 Unit Plan' was proposed, Toyota's total annual production of vehicles was approximately 150,000, meaning that it had already reached the optimum scale it had calculated for an assembly plant. Passenger car production, however, at only 40,000 units (27% of the total), remained far short of that level. Completed the year before in 1959, the Motomachi Plant was Japan's first factory devoted solely to passenger cars and had an initial monthly capacity of 5,000 units, still only half of the optimum scale. Thus, Toyota's efforts prior to 1960 were merely experiments in applying certain elements of the Ford system to low-volume production.

Production of the Crown benefited from realization of the 'Five Year Plan for the Modernization of Production Facilities and Equipment', instituted in 1951. The first phase of the organization of this plan, an observation tour of Ford's River Rouge Plant headed up by then managing director Eiji Toyoda between July and October of 1950, was succeeded by the second phase undertaken from October through January of the following year. The goal of the plan was to double monthly capacity to 3,000 units without any increase in personnel (total employment in 1951 stood at 5,685). Just as the plan got started, however, the Ministry of International Trade and Industry's 1952 'Fundamental Policy Concerning the Introduction of Foreign Capital to Passenger Car Production' led to the appearance of domestically produced but foreign designed autos such as the Nissan/Austin A40, the Hino/Renault 4VC, and the Isuzu/Hillman Minx. In response, the plan was revised in 1953 and implemented in two stages.[9]

In the first of these, replacement of out-dated equipment was carried out. The latter stage addressed the problem of rationalization of materials transport within the factory, since Eiji, in a report of his visit to River Rouge, had noted that 'what most attracted [my] interest was materials handling.'[10] A 'Transport Countermeasures Committee' was set up and a variety of conveyors and hoists were introduced at each step of the production process. Also, Taiichi Ohno's 'supermarket method', which allowed parts to be claimed in lots of five and eliminated intermediate warehousing was started in 1949.[11]

Monthly production of 3,000 units, achieved in 1955, was along the following lines:

a) Truck chassis assembly line, with capacity for 1,000 standard trucks and 1,000 small trucks. Three consignment assemblers, Toyoda Automatic Loom Works, Toyota Auto Body, and Araco, were called upon to supply the monthly requisite 2,000 bodies. (These were joined by Kanto Auto Works and Central Motor in 1956, and by Gifu Auto Body Industries in 1959.)

b) The passenger car chassis assembly line, with capacity for 1,000 Crowns, and the auto body assembly line, with capacity for 500 Crown bodies. The extra 500 Crown bodies were supplied by the consignment assembler Kanto Auto Works. (Central Motor was added in 1957.)

By adopting the tact system (where work moves intermittently) on these two assembly lines, and by making lots of five units, Toyota was able to reach production of 60 Crowns per day.[12]

As noted by Maxcy and Silberston, because assembly lines have a wide range of use compared to other equipment, and because they allow rationalization of complex movements of materials within the factory, they can be introduced at a rather low-level scale of production and still provide dramatically improved efficiency.[13] Although one should not underestimate the impact of the tact system, Toyota had yet to develop an automatic conveyor-type assembly line at the juncture of this 'Five-Year Plan'.

The six consignment assemblers were allotted the various complex bodies, and here too assembly lines geared to Toyota were introduced. At the time of Toyota Auto Body's construction in 1957 of its new Kariya Plant, there was considerable discussion over whether to install separate lines devoted to the production of each model, or whether to operate one mixed line. In the end, since 'specialized lines would require additional lines for every new model', the decision was made to integrate production of the three different truck models on a line with a 4.5-minute cycle time, allowing monthly production of 3,500 units. The construction of this new factory served as a test run for Toyota's Motomachi Plant, and 'with Toyota's unfailing guidance, many of their ideas were put into practice using our company's equipment.'[14] Central Motor proved to be the most ambitious and, in deference to Toyota's desire expressed at the signing of the consignment contract for a 'company that will turn on a dime', developed in 1957 a 'multi-model single line production method'. This produced three models initially, and employed a tact system completely driven by assembly jigs to yield 150 units monthly. In this instance, Toyota provided for the introduction of jig production technology from Kanto Auto Works.[15]

In this manner, technology exchanges were conducted among the consignment assemblers, and progress was made in the direction of forging their mixed line production skills.

Additionally, the product development abilities of the consignment assemblers deserve attention. Kanto Auto Works was entrusted with the development of the Crown commercial-type bodies, and a series including a taxi model, a pickup, and a van was marketed simultaneously in 1955.[16] And development of the Corona, released in 1957 in response to the rapidly expanding demand for compact taxis, was also undertaken by Kanto at Toyota's request. Here, Kanto was able to put many years of research and testing to work, coming up with Japan's first monocoque bodyshell to which it fitted existing Toyota parts.[17]

Toyota, at the core of these arrangements, had precious few managerial resources to spare. Its response, then, to an unstable auto market still in its infancy was to create a network of support.

2.3.2. *The 30,000 Unit Plan*

The 30,000 Unit Plan that had begun in 1960 culminated in 1966 with the achievement of production of 50,000 units per month; it was a period of close imitation and implementation of the Ford system for the production of passenger autos for Toyota. The Corona and the Publica both topped 100,000 units in 1965, with the Crown notching up 80,000. This was because these three models were finally being produced on a scale that justified an automated assembly line.

The first phase of construction of the Motomachi Plant saw the 1959 initiation of the Crown production line. This was followed by the second phase and the Corona line in 1960, and then by the third phase, which added the Publica line in 1962, for a total of three specialized, automated assembly lines. Then, in September 1962, the Crown and Corona lines became the first in the history of Japan's auto industry to adopt two shifts on a full-time basis. These were joined in August of the next year by the Publica, giving the Motomachi Plant three specialized lines and total monthly capacity of more than 30,000 units.[18] Meanwhile, the Corona model change concurrent with the second phase was taken advantage of to move production back to Toyota from Kanto Auto Works, leaving only the assembly of commercial-type Coronas there.

Still, a major part of the framework of the 30,000 Unit Plan was 'to expand dedicated facilities of outside suppliers' in order to 'overcome existing limits of factory space, personnel, production management and other factors, and to reduce costs.'[19] During this period the consignment assemblers were putting new facilities into operation one after another. Central Motor's new Sagamihara Plant, opened in 1960 and equipped with automatic conveyors, continued the shift to the 'multi-model single line' and, after experiencing a number of difficulties, finally reached monthly production of 1,000 units five years later.[20] Kanto Auto Works's Fukaura Plant, opened that same year, had a moving line with a 3-minute cycle time that allowed production of more than 3,000 various commercial-type autos per month. Subsequent improvements, including introduction of a 2.1-minute cycle time, and the running of two shifts brought monthly production to 9,500 units by 1970.[21] The year 1964 saw the opening of Toyota Auto Body's Fujimatsu Plant, which it built to specialize in truck production after taking over assembly of the Dyna small truck from Kanto in 1963. 'In an unusual move, the company inherited all production equipment from Kanto Auto Works', and it was at this point that Kanto ceased to manufacture trucks.[22]

It was in this manner that mixed moving assembly lines appeared in their mature form at the consignment assemblers.

Thus, the 30,000 Unit Plan centred on the Motomachi Plant and involved another three new assembly plants as well as two major changes in responsibility for the production of particular models. With regard to the passenger automobile assembler network, then, a clear distinction became apparent between (a), the dedicated (single-model) assembly lines for standard-type cars at Toyota, and (b), the multi-model moving lines for commercial models accepted by Kanto Auto Works and Central Motor. For itself, Toyota was pursuing economies of scale in Ford-style assembly plants.

2.3.3. Full Line, Wide Selection

The year 1966, when Toyota achieved annual production of 50,000 units, was a turning point in the company's history. The year's slogan was 'Quality Assurance Throughout Toyota', and it was decided to spread Toyota TQC activities to consignment assemblers and parts suppliers.[23] It was with the goal of internal TQC in mind that the General Planning Department was created from the former QC Promotion Department, but in accordance with wider policy and planning, the new Quality Assurance Department was put in charge of education and training, and a new Purchasing Management Department was set up within the purchasing organization.[24] Also, in that December the *kanban* method of parts delivery/supply was made applicable to all suppliers, and in 1970 the Production Investigation Department was established within the production management organization in order to provide staff to supervise the introduction of the *kanban* system.[25] The group of companies centred on Toyota began in this way to become linked at the factory management level.

Regarding the assembler network, it was of great significance that the beginning of this year (1966) saw the start of the 'wide selection', which initiated a ten-day cycle from the receipt of an order to the shipping of a finished car.

By 'wide selection' was meant availability of many different varieties of a particular model. The 1966 Crown, for example, was assembled in 48 different combinations of engine, body and transmission. With the inclusion of different interiors, exteriors, and paint colours, a total of 322 finished versions were possible. This expanded to 556 combinations (of which 88 were actually produced) and 54,912 possible finished versions (of which 667 were realized) in 1969.[26] The 'wide selection' offered by GM as part of its full-line policy of the 1920s involved merely the bolting of varying bodies on the same type of chassis and, in the case of the Chevrolet, the best selling model, five types were available; this system continues today in America.[27] By comparison, Toyota can be said to have embarked on an extremely ambitious programme of product diversification. According to Toyota, it was this diversification that 'demonstrated [the passenger car's] appeal as a general-purpose family car to the average person' in the exploding popular market outside that was added to the one that had formerly existed for taxis and other vehicles for commercial use.[28]

This bold diversification carried with it the very high risks of anticipatory production, leading Toyota to shorten as far as possible the cycle of production planning and response to changes in demand. This led in turn to the need for mass production of many different versions, meaning increased flexibility of the assembly line and its approach to custom order production.

It was at this point that Toyota finally introduced for the Crown the forecasting programme and ten-day production plans[29] that GM had developed in the 1920s and which are still widely used in America today. These were applied to the Corona and Publica in the following year, 1967, and to the Corolla in 1969.[30] There was a huge difference, however, between the scale of Toyota's diversification and that of GM. It was for this reason that, between 1972 and March 1974, Toyota was forced to develop its 'New Order System' which allowed daily modification of production plans in order to respond to actual changes in demand that could not be addressed with a ten-day system.[31]

The difference in use of the assembly line itself, as we shall see, was of even greater importance. In the late 1960s, during the phase of preparation for the liberalization of the auto industry, Toyota saw the two-million-unit production mark as a criterion for an open economy. The company's response was extremely dynamic, as demonstrated by the introduction of the Corolla in 1966, the Century in 1967, the Mark II and the Sprinter in 1968, and the Celica and the Carina in 1970, a total of six new model lines. These were accompanied by the formation of business tie-ups with Hino Motors in 1966 and with Daihatsu Motor in 1968. The production allocation involved in two-million-unit production was contemplated from the spring to the autumn of 1968 inclusive by the General Planning Department, and the necessary decisions were made.[32] Table 2.3 shows the results as of 1972, when the two-million mark was reached.

Toyota opened its Takaoka Plant in 1966 with the No. 1 assembly line, to which was added the No. 2 A line in 1968 and the No. 2 B line in 1969. Further, the Tsutsumi Plant began operation in 1970 with its own No. 1 and No. 2 assembly lines to produce a total of five new lines.[33] Kanto Auto Works also opened its 12,000 unit per month Higashi Fuji Plant in 1967, and Toyota Auto Body began the manufacture of passenger automobiles that same year when it opened the Fujimatsu Passenger Auto Plant, also with a monthly capacity of 12,000 units. In the area of trucks and buses, Toyoda Automatic Loom Works' Nagakusa Plant commenced operations in 1967 as well, with two more assembly lines, to produce an additional four lines.[34] The two companies with which Toyota had established business tie-ups (Hino and Daihatsu) each added a Toyota-dedicated assem-

TABLE 2.3 *Assembly Plants of TOYOTA Passenger Cars (1972)*

| | TOYOTA | | | | | TOYODA A. LOOM WORKS | TOYOTA AUTO BODY | KANTO AUTO WORKS | CENTRAL MOTOR | HINO MOTORS | DAIHATSU MOTOR | ARACO | GIFU AUTO BODY |
	HON-SHA	MOTO-MACHI	TAKAOKA	TSUTSUMI	TAHARA								
Century								•'64–					
Crown	•'55–							•'55-'69–	○'57–				
Mark II	•'68–						•'68–	•'68-'70–	○'68–				
Corona	•'60–						•'65–	•'57-'70–	○'64–				
Celica				•'70–									
Carlina			•'72–	•'70–			•'74–						
Corolla			•'66–			○'71–		•'67-'69–					
Sprinter			•'68–					•'74–					
Publica						○'67					•'67-'69–	•'69–	

note '67– year started.   • standard(sedan) type.   ○ commercial type bly line, bringing the overall total to eleven new assembly lines. Thus was formed a unique assembler network involving what might be termed 'multi-locational production of standard-type cars and single-locational production of multiple models'.

In the 1920s GM was using its assembly lines in the same way that Ford was, with each of its five divisions producing its model on a dedicated line. To this day, the American traditional method of production stands in sharp contradistinction to Toyota's assembler network.[35]

2.4. Administration of the Assembler Network

2.4.1. The Mixed Assembly Line

It was the Tsutsumi Plant, opened in 1970, that allowed Toyota to clear the two-million-unit hurdle. The body section of the new plant employed a special attachment jig that could be freely disengaged and connected, in a new process known as the 'gate line method'. This new line was used to assemble both Carinas and Celicas simultaneously, and was revolutionary in the sense that the ratio of these models could be adjusted in response to demand.[36] The origin of this 'gate line method' can be traced back to Kanto Auto Works' response in 1953 to orders for 300 units per month, when it experimentally developed Japan's first pushcar-type total assembly jig. A number of refinements were made subsequently, such that by 1956 it had evolved into what was called the 'block assembly'. This idea was relayed to Central Motor, Toyota Auto Body, and other consignment assemblers, and remains the basis for Kanto's mixed assembly lines to the present day.[37] Thus, Toyota's 'full line, wide selection' depended on the skills built up by the consignment assemblers and absorbed by Toyota itself, then giving Toyota assembly plants the same capabilities.

Let us now examine the question of the function carried out by the mixed assembly line within the assembler network.

2.4.2. Managerial Adjustment in Response to Changes in Demand

The new assembler network is able to respond to and absorb short-term changes in demand, such as those engendered by seasonal variations and the vicissitudes of hit models.

The assembler network as of 1982 in its mature form, centring on the Tsutsumi Plant, is laid out in Figure 2.2. Fluctuations in demand among various models can be absorbed through multilateral adjustment among:

	TOYOTA Tsutsumi plant	
	Assembly line No. 1	Assembly line No. 2
TOYOTA AUTO BODY Co. Fujimatsu plant	Corona	
CENTRAL MOTOR CO. Honsha plant	Carina	Carina
TOYOTA. Tahara plant	Celica C/P	Celica C/P–CB Camry Vista

Fig. 2.2. *Production adjustment at TOYOTA Tsutsumi Plant*

a) multiple models on a single assembly line; b) two dedicated assembly lines; c) other Toyota assembly plants; and d) Toyota's consignment assemblers.[38]

Thus, using its assembler network, Toyota is able to fine-tune a multilateral response to fluctuations in demand that would force a typical Ford system specialist assembly plant to dramatically slow down or speed up the pace of operations and lay off or rehire workers.

This kind of regulation is undertaken mainly by Toyota's Production Management Department, set up in 1960. Under the 'Ten-Day Order System' initiated in 1966, the Production Management Division decided on monthly production plans, or line-specific plans as they were also known, which were used to allocate the following month's production figures for every assembly line in both Toyota's and the consignment assemblers' plants. Accordingly, the ten-day plans provided instructions for final assembly, including bodies, options, interiors, exteriors, and colours. The 'New Order System', established in 1974, replaced the ten-day plans for final assembly with plans that were modified on a daily basis in order to respond directly to customer orders.[39] That is to say, with respect to the assembler network, production is regulated centrally and hierarchically by Toyota's Production Management Department, and this regulation is adjusted monthly at the 'full line' level and daily (formerly every ten days) at the 'wide selection' level.

The new assembler network has also allowed for the reallocation of production in response to long-term shifts in demand, such as the explosion in the market for personal automobiles and the trend towards more up-scale models. It has also provided for a gradual but continuous risk-sharing approach to capital investment throughout the production process.

As previously noted in Table 2.3, Toyota passed through three periods of major shifts/additions of production responsibilities/locations on its way to realizing annual production of two-million units; the '30,000 Unit Plan' in 1960; 1964 to 1965 just before reaching monthly production of 50,000; and in 1968 with the achievement of one-million units. For the Corona, this meant a total of five transfers of control to Kanto Auto Works from 1957, to the Toyota Motomachi Plant from 1960, to Central Motor from 1964, to Toyota Auto Body from 1965, and to the Toyota Tsutsumi Plant from 1967. Five transfers of control were also registered in respect of the Publica, to the Toyota Motomachi Plant from 1961, to the Takaoka Plant, Toyoda Automatic Loom Works, and Hino Motors, all from 1967, and to Daihatsu Motor from 1969.

From its establishment in 1966 onward, the General Planning Department was responsible for the planning of these transfers within the assembler network. Thus it was that long-term production responsibilities became centrally regulated.

'In aiming to become a two-million-unit producer . . . it was imperative for the entire Toyota Group to consider the efficient organization of production'. Hence, 'The results of the deliberations carried out from the spring of 1968 through the fall of that year concerning the rational allocation of production . . . led to the allocation of production within the Toyota Group [with that specific aim in mind].' Further, 'This production allocation continues to be rationally refined in response to fluctuations in demand for the various models.'[40]

2.4.3. *Criteria for Managerial Adjustment*

In answer to the question of how managerial adjustment (regulation of production) was carried out, Table 2.4 shows the results of production allocation during a dynamic phase for Toyota, which saw rapid growth of new domestic demand (1973, as compared with 1965, shows growth of 299%), the country's first postwar decline in the auto business (production in 1974 is 78% of that in the previous year), and speedy recovery (1975 shows year-on expansion of 112%).

Even from before the war, Toyota had sought and repeatedly demonstrated 'mutual prosperity' through 'permanent business relations' with its outside suppliers.[41] The implementation of this philosophy can be confirmed by referring to the results of managerial adjustment in Table 2.4, which shows changes in the numbers of units produced. It can be

TABLE 2.4. *Trends of Production Allocation*

	1965–1973 (1965 = 100)	1973–1974 (1973 = 100)	1974–1975 (1974 = 100)
TOYOTA total	483.2	91.6	110.5
passenger car	691.1	91.0	115.5
bus & truck	280.0	93.2	98.6
Assemblers total	534.2	92.9	111.1
TOYODA A. LOOM WORKS	7871.0	95.6	116.9
TOYOTA AUTO BODY	356.2	88.6	96.0
KANTO AUTO WORKS	430.0	84.7	94.7
HINO MOTORS	–	102.7	112.8
DAIHATSU MOTOR	–	102.7	121.9
CENTRAL MOTOR	468.1	100.1	115.2
ARACO	353.9	128.8	157.7
GIFU AUTO BODY	883.3	83.2	150.9

observed that Toyota chose the policy of distributing the effects of rapid growth, swift decline, and speedy recovery virtually equally throughout the supplier network (including its own plants).

2.5. Conclusion

Toyota unceasingly sought the introduction and smooth operation of the moving assembly line, the nucleus of the Ford system, based on the presumption of its assembler network. The intermediate result was bipolar, with Toyota adopting the American-style specialized assembly line and its consignment assemblers using the mixed line. In meeting the new demand accompanying motorization in Japan, however, with a diversity of production on a different scale from that in America, Toyota itself was also forced to introduce mixed production lines. Through the consignment assemblers, then, the so-called Toyota system reached maturation in the early 1970s, setting the standard for Japan's version of the Ford system.

Incidentally, according to Chandler, during the implementation of the full line policy in the American automobile industry at GM in 1921 and at Ford and Chrysler following World War II, the system of product divisions was adopted. This same conclusion is arrived at by Rumelt as well.[42] Toyota, on the other hand, since its founding in 1937, has continued to expand its system of functional organization right down to the present day, never giving serious consideration to product divisions. What could have given rise to this difference between Japan and the United States? Any number of factors, such as differences in the geographical scale of markets, differences in the pace of differentiation, and differences in the rate of internal manufacture of parts can be cited. As this paper has elucidated, however, the more important reason is that the dynamic operation of the assembler network requires a single, central managerial pivot (the General Planning department and the Production Management Department at Toyota); it does not lend itself to decentralized supervision. Starting with a lack of managerial resources on the part of the core enterprise, Toyota, the assembler network continued to develop in order to disseminate the risk involved in capital investment, and exhibited the continuing capacity to absorb the fluctuations in demand accompanying mass production of differentiated models. In this way, the network can be said to have softened negative effects on employment, parts suppliers, and regional society.[43]

NOTES

1. Compiled from: Toyota Motor Company, *Sozo kagiri naku—Toyota Jidosha 50 nenshi* [Creation without limit—A 50-year history of Toyota Motor] (1987), 97, and Allen Nevins and F. E. Hill, *Ford, Decline and Rebirth 1933–62* (New York, 1962), 478. Tractor production is omitted for Ford.

2. For current analysis of the parts supply system in the Japanese automotive industry, see: Kazuo Wada, 'Jidosha sangyo ni okeru kaisoteki kigyo-kan soshiki no keisei—Toyota Jidosha no jirei' [The formation of hierarchical interfirm organization in the automotive industry: The case of Toyota Motor], *Keieishigaku* [Japan Business History Review], 26/2 (1991); Banri Asanuma, 'Jidosha sangyo ni okeru buhin torihiki no kozo—chosei to kakushinteki tekio no mekanizumu' [The structure of parts transactions in the automotive industry: Adjustment and the innovative response mechanism], *Kikan gendai keizaigaku* [Quarterly Modern Economics] (summer 1984); and Haruhito Shiomi, 'Kigyo gurupu no kanriteki togo—Nihon jidosha sangyo ni okeru buhin torihiki no jissho bunseki' [Management control of enterprise groups; Demonstrative analysis of parts transactions in the Japanese automotive industry], *Oikonomika*, 22/1 (1985).

3. Of these assemblers, Hino Motors and Daihatsu Motor manufacture their own lines of trucks, buses, or passenger cars; the other six are so-called 'body makers' that Toyota affiliates.

    Each of these assemblers, however, has published its own corporate history: Toyoda Automatic Loom Works published its *40 nenshi* [A 40-year history] in 1967; Toyota Auto Body issued *Toyota Shatai 20 nenshi* [A 20-year history of Toyota Auto Body] in 1965, *Toyota Shatai 30 nenshi* [A 30-year history of Toyota Auto Body] in 1975, and *Toyota Shatai 40 nenshi* [A 40-year history of

Toyota Auto Body] in 1985; Kanto Auto Works released *Kanto Jidosha Kogyo 30 nenshi* [A 30-year history of Kanto Auto Works] in 1978 and *Kanto Jidosha Kogyo 40 nenshi* [A 40-year history of Kanto Auto Works] in 1988; Central Motor published *20 nen no ayumi* [A 20-year journey] in 1961, *30 nen no ayumi* [A 30-year journey] in 1980, and *50 nen no ayumi* [A 50-year journey] in 1991; Hino Motors produced *Hino Jidosha Kogyo 40 nenshi* [A 40-year history of Hino Motors] in 1982; Daihatsu Motor issued *60 nenshi* [A 60-year history] in 1967 and *Daihatsu 70 nen shoshi* [A concise 70-year history of Daihatsu] in 1977; Araco released *Arakawa Shatai 25 nenshi* [A 25-year history of Arakawa Auto Body] in 1972, *Kiseki 10 nen—35 shunen shoshi* [Tracks of 10 Years—35th anniversary concise history] in 1977, and *Arakawa Shatai no 40 nen* [40 Years of Arakawa Auto Body] in 1982; finally, Gifu Auto Body Industries published *Gifu Shatai Kogyo 40 nenshi* [A 40-year history of Gifu Auto Body Industries] in 1981.

4. *Toyota Jidosha 30 nenshi* [A 30-year history of Toyota Motor] (1967).
5. Each of the facilities that make up Toyota's assembler network is equipped with presses, body manufacturing facilities, a paint shop, and a final assembly.
6. Compiled from: *Toyota Jidosha 20 nenshi* [A 20-year history of Toyota Motor] (1985); *Toyota Jidosha 30 nenshi* (1967); *Toyota Jidosha 40 nenshi* (1978); *Sozo Kagiri naku—Toyota Jidosha 50 nenshi* (1987); and the corporate histories listed in n. 3.
7. *Sozo kagiri naku*, 362–3.
8. George Maxcy and Aubrey Silberston, *The Motor Industry* (London, 1959), 75–86.
9. *Toyota Jidosha 20 nenshi*, 360–8, 409–20; *Toyota Jidousha 30 nenshi*, 327–40.
10. *Toyota Jidousha 30 nenshi*, 328. Also, Eiji Toyoda, 'Amerika dayori' [Tidings from America], *Toyota Shimbun* [Toyota Newspaper], 2 (29 Sept. 1950).
11. *Toyota Jidosha 30 nenshi*, 329–30, and *Toyota Jidosha 20 nenshi*, 490.
12. *Toyota Shimbun*, No. 130, (12 Oct. 1954), introduces 'Japan's first modern assembly plant'. This included rationalized transport of parts by a variey of conveyors and a testing line for finished cars. Referred to in *Toyota Jidosha 20 nenshi*, pp. 409 and 489–90.
13. Maxcy and Silberston, *The Motor Industry*, 79.
14. *Toyota Shatai 30 nenshi*, 90, 93–4. The new 4.5-minute tact was a full one-third closer than what the company had been using up to that time. This shortening was supported by high-precision assembly jigs.
15. *30 Nen no ayumi*, 77, 82, 113.
16. *Toyota Jidosha 30 nenshi*, 368–9, and *Kanto Jidosha Kogyo 30 nenshi* [A 30-year history of Kanto Auto Works], 100–1.
17. Kanto Auto Works was founded primarily by a group of aeronautical engineers who had worked for Nakajima Aircraft Co., Ltd prior to World War II, and development of the monocoque bodyshell exploited this latent resource. In this case, Kanto received a set of five specifications from Toyota that it was to incorporate in its design, making it the first 'drawing approved maker' among the consignment assemblers. *Kanto Jidosha Kogyo 40 nenshi*, 54–60. According to the same source, by 1978 Kanto had developed ten different automobiles, six of which were included in Toyota's 'wide selection' and produced in high volumes (ibid. 73).

Similar examples can be found with Central Motor's development of the 1963 Publica Convertible (*40 nen no ayumi* [A 40-year journey], 32–3) and the development by Toyota Auto Body of the 1965 Corona Hardtop, Japan's first-ever hardtop model, and the 1968 Mark II Hardtop (*Toyota Shatai 30 nenshi*, 147–53).

18. *Toyota Shimbun*, 474 (15 Sept. 1962); 522 (31 Aug. 1963). Also interviews.
19. *Toyota Jidosha 30 nenshi*, 485.
20. *40 nen no ayumi*, 28, 33. Central adheres to the multi-model single line even today, but faced a variety of problems when it first switched from the tact system to the automatic conveyor system; production remained stuck at about twice the level of the old system, at about 400 units per month. This was because 'the assembly process failed to keep pace with the conveyor and improperly fastened parts' were a frequent occurrence. 'The efforts of all concerned to address this problem [eventually] proved successful' and four years later 'the rhythm was grasped' (ibid. 32, and *30 nen no ayumi*, 122–3).
21. *Kanto Jidosha Kogyo 40 nenshi*, 107–8. Renovation of the line in 1969 allowed a 2.5-minute tact (two-shift capacity of 8,000 units), with the 2.1-minute tact implemented in 1970 leading to 9,500 units per month (ibid. 116–17).
22. *Toyota shatai 30 nenshi*, 110.
23. Concerning the diffusion of Toyota's TQC activities to suppliers, see Wada, 'Jidosha sangyo ni okeru kaisoteki kigyo-kan soshiki no keisei'.
24. Shoichiro Toyoda, 'shiire-saki wo fukumeta zenshateki hinshitsu kanri no donyu to suishin' [The supplier inclusive company-wide introduction and promotion of quality control], *Hinshitsu kanri* [Quality Control], 32, 1 (Jan. 1981), 41.
25. *Sozo kagiri naku*, 377, 586–7.
26. Toyota Motor Sales, *Motarizeishon to tomo ni* [Along with notorization] (1970), 369. Also, Susumu Uchikawa, 'Kore kara no seisan taisei' [The future production system], *Toyota Manejimento* [Toyota Management], 14/2 (Feb. 1972), 41.
27. Haruhito Shiomi, 'GM sha no fururain seisaku ni okeru seisan kozo' [The production structure of GM's full line policy], *Oikonomika*, 12/1 (June 1975), 78–9.
28. Toyota Motor Sales, *Motarizeishon to tomo ni*, 366.
29. Shiomi, '*Kigyo gurupu no kanriteki togo*', p. 21.
30. *Motarizeishon to tomo ni*, p. 370.
31. *Sozo kagiri naku*, pp. 589–90.
32. *Toyota Jidosha 40 nenshi*, 328.
33. Ibid. 286–9, 328–31. Supplemented by interviews. The Takaoka Plant renewed a cycle-time of one minute, Japan's highest, in 1970. Each assembly line produced 20,000 units per month on two shifts, and this figure was used thereafter by Toyota as a criteria for investment.
34. *Kanto Jidosha Kogyo 40 nenshi*, 111–17; Toyota Automatic Loom Works, *40 nenshi*, 504; *Toyota Shatai 30 nenshi*, 168–75. Note also that the Fujimatsu Passenger Auto Plant shifted to two lines in August of 1974, allowing specialized production of Coronas and Mark IIs, both of which had been assembled together on a mixed line since 1968. Thought was originally given, however, to giving the two lines interchangeability (ibid. 175).
35. Shiomi, 'The Production Structure of GM's Full Line Policy', 83.

36. *Toyota Jidosha 40 nenshi*, 331–2.
37. *Kanto Jidosha Kogyo 40 nenshi*, 50, 75–6. The more fundamental historical origin of Toyota's mixed assembly lines, then, can be traced back through Kanto Auto Works to Nakajima Aircraft, providing a link in both technology and personnel with the prewar aircraft industry.
38. Kan'ichi Kitada, 'Juyo hendo, tayoka ni tai suru kojo no taio' [Factory response to changes in demand and diversification], *Toyota Manejimento* 25/9 (Sept. 1982), 22.
39. Concerning Toyota's production and sales adjustment systems, see Toyota Motors Production Management Division, 'Production Management'. Also, Rintaro Muramatsu (ed.), 'Jidosha no seizo kanri', (*Jidosha kogaku zensho 18*) [Production management of automobiles (vol. 18 of the complete series on automotive engineering)] (1980), 84–8.
40. *Kanto Jidosha Kogyo 40 nenshi*, 328.
41. Wada 'Jidosha sangyo ni okeru kaisoteki kigyo-kat soshiki no keisei', 4–6.
42. Alfred D. Chandler, Jr., *Strategy and Structure* (Cambridge, 1962), 373; Richard P. Rumelt, *Strategy, Structure, and Economic Performance* (Boston, 1974), 183 and 191.
43. In the American system of product divisions, according to O. Williamson, business decision-making and such basic functions as purchasing, production, and sales, along with managerial adjustment of financial dealings, are decentralized among the various divisions, with general headquarters overseeing strategic decisions and allocating resources. The key to this type of organizational structure lies in the concentration of control over cash flow with top management, and in their desire to secure high internal efficiencies of capital. See Oliver E. Williamson, *Markets and Hierarchies* (New York, 1975), ch. 8.

    This type of concentration is not found in the Toyota assembler network, as each consignment assembler undertakes its own individual administration of funds. Conversely, the concentration of decision-making in Toyota's assembler network has increased, such that the entire organization operates as functional divisions.

    Regarding so-called 'group dynamics' in Japanese industry, the maturation of the process of operational adjustment around a central pivot is attracting much attention. Concerning this point, see *The Future of the Automobile, The Report of MIT's International Automobile Program* (Cambridge, 1984), 146–7.

3

Planning and Executing 'Automation' at Ford Motor Company, 1945–65: *The Cleveland Engine Plant and Its Consequences*

DAVID A. HOUNSHELL

Since the late 1970s, as Americans watched manufacturing jobs flow out of the United States while imports of both raw materials and finished products flowed into the country, ordinary citizens, scholars, policy analysts, and politicians have put forth a variety of explanations to account for these trends. A large part of this theorizing has centred on the automobile industry, one of the great growth engines of the twentieth century. While the US automobile industry was laying off more and more workers, imports of Japanese-made automobiles rose dramatically in spite of such barriers as the Pacific Ocean, tariffs, and 'voluntary' import quotas. From among a host of theories offered to explain why the Japanese car makers were taking market share away from US manufacturers emerged a near consensus that the Japanese manufacturers were simply more nimble at manufacturing cars of better design and quality than their US counterparts. Japanese carmakers had developed production systems that were more flexible than those of US automobile companies, which had been committed too long to 'Fordist' production ideals.

In 1984 Michael J. Piore and Charles F. Sabel published their highly influential book, *The Second Industrial Divide: Possibilities for Prosperity*, in which they argued that the Fordist production paradigm was drawing to a close.[1] A new order—'flexible specialization'—provided the only path to prosperity, not just in the automobile industry but in all manufacturing. Little wonder that the German edition of their book bore the title *Das Ende der Massenproduktion*—the 'End of Mass Production'.[2] Some analyses of

I am indebted to James P. Womack, David Jardini, Steven Klepper, and Takamoto Sugisaki for their helpful criticism of an earlier draft of this paper. I thank Darleen Flaherty, Archivist, Ford Industrial Archives, for her enormous help in facilitating my research at Ford Motor Company.

the US automobile industry linked the rigidity of US automobile manufacturing to design regimes that had been conditioned by Detroit's commitment to automation in the post World War II era. MIT's Commission on Industrial Productivity was among the leaders of this assessment. Its summary report in *Made In America: Regaining the Productive Edge* recounted the history of product and process development in the US automobile industry and argued that 'By the late 1940s, the US auto industry had adopted the structure and strategy it would pursue for nearly 40 years.'[3] The report went on:

Well-defined tasks eased the burden of automating, since a dedicated machine could be assigned to each specific task. This policy led to inflexible automation, so that by the 1960s, even a small change in the dimensions of an engine would require millions of dollars for new tooling. Until the mid-1960s this 'hard' automation steadily boosted productivity by 5 per cent per year, about double the increase for the rest of manufacturing. Once most of the defined tasks had been automated, however, productivity growth dropped back to the rather low average of 2.5 per cent per year found in other US manufacturing sectors. Wages, meanwhile, continued to rise at a 5 per cent annual rate. The high cost of design changes also led to reluctance to diversity product offerings when European automakers introduced compacts in the late 1950s.[4]

A year later, MIT's International Motor Vehicle Program's best selling book, *The Machine That Changed the World*, merely repeated this assessment of American automobile manufacture and contrasted it with the 'lean production' practised in Japan.[5] Once a signal of American manufacturing might, 'Detroit automation' became an epithet—a code for outmoded, inefficient, and inflexible methods.

This paper examines the development of 'Detroit automation' as it was carried out at the Ford Motor Company between 1945 and 1965. Contemporaries saw Ford as one of the nation's leaders in automation of manufacturing processes, and the company's Cleveland Engine Plant, opened in 1951, was viewed as the pioneering plant of Detroit automation. Ironically, Ford's automation programme helped to set off a wave of panic that gripped most blue-collar and a good many white-collar workers in the United States during the 1950s and early 1960s. These workers feared loss of their jobs through automation, not the eventual decline of the US automobile industry.

Why did Ford automate? What was the organizational context for its automation programme? How did the company carry out this programme? Did the company change the programme's objectives over time? Was the programme seen within Ford as a sharp break with the company's manufacturing traditions or viewed as a continuation of them? Did Ford's managers foresee that their work in automation would eventually be interpreted as providing one of the underlying causes of the decline of the American automobile industry? These are some of the questions this paper seeks to address.

No single factor gave rise to Ford's automation programme. It occurred in a rich and rapidly changing context that defies monocausal analysis. Above all, the programme must be seen as a manifestation of the company's reinvention of itself—and its explicit emulation of General Motors—under the new leadership of Henry Ford II and his mentor, Ernest Breech. Labour problems at Ford and throughout the automobile industry also influenced the programme's course, although these problems were not paramount. Automation must also be understood as a programme that was part and parcel of steps taken in product design and innovation. The robust nature of the post World War II market for automobiles and of the economy in general also shaped the speed and degree of implementation. Finally, automation at Ford must be seen as an innovation within the Fordist production paradigm established in the second decade of the twentieth century—an intensification of mass production methods—rather than as a revolutionary break with the past.

3.1. Ford Motor Company's Reorganization

That Ford Motor Company would come to be viewed as a pioneer of new manufacturing technology in the 1950s is something of a miracle considering the state of the company in 1945 when Henry Ford II, the 28-year-old grandson of Henry Ford, wrested control of the seriously deteriorated company from the scheming Harry Bennett and determined to right the listing ship. At its height, Ford Motor Company (FMC) had dominated the US automobile industry, taking some 55 per cent of the market. Yet the decline of the Model T in the mid-1920s signalled the decline of FMC. Following a failed attempt to revive the Model T by dressing it up, the stubborn Henry Ford had ended the car's production in 1927 after some 15 million had been manufactured. He vowed to introduce a thoroughly 'up-to-date' car, which after a virtual year-long shutdown of FMC, emerged as the Model A. Ford had kept his word. The Model A proved to be enormously popular, and the company regained market share, garnering 42 per cent of the US market in 1930. But the success of the A proved to be short-lived. By 1933, FMC claimed barely 22 per cent of the seriously depressed US market. Henry Ford's response—the V-8—brought the figure back up to 30 per cent in 1935. But the company, now showing the full effects of its deteriorating organization as its founder grew more senile and as Bennett gained power, declined steadily. By 1941, when its manufacturing plant was converted to production of matériel for

World War II, FMC enjoyed less than 20 per cent of the US market. (See Figure 3.1.)

During the war, FMC manufactured tanks, jeeps, trucks, aircraft engines, amphibious vehicles, gun mounts, and B-24 bombers. At war's end Ford stood eighth among US corporations in terms of the value of products it had manufactured for the conflict. During this period, however, the Roosevelt Administration had grown so concerned about the stability of the company's leadership that it had seriously entertained the idea of nationalizing Ford Motor Company. The death in 1943 of the company's 50-year-old president, Edsel Ford, son of the 80-year-old founder, and the company's failures in meeting early production targets for the mass production of the B-24 Liberator bomber provided the immediate context for the Administration's considerations. Edsel's struggle with stomach cancer and subsequent death left a power vacuum in which intrigue and chaos reigned supreme. Although the newsreels showed Henry Ford once again taking command of his company for the nation's safety and welfare, in fact, Henry Ford was rapidly failing. So was his company.

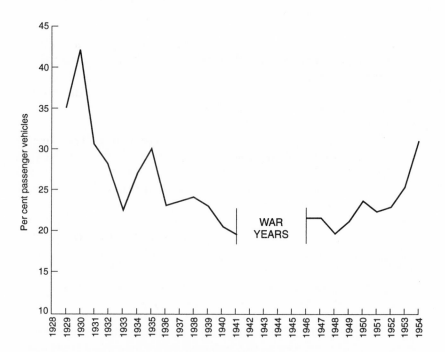

Fig. 3.1. *Ford's share of the US automobile market.*

Source: Ford Motor Company Annual Reports, 1947, 1954.

Instead of system, order, and clear managerial hierarchies, FMC relied upon management by fear and intimidation, which served to undermine whatever manufacturing capabilities remained in the company. The struggles for control of the company possessed all the drama of a *Macbeth* or a *Julius Caesar*. Following Edsel's death, the protagonist Henry Ford II, namesake and eldest grandson of the company's founder, was relieved of his post in the US navy and positioned at Ford's sprawling River Rouge factory. Ostensibly assigned to work out how to improve the plant's miserable performance in fulfilling FMC's various government armaments contracts, the heir-apparent knew he had a more important task. He needed to discover who were his friends—and enemies—in the company. The drama did not end until his mother, Eleanor Ford, backed by her mother-in-law and wife of Henry Ford, threatened to sell the shares of the privately held FMC she had inherited from Edsel unless her eldest son was given control of the company by the increasingly incapacitated Henry Ford. The first Henry Ford finally abdicated 21 September 1945.

Henry Ford II (hereinafter alluded to simply as Ford) began immediately to rebuild the company. Within minutes of acceding to power, Ford fired Harry Bennett and pledged to end the reign of terror that had prevailed. He assembled a temporary team of managers to help him cope with the enormous problems of reconversion to peacetime production. One of the team, John R. Davis who had been banished to California by Harry Bennett, recounted the state of Ford Motor Company at that time: 'when young Henry came in here the Company was not only dying, it was already dead, and *rigor mortis* was setting in.'[6] Ford also hired John S. Bugas, the head of the FBI's Detroit Office, to take charge of Industrial (i.e. Labour) Relations. At the same time young Ford sought help for the long-term governance and strategy of the firm.

Ford's uncle, Ernest Kanzler, a successful banker who had tried unsuccessfully to turn the FMC into a modern corporation near the end of the Model T era,[7] provided him with critically important guidance. A director of the General Motors-affiliated Bendix Aviation Corporation, Kanzler guided Ford to Bendix's president, Ernest Breech, to help rebuild the company. Trained formally in accounting, Breech had been with General Motors (GM) since 1925 when GM acquired the Yellow Cab Manufacturing Company for which he worked. By 1939 he had become a vice-president of GM, having done stellar work in the company's North American Aviation Division. He then became president of Bendix—and thus a senior vice-president of GM—during the war. Breech had earned a reputation in GM for his abilities as a manager of turn-arounds—taking poorly operating businesses and making them into efficient and profitable enterprises. At age 49, Breech had recently seen GM's board of directors pass over him when it named Charles E. Wilson to the presidency. Initially Breech had turned down Ford's invitation to become the number two man at Ford Motor Company but agreed to consult with Ford only to soften the rejection and to retain the company as an important customer of Bendix. But after examining the crude way in which the company kept its books ('about as good as a small tool shop would have') and realizing that FMC was losing $10 million per month, Breech signed on.[8] He recognized a good challenge—and opportunity—when he saw one.

Henry Ford II hired Breech to help him transform the Ford Motor Company into a General Motors-type organization. In a single stroke he also brought to FMC ten Army Air Force officers destined to be called 'the Whiz Kids' because of their statistical and organizational wizardry gained during the war in the Office of Statistical Control. Two members would go on to become president of the company and four others vice-president.[9] Ford, Breech, and the Whiz Kids worked furiously simply to acquire a sense of the company's baroque organization because the company's founder had contemptuously avoided organization charts, titles, and other trappings of managerial organization and hierarchies. During this period Ford's team turned to Peter F. Drucker's *Concept of the Corporation* as a kind of Bible for true believers in the miracles of what has become known as the M-form of the modern corporation—that is, the decen-tralized, multi-divisional firm.[10] GM had invited Drucker to study its management, and out of this experience he wrote *The Concept of the Corpo-ration*. Published in 1946, this now-classic book reified not only GM's organizational structure and GM itself but also modern managerial capi-talism. Drucker's book served as the inspiration for Ford and the Whiz Kids; Breech, who knew the book and brought copies with him when he joined Ford, did not need to resort to it, for he had first-hand experience with the GM system.[11] Breech also began to recruit others from GM to round out the management team that he and Henry Ford II were assembling.

Breech hired three men to fill top executive positions at Ford Motor Company, all with GM experience and thus critical in the restructuring of the company along GM lines. Lewis D. Crusoe had worked with Breech at Bendix and would serve initially as vice-president of operations (i.e., Breech's assistant), then as head of the division of planning and control, then as controller, and by April 1947 as vice-president of finance. He would then become the first head of the newly created Ford Division of the decentralizing FMC. Crusoe's experiences as an employee of Fisher Body Company (later a division of GM) and assistant treasurer of GM gave him solid grounding in GM's system of financial controls and accounting. Breech believed that in addition to finance, engineering constituted one of the weakest areas of FMC. To strengthen this function, he hired Harold T. Youngren, who had worked as an engineer in GM's Oldsmobile Division and was then serving as chief engineer of the Borg–Warner Corporation, a major supplier to GM and the automobile industry. Breech also recruited a new vice-president for manufacturing, Delmar S. Harder, who was serving comfortably as the president of the E. W. Bliss Company, the nation's leading manufacturer of stamping presses, one of the most important pieces of equipment in automotive manufacture.

Like Breech and Crusoe, Harder had also begun his career with the Yellow Cab Company in 1912. Two years later he moved to Chevrolet, where he eventually became superintendent of that company's New York plant. After Chevrolet became part of GM, Harder was moved to GM's tractor division in Janesville, Wisconsin, but he soon joined William C. Durant's new automobile company, serving as superintendent of Durant Motor's Oakland, California, plant. In 1928, Harder took a position as the manager of the E. G. Budd Company's Detroit plant, and four years later he assumed supervision of Budd's Philadelphia plant as well. (Budd supplied bodies to a number of automobile companies, including Ford.) But in 1934, Harder returned to GM when he became the general manager of the Fisher Body Division's fabricating section, a position in which he remained until 1945, when he took the presidency of Bliss.[12]

Thus Harder possessed a wide variety of experience in automobile manufacture, but his expertise clearly resided in the area of body production and sheet steel stamping technology. After 'automation' became a buzzword in the 1950s and definitely associated with Harder's work at FMC, he would claim that he had first coined the word in 1936 while at Fisher Body to describe automatic loading and unloading of work pieces into and out of stamping machinery.[13] With eleven years of recent experience in a GM division, Harder also knew a good deal—though certainly not everything—about the GM system of management. He would remain with FMC until his retirement in 1961. During his tenure he would preside over not only the decentralization of FMC's manufacturing operations, but more importantly the dramatic expansion of the company's productive capacities, which included thirty-three new manufacturing and assembly plants totalling some forty million square feet of floor space (an area more than twice the size of the company's famous Rouge Plant). During his tenure, the company also acquired some 33,000 machine tools, many of which became part of the automation programme that Harder initiated.[14]

Harder found the manufacturing base of FMC seriously eroded when he assumed his new position. Charles H. Patterson, later vice-president of manufacturing who had spent most of the war years at the company's Willow Run bomber factory, returned to the Rouge Plant in 1945 and was 'shocked' to find it 'wrecked' and in a 'state of deterioration'.[15] Once a show-piece of pure Fordist production methods where iron ore came in one end and finished automobiles rolled out the other, the Rouge Plant had been allowed to run down during the war because of the struggles for power in the company. As one of the first published annual reports of the company put the matter, 'for five of the past ten years the Company has used up plant, property, and equipment faster than it has replaced it, leaving a substantial capital deficit. . . .'[16] Judging by the situation at the Rouge, Patterson noted, FMC 'was a gone goose'.[17]

3.2. Ford's Postwar Labour Situation

If the company had used up capital during the Bennett regime, it had also exhausted the goodwill of its workers. The image of Bennett's goons in the Ford Service Department beating up Walter Reuther and other organizers from the United Auto Workers in the infamous 'Battle of the Overpass' in May 1937 is deeply ingrained in the annals of organized labour in the United States. Yet after a long and bitter struggle within labour's own ranks, Reuther and the UAW had ultimately triumphed. Following a bloody but successful strike at the Rouge Plant in April 1941 the union won a contract more generous than the contracts its sit-down strikes of 1937 had yielded from Chrysler and GM.[18]

Although FMC had capitulated to its 78,000 Rouge workers, under the continued Bennett regime it had not altered its management practices. Not long after moving from GM to Ford, Ernest Breech commissioned a survey of all FMC employees to find out how they viewed the company. The results showed overwhelmingly that Ford employees were not happy with the way they had been treated by their managers, supervisors, and foremen. Breech had observed the rough treatment of workers at the Rouge and found it abhorrent. 'I don't think that any company can operate in a competitive system,' he told FMC's superintendents and division heads, 'when all their relationships are predicated on fear . . . [and they cannot] build their products as cheaply as they could build them if they had good human relations.'[19] Breech pushed hard to end management by fear at Ford.[20]

During the war, the company had experienced some 750 unauthorized work stoppages. Even before taking the reins of the company, Henry Ford II had witnessed the effects of these stoppages. Thus Ford also knew that in the postwar era, he and his company must address issues of labour management. In early 1946, while negotiations between Ford Motor Company and the UAW were taking place, he delivered a speech to the Society of Automotive Engineers entitled 'The Challenge of Human Engineering'. Widely hailed by UAW officials for the industrial statesmanship displayed in this speech, Ford acknowledged that 'Labour Unions are here to stay' and promised that Ford Motor Company had 'no desire to "break the unions. . . ."'[21] Yet he demanded that the UAW take responsibility for curbing the unauthorized strikes of its members. Ford's challenge echoed what his Industrial Relations head, John Bugas, was proposing in the negotiations—initially a provision to ensure employer rights by fining the union $5 per day for each worker who engaged in unauthorized work stoppages. This specific proposal fell by the wayside before the new contract was signed in late February 1946. The union, however, acknowledged the company's need for 'security' against work stoppages and granted the company the right to discipline (including discharging) those who engaged in work stoppages. It also gave Ford the unquestioned right to set and enforce production standards.[22] Yet in spite of its provisions and a pay hike of 15 per cent, the contract did not stop all unauthorized work actions from occurring at Ford.

The following year FMC and the UAW began early contract negotiations not long after the company, aided by the passage of the Taft–Hartley Act, crushed the 4,000-strong foreman's union, the Foreman's Association of America. FMC had recognized the FAA in 1942 after the firm recognized the UAW, but the FAA had had little support from the UAW itself.[23] In spite of the lost man-days from the foremen's failed strike, FMC experienced markedly reduced lost time in 1947 as compared to 1946 (and significantly less than during any of the last five years of the Bennett regime).

Industrial relations at FMC continued to improve during 1948, when only 5,620 man-days were lost through non-authorized work actions, but in May 1949 FMC experienced a debilitating strike at its Rouge Plant and at the much smaller Lincoln Plant. The 24-day strike erupted over the issue of production standards. The union charged the company with a 'speedup', while the firm insisted on its contractual right to set—and enforce—production standards. Costing the company more than 1.7 million man-days at a time when making cars was critical to the company's profitability and market position, the strike greatly upset FMC's top executives. Even before the strike, the company had been unable to make enough cars to satisfy the extraordinary postwar demand for its products. The results of the strike proved all the more disconcerting to them, as is suggested only cryptically in the annual report for 1949:

The major issue of the dispute was finally settled by arbitration, as the Company had proposed repeatedly both prior to and during the strike. The arbitration award did not impair the Company's right to establish and enforce reasonable production standards, but did establish certain general principles to be followed in establishing and applying production standards on assembly operations.[24]

Any encroachment on the company's prerogatives *vis-à-vis* production standards posed a serious threat to this production-driven company.

Labour historian Nelson Lichtenstein has suggested that Ford worked actively to 'recapture' authority on the shop floor by decentralizing pro-

duction—by moving production away from the Rouge Plant, which har-
boured the most militant segments of the Ford labour force—and by
implementing automation as a way of deskilling and displacing workers
at the Rouge. He also argues that Ford was aided by the UAW itself,
which focused its efforts on winning concessions of money and fringe
benefits from Ford, and from the Cold-War-bred fears of communism,
which undermined UAW Local 600 with its militant communist
leadership.[25]

Without question, as will be examined below, Ford both decentralized
and automated much of its production. But the primary evidence left by
top executives shows no overt strategy to undermine labour at the Rouge
as the primary motivating factor in decentralization and automation.
Although concerns about labour power were unquestionably part of top
management's calculus, FMC's steps toward decentralization of production and automation can be explained more fully by other factors at work
at the company in the period after 1945. One of these factors was the
degree of integration of FMC.

3.3. Vertical Integration and Product Planning, 1946–52

Almost as soon as manufacturing expert Del Harder arrived at FMC and
began building his staff, many of whom he recruited from General
Motors, he identified a fundamental problem with Ford's manufacturing
system—the lack of integration. Although the Rouge had once been the
most integrated manufacturing plant in the automobile industry, Ford
Motor Company had since the changeover from the Model A to the V-8 in
1932 come to rely increasingly upon external suppliers for many of its
major components.[26] Harder concluded that Ford could never succeed
with this strategy. As Harder's assistant, John Dykstra, a former GM
production expert, remarked, 'Ford purchased bodies, stampings, axles,
drive shafts, transmissions; it had no manufacturing facilities except at the
Rouge. . . . Ford had bought too much outside. If the company was to be
competitive, it must become more integrated.'[27] Thus, pursuit of more
extensive integration of components manufacture became a priority at
Ford after 1946 and shaped the company's decisions that lay behind the
Cleveland Engine Plant. Another and closely related priority also contributed to the company's actions: product planning.

While seeking to get the FMC back into production of automobiles after
the war, Henry Ford II and Ernest Breech looked to future product design
as yet another way to revive the beleaguered company. Ford deferred to
Breech and the company's new chief engineer, Harold Youngren. Breech,
as well as Ford, had been captivated by the possibility of introducing a

small economy car, both to serve perceived wants in the market-place and, as important, to help achieve a greater degree of organizational decentralization along the lines of GM. They created a Light Car Division in April 1946, hoping that FMC might be able to introduce such an entirely new economy vehicle in 1947, to be produced and sold by an autonomous division of the company. But after Youngren officially joined the company on 1 August 1946, they abandoned the idea of a light vehicle for 1947, largely because of design problems with the Ford automobile (the mainstay of the company's product line) and its cousins, the Mercury and the Lincoln.[28]

Breech believed that the design of FMC's cars was poor. By August 1946 he had concluded that the company needed to start from scratch. For the time being, the Policy Committee—the 'team' that Ford and Breech had assembled to recreate Ford Motor Company[29]—determined to use the face-lift planned for the Ford as the company's Mercury product and to stick with the existing Ford until an entirely new Ford car could be brought out through a crash development programme. By December 1946, Youngren's staff had readied various clay models for the Policy Committee's review and selection. One of the models met with the committee's praise and approval. Two months later Harder promised that this new Ford would be in full production by February of 1948. But problems in securing machine tools and various production materials (which, in general, plagued all the automobile companies during 1947) slowed the new car's public introduction until June 1948. Even with the delay, FMC was the first of the Big Three to come out with a thoroughly redesigned, popularly priced automobile for the booming postwar market.[30]

The car's design proved to be sensational. Introducing the car at New York's Waldorf Astoria Hotel, however, the company warned that because of the pent-up demand for automobiles and the problems of Ford's decayed manufacturing infrastructure, the average buyer might face a two-year wait for a new Ford.[31] Writing in his president's letter for the 1947 Annual Report—dated in August 1948—Henry Ford II claimed that the 1949 Ford ranked with another great changeover in the Ford Motor Company's history—from the Model T to the Model A.[32]

Outwardly the new Ford car appeared revolutionary; it contained a new chassis and a new body without any running boards. But the 1949 Ford was not entirely new. The power plant continued to be the standard Ford V-8 that had kept the company afloat during the 1930s. It was manufactured in the company's Motor Building at the Rouge Plant, which, as will be discussed in another section of this paper, Harder judged to be obsolete.

While the 1949 Ford was being designed and readied for production, the issue of the light car continued to simmer. Some top executives' fascination with the light car made planning for the post-1949 model year very difficult.[33] In part because of the confusion over the light car and in part because of the need to figure out what other products lay beyond the 1949 Ford, Henry Ford II and Ernest Breech created what eventually became known as the Forward Product Planning Committee. This body reported organizationally to the Policy Committee. Its fundamental mission was to determine both short-run (next model year) and long-range (four model years hence) characteristics for FMC's products. As part of the Forward Product Planning process, discussion of engines for future models became a driving force in process innovation at Ford. By mid-1948, the implications for products planned through the 1951 calendar year had become clear, such that Ford asked his Finance Department to conduct a thorough, cross-company study of FMC's proposed engine program. The study required the concurrence of the three departments involved most heavily in the engine program—Engineering, Manufacturing, and Finance.

Under the leadership of Finance Vice-President Lewis Crusoe, the forward engine programme review made it clear that, even though the Light Vehicle Division had ostensibly been killed in September 1946, the engine for the small car continued under development—as though its production were inevitable. The reason for this was because the engine—dubbed the XA-1—possessed a block that was common with the six-cylinder overhead valve engine—called the XB-1—planned to replace the V-8 in Ford automobiles.[34] Therefore, in planning for a thoroughly re-engineered Ford (which was eventually first shown as the 1952 Ford in the second quarter of 1951), production of the XA-1 always seemed to make sense because of the inherent scale economies in the engine foundry and machining operations. Not until Henry Ford II and Ernest Breech created the semi-autonomous Ford Division in February 1949 and named Crusoe to head it was the matter of the light car removed from the forward engine planning process.

Crusoe was adamantly opposed to the idea of a light car. He was also intent on establishing the independence of his new division. Thus very early in the Ford Division's history, he eliminated any discussion of a small car in his division's plans (although he would periodically have to slay this dragon that dogged him from above). Once freed of the light car, planning then could proceed on finding a way in which the XB-1 six-cylinder Ford engine could be produced.[35]

Although successful in establishing enough autonomy in his new Ford Division to deny the light car, Crusoe failed when it came to the sphere of production. He tried hard to convice Ford, Breech, and others that his division should manage its own production facilities. The company's initiatives (discussed more fully below) in building a stamping plant outside Detroit and the apparent necessity of building at least one new engine plant beyond the Rouge seemed like the perfect way to create real autonomy for the new division.[36] But the strong feelings of the company's marketing people that the 1952 Ford should be sold with an optional V-8 (presumably that would continue to be produced at the Rouge Motor Building along with the engines of other FMC products such as trucks, tractors, and Mercuries) seriously undermined Crusoe's case for having his own engine plant.[37] Organizational responsibility for the proposed engine plant became highly contested terrain. Crusoe was not the only division manager who wanted to make his own engines. The general manager of the Lincoln-Mercury Division (a division that Henry Ford II had created even before Ernest Breech joined the company) had earlier proposed that the Lincoln-Mercury Division be split up and that both the new Lincoln Division and the Mercury Division be responsible for their own production facilities, thus carrying the principles of decentralization of management to their logical conclusion.

To explore the implications of the fissioning of the Lincoln-Mercury Division and the more general questions of production facilities for the new products being planned within the company, Breech and Ford created a Special Facilities Committee and placed Del Harder at its head.[38] In announcing the appointment of this special committee, Breech stated how bound together the issues were:

The consideration of forward product planning by the Ford Motor Company is unalterably involved with the consideration of the available manufacturing facilities. The number of lines of cars that it may see fit to manufacture in the future involves not only our dealer organization, and its adequacy to handle certain lines of cars, but the adequacy of our manufacturing facilities. Not only is the adequacy of the manufacturing facilities an important consideration but, likewise, location of those facilities is of equal importance.[39]

Thus, new product design—although tangled up with issues of organizational responsibility in the decentralization of FMC, its commitment to achieve greater vertical integration, and the decentralization of production away from the Rouge—drove the decision to build a new engine plant. As FMC's executives saw the matter, a new line of engines was being created for a new line of products; thus a new plant was needed. Their reasoning was shaped by the knowledge of their company's inability to manufacture all the '49 Fords it could sell in spite of its plants running at capacity.[40] The Rouge Plant's Motor Building was fully engaged. How could the manufacturing units be expected to plan for the production of entirely new engines targeted for the '52 models while also meeting current demand? Although Ford liked to compare FMC's major model changes during this period with the great changeover from the T to the A, he would not consider shutting down the company as his late grandfather had done in 1927 when he changed over to the Model A.[41]

Thus the idea of a completely new engine plant in which to build a completely new engine that could gain market acceptance while the old V-8 was being phased out proved to be highly appealing.

3.4. Integration, Decentralization of Production, and Automation

Along with new product design, the forces of integration and manufacturing decentralization pushed the Ford Motor Company to build what became the Cleveland Engine Plant. A critical prelude to this highly touted factory, however, was the company's new stamping plant built in Buffalo, New York. Del Harder's deep experience with the Budd Company, Fisher Body, and Bliss explains why FMC's first major step in the decentralization of production lay with body parts production. It also illuminates why the firm's first significant step in automation occurred in its new stamping plant.

Authorized in mid-1949 and completed in 1950, the Buffalo Stamping Plant was built to lessen FMC's dependence upon outside suppliers,[42] to shift some production out of the Rouge Plant, to provide the firm with capabilities to handle new models of automobiles, to lower production costs, and to handle expected capacity increases. No single factor predominated in FMC's decision to build the plant in Buffalo. Yet the former GM executives who now managed much of FMC's business were steeped in GM's response to the problem it had encountered during the 1930s in south-eastern Michigan, namely the sit-down strikes in Flint that had forced the company to recognize the UAW. GM had undertaken a programme of dispersing plants to counter labour power and, ostensibly, to reduce transport congestion in the Flint/Detroit region.[43] As Harder, Breech *et al.* were well aware, GM had satisfactorily relocated some of its production to Buffalo, New York.[44] The rubber industry had pursued much the same strategy as GM to deal with sit-down strikes it encountered in the 1936–8 period.[45]

Plans for the Buffalo plant, whose cost was estimated at $35 million, also called for a major overhaul of the Rouge's stamping plant—at more than $11 million—and shifting the mix of products made there. Harder's department estimated that some 3,600 workers would be added to the company's payroll in Buffalo, while almost 3,000 would be laid off at the Rouge. Thus, for a total increase of about 600 workers, or fewer than 7 per cent of the total labour force in the pressed steel area of the Rouge, FMC would be gaining an additional 33 per cent of its pressed steel needs. When netted out, this plan would save the company roughly $7.2 million per year—or a 27 per cent return on investment.[46]

The Buffalo Stamping Plant not only promised to save money, but it also served to inspire confidence in the work being done in the Automation Department, a unit that Harder had established not long after assuming his duties at FMC. In both naming this department and defining its mission, Harder had laid much of the groundwork for the automation wave that passed through American manufacturing in the 1950s. When he created the Automation Department in April 1947, Harder defined 'automation' as 'the automatic handling of parts between progressive production processes'.[47] By October 1948 the department employed about fifty staff members who were responsible for identifying 'parts-production lines where it is obvious that only a few operators are needed to control the [production] equipment, where the majority [of] work [is] feeding and unloading, turning parts over and stacking them'. Three criteria were paramount in the department's work. First, output had to be increased, and second, 'probably [sic] cost of the device will not exceed $3,000 per man transferred [i.e., displaced by a particular piece of automation equipment]'. (The average weekly earnings for a US auto worker in 1947 were roughly $55, which, assuming a fifty-week work year, meant annual wages of $2,750.[48]) Third—and determined in large measure by the second criterion—'automation *tooling* must be amortized in one year because of obsolescence by model changes.' *American Machinist* reported in late 1948 that during its eighteen months of existence, the department had 'produced approved designs for over 599 devices [that were expected] to cost $3,000,000'.[49] Most of these devices were designed to load and unload stamping presses making body parts and components made from sheet steel. Thus, when the Buffalo Stamping Plant began operations, it contained hundreds of automation devices for loading and removing work from the huge presses and machinery to carry work pieces from one press or operation to the next.[50]

The automation work at FMC during Harder's early tenure was inspired by work done at GM—work about which Harder and the former GM production experts he hired no doubt knew. Some of FMC's automation devices were, in fact, produced under licence from GM.[51] During the 1930s, as GM experienced increasingly militant workers at some of its plants, the company, in addition to dispersing its production plants, turned increasingly to automatic machinery in an attempt to negate those troublesome workers.[52] Immediately after the war, GM installed transfer machines in its Buick engine factory.[53]

A 'transfer machine' was an assemblage of single or multiple spindle machine tools ('stations') that were linked together such that they could carry out several steps sequentially and automatically without human intervention. They were first developed and used in the American watch industry in the late nineteenth century, owing to the pioneering work of the Waltham Watch Company's Duane H. Church.[54] But transfer machines were not found in the American automobile industry until the 1930s, when Ford and GM began to experiment with them in conjunction with a few machine tool manufacturers. From 1923 to 1925 Frank G. Woollard and the British company, Morris Engines Ltd. of Coventry (later the engine division of Morris Motors), did large-scale development work on transfer machines for automobile engine blocks, gearbox castings, and flywheels. But, as Woollard recounted, 'in the state of the art at the time, they proved in practice to be over-complicated. They were, therefore, divided into individual machine units and the automatic system was abandoned, to be revived [in the USA] 20 years later.'[55] In spite of this British failure, transfer machines became the 'hot' item of immediate post World War II US machine tool and automobile industries.

FMC had also done a limited amount of experimentation during the 1930s with building and operating transfer machinery. According to William Abernathy, FMC tool designers built a two-station transfer machine in 1932 and four years later employed a seven-station transfer machine, manufactured by Baird, to machine part of the Ford car's drive train.[56] James Bright described FMC's transfer machine design of the mid-1930s as a type of 'functional specialization' because the transfer machines coupled machine tool operations performing common functions (e.g., drilling and tapping holes or rough boring and final milling of holes).[57] Whether anyone in Harder's Automation Department had been involved or not in the transfer machine work of the 1930s is not known. But FMC's leading production people definitely would have known that in the immediate postwar period the machine tool industry was making major commitments to the development of transfer machinery for the automobile industry.[58]

Not long after Harder arrived at FMC, he ordered a major rearrangement of the Motor Building at the Rouge Plant. This reconfiguration of the highly crowded Rouge engine plant entailed the installation of numerous new machine tools, including transfer machines.[59] The project convinced Harder that, whatever the particular outcome of the company's forward engine programme, FMC desperately needed new facilities in which to produce engines. Even with the rearrangement of old machinery and the installation of the latest machine tools the Motor Building was, in Harder's mind, simply obsolete.

3.5. Planning the Cleveland Engine Plant

Writing in the mid-1950s after conducting extensive interviews with Ford Motor Company's manufacturing executives, including Del Harder, James Bright recounted that, in late 1948, Ford's senior management asked the manufacturing engineering department to 'give us the most modern, efficient plant in the country' to make engines.[60] The department responded with two alternative plans. The first projected a plant using off-the-shelf materials handling systems and machine tools. The second described a plant that would feature the automation devices being designed by Harder's Automation Department—devices that would automatically load and unload work pieces into and out of production equipment (mostly transfer machines) and that would automatically carry out other functions, including inspection, gauging, and weighing, that hitherto had been done by workers.

Bright's account conforms roughly but not precisely to the manuscript records that Ford's top executives left from that period. When, under the aegis of the Forward Product Planning Committee the heads of the Engineering, Finance, and Manufacturing departments began the thorough review of the company's forward engine programme in the autumn of 1948, they established five 'first principles' to guide the formulation and execution of the programme. One principle simply stipulated which new engine would be built first (as already discussed), while the second stated that the daily capacity figures were derived from estimations of anticipated sales of the new models. Importantly, the third and fourth principles served to shape the thinking behind what became the Cleveland Engine Plant. The third principle specified that engines 'would be processed without limitation by present practice'. That is, the new engines would be made using pioneering production processes in an ideal plant. Fourthly, the manufacturing process developed to make the new engines 'would be made without regard to present facilities' (i.e., the Motor Building at the Rouge). For the purposes of establishing high and low investment information, the cost of this 'ideal' production process would then be estimated if installed in an entirely new plant (high) and if 'fitted into existing buildings with a minimum of outside plant' investment (low). The three executives' final principle was that FMC should aim at developing capacity to make 100 per cent of its engine requirements.[61] This goal thus departed from GM's strategy, which was to manufacture only part of its own components, thereby positioning the company to absorb swings in demand and to lower its risks of being shut down by its own workers.

As reported in December 1948, initial estimates suggested that the 'maximum' new plant scenario—the 'complete relocation' of engine manufacture requiring new foundry, new machining facilities, and new assembly areas totalling some 3.7 million square feet—would cost $157 million. The 'minimum' plan, costing $123 million, would involve the reworking of existing plant but would still require 1 million square feet of new manufacturing space to achieve the plan's targets.[62]

Roughly two months later, the Forward Product Planning Committee was still trying to work out what the company should do with its engine programme. Harder stated categorically that no matter what the sequence of new engine introductions would be, the company needed a new engine factory. He argued that the 'present motor building [at the Rouge] is badly overcrowded'. To improve its efficiency, more space was required. Moreover, additional space would be imperative to handle a changeover to new engine types 'without either a shut-down or serious interruption of engine production'.[63] Harder's plea must have worked, for at the same meeting, the committee resolved that a new plant should be built to manufacture the new overhead valve six with a standard daily output of between 2,000 and 3,000 engines.

Despite this early resolve, however, the top management at FMC continued to contemplate the forward engine programme and the need for a new engine plant. The matter was intimately bound up with other issues faced by the Policy Committee, which have already been discussed. The actions being taken by FMC's competitors—especially Chevrolet, the main target for the company—also conditioned the committee's thinking. Finally, in November 1949, the Facilities Committee[64] reported on the engine programme to FMC's Executive Committee, which had by this time become the highest level decision-making committee in the company. Harder, head of the Facilities Committee, made the presentation and distributed his committee's report to members of the Executive Committee. It considered four alternative plans but endorsed only one of them, which after substantial discussion was approved by the Executive Committee. The plan called for the demolition of the Rouge Plant's old Motor Building, the relocation and upgrading of its machinery to the Rouge's Parts and Accessories Building, and the production of 4,135 Ford V-8 engines per day there. The report also projected the construction of two new engine plants, one to make the new overhead valve Ford 6 (XB-1) at the rate of 2,205 per day, and the other to manufacture new overhead valve V-8s for the Mercury (1,973 per day) and the Lincoln (376 per day). (The Lincoln-Mercury Division had recently killed the idea of adopting a bored-out six for the '52 Mercury, convincing executives that a new, hot overhead valve V-8 was necessary to meet the competition.) The existing foundry at the Rouge would continue to supply the relocated engine plant there, but Harder's committee called for the construction or purchase of a foundry to supply the castings for the two new engine plants. Thus, this plan allowed for control over the two new engine plants by the Ford and Lincoln-Mercury product divisions, respectively, but it posed the question about who would manage the foundry.[65]

The ink had barely dried on the Executive Committee's approval, however, when Robert S. McNamara, one of the Whiz Kids who was now FMC's Controller and who had been a member of the Facilities Committee, put forth an alternative idea—'Plan A.' It effectively questioned the assumptions under which the Facilities Committee had formulated its report. McNamara's plan preserved the assumptions of organizational separation (i.e., 'engine machining and assembly must be segregated by end product division without regard to engine interchangeability or relative operating costs of different plant sizes') but shifted the production of the new Ford 6 back to the Rouge. It thus violated one of the committee's explicit criteria, which was to lower the Rouge's production to 50 per cent of the company's total engine needs.[66] McNamara suggested that the company simply purchase castings for the new Ford 6 as well as those for the new Lincoln V-8. With some minor rearranging, he argued, the Rouge's foundry could produce the castings for the new Mercury V-8. He also suggested other alternatives to the approved plan, including using the company's Detroit-Lincoln plant for engine manufacture. Plan A, McNamara maintained, would save the company some $46 million in investment and manufacturing costs.[67] McNamara's proposal threw Ford's top executives into a spin and put Harder very much on the defensive.

The Executive Committee met in early December 1949 to reconsider its earlier decision in light of McNamara's Plan A. It summoned a large number of interested parties and the members of the Facilities Committee, including McNamara, to attend the meeting. Harder first presented his department's assessment of costs associated with using the Detroit-Lincoln plant versus building a new engine plant, which demolished McNamara's assumptions. The committee agreed with Harder that further pursuit of the Detroit-Lincoln plant idea would be 'inadvisable'.[68]

Harder then distributed—and read—a new memorandum to the Executive Committee in which he reassured the committee members that the plan they had approved a month earlier had been well formulated. The Facilities Committee, Harder stressed, had carefully developed twenty alternative plans, of which he had presented only the four most attractive ones. He reiterated that the committee's decision to build two new engine plants was sound. Presenting locational data for the two new plants, Harder then asked the committee to make a decision about where the plants would be built. The choices included Detroit, Chicago, and Cleveland and combinations thereof. With these choices Harder also presented cost estimations that included site costs, tax burdens, and freight rates but did not include wage rates and 'labour efficiency' figures in each of the cities.[69] As the minutes of the meeting note, discussion then 'ensued'.[70]

When the meeting adjourned, the committee had determined to build only one engine plant (complete with foundry) and to locate it in Cleveland. But unlike either the previously approved plan or McNamara's Plan A, the Cleveland plant would be built to manufacture 4,000 to 4,500 engines a day—essentially the combined daily output projected for the two previously approved plants. Scale thus became the overriding factor in the Cleveland plant. 'The actual mix of engines to be manufactured in the new plant . . .', note the minutes, 'would be the subject of further study'. The decision to build a single plant unquestionably threw out the idea of the product divisions manufacturing their own engines and with it the principle of divisional control of manufacturing assets. The Executive Committee immediately deferred 'consideration of the question of whether or not to make the Buffalo Stamping Plant a part of the Ford Division'.[71] Ultimately, the company's decisions about production would lead to the vesting of organizational responsibility for both the Buffalo Stamping Plant and the Cleveland Engine Plant in a 'group' of the manufacturing department named the Engine and Pressed Steel Group.[72] FMC thus remained production-oriented, rather than product-oriented. The Executive Committee also approved Harder's recommendation to raze the Rouge's old Motor Building and to relocate and upgrade engine production to the site's Parts and Accessories Building.

Immediately, Harder's department moved into action. By mid-January, the company had purchased the Cleveland site—200 acres in Brookpark, Ohio, a community adjacent to the Cleveland Airport. Given that not even Harder yet knew exactly what would be made at the new plant except 2,000 Ford six-cylinder overhead valve engines per day, FMC's press release was notably vague in its details about the Cleveland product. It simply said the new plant would make about 4,000 engines per day.[73] Interestingly, the company also said nothing about 'automation' or advanced manufacturing technology in the release. The company broke ground for the engine plant in May 1950 and for the foundry a month later. In December 1951, production of the six-cylinder Ford engines began (using purchased castings because the new foundry was still not finished). Production at the 1.25-million-square-foot foundry started on a pilot basis in June 1952 and reached the volume needed for the Ford 6 shortly thereafter. In October 1953, the Cleveland Engine Plant began making the new Mercury overhead valve V-8.[74] By this time the company had already committed itself to investing in yet a second engine plant at the Brookpark site, this one to make a new Ford V-8 overhead valve engine whose introduction was anticipated for the 1956 model year.[75]

3.6. The Cleveland Engine Plant and Its Sisters

When it opened, the Cleveland Engine Plant was hailed in the technical literature as 'a veritable triumph of production planning, presenting the latest expression of the Ford philosophy of Automation—automatic handling of work in process and its transport from receiving, through the automatic transfer machines, and finally to the assembly, testing, and shipment of engines'.[76] (See Figures 3.2–3.6.) Almost all the machine tools in it were automatic, multi-station transfer machines manufactured by machine tool firms, including Ex-Cell-O, Cross, LeBlond, Snyder, Sheffield, Sundstrand, and W. F. & John Barnes. These transfer machines were state-of-the-art for the industry but by no means exclusive to FMC. What made the Cleveland Engine Plant so interesting to contemporaries was the way in which Ford had linked these transfer machines together by automatic conveying equipment such that little human assistance was necessary for conveying, loading, or unloading work in process. Also numerous labour-and skill-intensive processes, such as weighing, dynamic balancing, and gauging, had been automated and integrated into the production sequence. Handling devices allowed parallel or multiple machining tasks to be fed automatically, thus achieving what process engineers call 'balance' in machining capabilities. All these automatic

Fig. 3.2. *Block 'cradle', Cleveland Engine Plant, 1952. This photograph shows a good example of the many devices that FMC's 'Automation Department' custom-built to handle work automatically. The cradle both conveys the engine block at a ninety degree angle and turns it over to position it for further machining.*

Source: Ford Motor Company.

Fig. 3.3. *Block machining line, Cleveland Engine Plant, 1952. Note how the conveyor is able to move the blocks at right angles.*

Source: Ford Motor Company.

conveyance devices were controlled by a hard-wired control panel of telephone-type relays—an electromechanical 'brain'. The system included a series of interlocks to ensure correct conveyance and feeding of parallel lines. With such a tightly coupled system, as James Bright quoted an unnamed Ford employee, 'Everything had to work or nothing worked.'[77] Assembly operations at the plant also relied extensively upon automated conveyance systems, which with their 'power and free' capability (i.e., provision to take up slack at any point in the line without having to slow down or stop the entire line) allowed for greater flexibility than the automated equipment in the machining areas.

The original authorization of the Cleveland Engine Plant called for expenditures of $47 million for the building, machinery, and tooling of the plant and another $33 million for the foundry. By May 1951, the company had to appropriate another $16 million for the engine plant and $7 million for the foundry. Many of the executives who worried about finances expressed concern about the lack of detail in Harder's requests

Fig. 3.4. *V-8 block transfer line, Cleveland Engine Plant, 1954.*
Source: Ford Motor Company.

and the large sums of money being expended on the engine program.[78] Bright reported in 1957 that the 'Ford automation program was criticized both inside and outside the company during 1950–1953. . . .'[79] But Harder and his team stuck to their plan, both at Cleveland and at the Rouge. The Controller's Office estimated in May 1951 that tooling up to produce the new engine for the '52 and '53 automobiles would run in excess of $200 million.[80] The test of Harder's strategy came, however, when the Cleveland plant started up and when cost data began to be analysed.

Surviving manuscripts contain virtually no evidence that the start-up of the Cleveland Engine Plant posed enormous difficulties. Bright's 1957 study suggests implicitly, however, that problems plagued the early days of the new plant. The most critical problem in the highly linked factory was down time, and thus the 'timing and/or elimination of maintenance downtime became a critical part of operating success'. Bright further noted, 'It took time and money to get rid of the "bugs". . . .'[81] Initial cost studies of the Cleveland Plant showed that unit manufacturing costs were higher for the new engine than anticipated. Planned unit cost for the

Fig. 3.5. *One of the 'brains' of the Cleveland Engine Plant, 1952. The plant was controlled by electromechanical circuitry.*

Source: Ford Motor Company.

Fig. 3.6. *Assembly line, Cleveland Engine Plant, 1952. Note that the arms allow the assembler to position the engine in different positions to reduce worker fatigue.*

Source: Ford Motor Company.

engine at the 383,159 'standard volume' of engines was $157, but the measured cost in March 1952 was $196 at an 'annual volume' of 182,256 engines. When adjusted for scale effects, the cost was still higher at $182 than what had been planned.[82] By May 1952, the plant had lowered its unit cost to $172 (corrected for scale effects), and by September the figure had reached $159 (while the newly calculated planned cost had been lowered to $138). In a detailed accounting of the Engine and Pressed Steel Division's costs prepared by McNamara's Controller's Office in November 1952, the automation programme came in for implicit criticism as McNamara's numbers pointed at the 'Efficiency Less than Planned at Cleveland Engine Plant' as the principal culprit in real costs exceeding planned costs. Actual overhead costs exceeded planned overhead costs by almost 22 per cent, and actual labour costs were 30 per cent over planned costs.[83] (It is interesting to note that planned labour costs were only $13 per unit,[84] and this comparatively small figure would eventually occasion a member of the Engine and Foundry Division to write that 'direct labour in today's automated plants offers limited opportunities for major costs reductions.'[85])

Measured engine costs, as calculated by McNamara's office, continued to dog the Engine and Foundry Division, especially after the Cleveland Engine Plant started up the production of the overhead valve V-8 for the Mercury and cost figures began to come in. The division had promised that, using automation, the new engine would cost $25 to $30 less than the standard-bearer V-8 engine ('L' Head). Instead, it came in costing some $23 more. As one member of the Controller's Office wrote to another in early 1954, 'the whole subject of the cost differential on engines has been such a touchy one for several years. . . .'[86] Controversy between Harder's and McNamara's respective offices about costs at the Cleveland facility might have occasioned an unusual visit to the facility by Ford's board of directors in October 1954 to judge the success or failure for themselves.[87]

In briefing the board members before their actual tour of the Cleveland complex, the Engine and Foundry Division took pains to argue that FMC had made a wise investment in automation. The division prepared charts showing that the Cleveland Engine Plant had moved squarely into the black in terms of its profit position since the plant was now producing both the Ford-6 and the Mercury V-8 overhead valve engines and was now projecting a return on investment of 25 per cent. To impress the board, the division marshalled other evidence. One chart, labelled 'Advantages of Automation', compared direct labour minutes for machining eight-cylinder blocks and heads with and without automation. The figures for blocks were 20 and 41 labour minutes, respectively, and for heads 14 and 26 labour minutes, respectively. Thus automation reduced direct labour minutes by 49 per cent. The same chart also showed the square feet of factory space required for block and head machining with and without automation, and again automation showed advantages of 17 per cent. Similar figures were presented for crankshaft machining operations using FMC's new shell-moulded casting for crankshafts rather than the industry-standard forged crankshaft. Output per hour of a Cleveland Engine Plant machine was 35 crankshafts compared to the competition's 15 per hour. The division even promised that this number would reach 60 per hour in the new engine plant, which was nearing completion, 'by installing completely automatic loading devices'. Thus, automation of crankshaft machining reduced the number of machines by 86 (the total number was not given), facility investment by $4.8 million, machining time by 20 per cent, and machining costs by $3.2 million annually. The

Engine and Foundry Division's presentation never took on the questions about total costs raised by the Controller's Office.[88]

Whether the Engine and Foundry Division deliberately deceived Ford's board of directors or whether the board had truly been concerned with the performance of the Cleveland Engine Plant is unclear from the evidence. But clearly the board had already committed major funds to the construction of a second new, automated engine plant to be located on the other side of the Cleveland foundry and to manufacture overhead valve V-8 engines for Ford cars and trucks with several different cubic inch displacements. Moreover, in April 1955—six months after the board visited the Cleveland plant—the Executive Committee approved yet another engine plant to manufacture hotter engines for the Mercury and Lincoln product lines, part of FMC's response to Detroit's 'horsepower race'.[89] Built at Lima, Ohio, this plant was originally projected to cost about $68 million, but by late 1955 its anticipated costs had risen to $86 million. They would go even higher by the time the plant was completed.[90] The Lima facility would advance the art of automation beyond the two Cleveland plants and by the early 1970s would contain extensive automated assembly processes as well as automated machining.[91]

FMC's need to construct the Lima Engine Plant stemmed from the rapid change in engine types that Detroit was moving through in the years after 1954. These changes also profoundly affected the two Cleveland Engine Plants and pointed up some of the problems of Detroit automation as pioneered at those plants. In early 1955, under Robert McNamara's watchful eye, the Ford Division sought to have the Dearborn Engine Plant, which had also made substantial commitments to automation (though not as thoroughly as Cleveland),[92] convert from the 272 cubic inch engine to a 312/331 cubic inch engine. At the same time, the division advocated shifting production of the Ford 6 from Cleveland Engine Plant no. 1 to Cleveland Engine Plant no. 2 and expanding the former plant's capacity for the overhead valve V-8. These additional V-8s would be only the larger bores produced there (292 cubic inches vs. 272 cubic inches but using the same casting) and would incorporate modified valve angles.[93]

These recommendations, which were approved, came at a heavy price. Changes at the Dearborn Engine Plant were projected to cost $65 million, while those at Cleveland Engine Plant no. 1 exceeded $43 million. The movement of the increasingly obsolete equipment built to manufacture the Ford 6 from Cleveland Engine Plant no. 1 to Cleveland Engine Plant no. 2 was projected to cost another $9 million, largely for an expansion to that plant to accommodate the extra machinery. By the time they had been completed, product changes approved in 1955 would cost FMC some $236 million for engine manufacturing plants alone.[94] The horsepower race exposed a major drawback to 'Detroit automation'.

3.7. The Call for Flexibility

Seeing almost immediately the impact of the horsepower race on their sphere, Harder and various manufacturing managers had begun to call for more flexibility in manufacturing technology. As early as 1954, the company's Manufacturing Engineering Office recognized that, in spite of all its benefits, automation 'has its limitations. It does not permit the same degree of flexibility as does the conventional type processing.'[95] D. J. Davis, whose name is often associated with Ford's automation programme because of his widely reprinted 1955 congressional testimony about Ford's automation programme, spoke on 'Trends in Manufacture' before the Armour Research Foundation in Chicago in early 1956 and 'stressed the serious problem of obsolescence costs resulting from complex specialized machine tools which are not adaptable to frequent product changes'. Davis called for development of more flexible production methods.[96]

A year later Davis reported to the Northern New Jersey Chapter of the American Society of Tool Engineers on 'New Manufacturing Horizons'. Again he noted that Ford had been 'plagued with high obsolescence in machine tools' and attributed much of it to the development and widespread adoption of the transfer machine. Ford Motor Company had pushed hard to lessen these problems by forcing machine tool manufacturers to design transfer machines based on FMC's 'building block' idea in which interchangeable standard units and components could be added to or taken away from a 'station' in a transfer machine. Thus, if a new machining step were added to a manufacturing process as a result of design changes, instead of having to rebuild the transfer machine to accommodate this new step, it could be achieved simply by plugging a new station into the transfer machine.[97] FMC's production planners called this system 'unitized automation'.[98]

Obsolescence of production equipment—especially in automated plants—continued to burden the company as its products shifted rapidly in the late 1950s with the institution and then abandonment of the Edsel Division and its products, the widening of the product line of the Ford Division under Robert McNamara's leadership (including the development of the highly successful compact car, the Falcon), and the continued horsepower race. In early 1958, the company's Manufacturing Engineering Office began to take a serious look at 'automation obsolescence', first in the company's stamping operations and then in machining plants.[99] Charles H. Patterson, who had supervised much of FMC's automation programme and who became a vice-president for manufacturing in the early 1960s, continued to preach for greater flexibility in automation. Efforts to achieve greater flexibility led to a new generation of automation installed first in FMC's new Windsor (Ontario) Engine Plant and then at the Lima Engine Plant. The need for greater flexibility was pointed up by the Lima Engine plant, which by 1967 had produced three different six-cylinder engines and two different V-8s in its relatively short lifetime (less than a decade), largely by ripping up old lines and installing new ones.[100]

Even with greater flexibility in accommodating product changes in the automated engine plants, these plants still posed problems for production planners in that their output was tightly bound by the 'balancing' imperative. That is, a carefully balanced plant could not easily increase output of a given engine by a few thousand or even ten thousand units per year because, once the plant was producing at balanced, rated capacity, increasing output above that capacity could only be achieved on a sustained basis by investing a great deal in new equipment to fit into the tightly coupled system. Production increases in a highly automated plant performing at or near capacity, in other words, had to be achieved on a step-function basis.[101]

3.8. Automation: Revolution or Evolution?

Was FMC's automation programme, as embodied in the Cleveland Engine Plant no. 1 and later plants, revolutionary or evolutionary? When the first plant opened, enthusiastic journalists in both the technical and popular press tended to hail it as revolutionary. *Business Week* wrote about the plant in an illustrated article entitled, 'Automation: A Factory Runs Itself'.[102] Even Ford's own employees got caught up in the revolutionary rhetoric. Ray Sullivan, Harder's assistant who had been placed in charge of the newly created Engine and Pressed Steel Division, told an audience in Cleveland at the premier of a new Ford-produced film on automation, 'Technique for Tomorrow', that 'Those plants of ours over in Brook Park are pioneering what amounts almost to an industrial revolution'.[103] Even the mild-mannered Ernest Breech, in addressing the Committee of One Hundred in Miami, Florida, on 'Some Responsibilities of American Business in the Free World', cited the Cleveland Engine Plant as a revolutionary instrument for the preservation of the free world:

We see great possibilities . . . in the development of automation. Essentially, 'automation' is the push-button factory. . . . For example, in our Cleveland Engine Plant's machining department . . . we have 41 in-line, transfer-type machines, comprising two basic lines, tied together into a single transfer machine 1,200 feet long. Manual handling of the block occurs only at one point in this long line—at the loading station. From there on, the entire operation is automatic. Materials are transferred from one machine to another and from one operation to another without the use of human hands or human back muscles. That's automation. It is more than a new technique; it represents an entirely new concept of production. And to me it is characteristic of our modern-day, free, democratic, capitalist economy.[104]

Business management expert Peter Drucker echoed Breech's assessment, calling automation 'a major economic and technological change, a change as great as Henry Ford ushered in with the first mass production plant fifty years ago'.[105]

FMC clearly sought all the attention given to its Cleveland Engine Plant and its 'Detroit automation'. The firm was unprepared, however, for the fire-storm of fear about the workerless factory that spread rapidly throughout the United States in response to the Cleveland Engine Plant and a host of other developments in the United States, including the increasing adoption of the electronic digital computer in white collar work.[106] From its revolutionary rhetoric the company quickly back-pedalled to a new position in which the techniques employed at the Cleveland Engine Plant were interpreted as being evolutionary and based on the same wholesome principles that Henry Ford had used at Highland Park in the birth of the Model T assembly line. D. J. Davis's testimony before the Subcommittee on Automation of the Joint Committee on the Economic Report in 1955 merely solidified the company's reformulated position on its automation program. Above all, Davis and FMC sought to assure the American public and especially American workers that auto-mation did not spell an end to their jobs but was yet another step in bringing bounty to the United States.[107]

Whether the Cleveland Engine Plant and the automated factories that followed at FMC were evolutionary or revolutionary is far less important a question than why this and its sister plants were built in the first place. This paper has attempted to show that the Cleveland Engine Plant was the product of several forces at work in the Ford Motor Company after Henry Ford surrendered the reins of his faltering company to his young grandson, Henry Ford II. Corporate reorganization, difficulties with labour, new product development, greater integration, decentralization of production away from the Rouge, transfer of production approaches from GM via the hiring of GM personnel, and a vast, unfilled demand for automobiles in the postwar years all contributed substantially to the making of the Cleveland Engine Plant and the entire Ford Motor Company's production system of the 1950s. The plant was not without its critics, both inside and outside the company. Well before the end of the decade, FMC's production engineers and executives realized that the once-touted system was not without its faults. The most important of these was its lack of flexibility in accommodating design changes occa-sioned by the horsepower race and the ever-changing desires of the auto-mobile consumer.

NOTES

1. New York, 1984.
2. Berlin, 1985.
3. Michael L. Dertouzos *et al.*, eds., *Made in America: Regaining the Productive Edge* (Cambridge, Mass., 1989), 176.
4. Ibid. 176–7.
5. James P. Womack *et al.*, *The Machine That Changed the World* (New York: Rawson Associates, 1990).
6. Allan Nevins and Frank Ernest Hill, *Ford: Decline and Rebirth, 1933–62* (New York, 1962), 294.
7. See David A. Hounshell, *From the American System to Mass Production, 1800– 1932* (Baltimore, 1984), 277–8.
8. Nevins and Hill, *Ford: Decline and Rebirth*, 315.
9. For a journalistic and none-too-well documented history of this group, see John A. Byrne, *The Whiz Kids: Ten Founding Fathers of American Business and the Legacy They Left Us* (New York, 1993).
10. New York, 1946.
11. Nonetheless, the parallels between Drucker's text and the presentation that Breech gave at the company's second Management Meeting, 13 May 1947, are quite striking. See Report of Management Meeting, May 1947, Ford Industrial Archives, Dearborn, Michigan, 65-71:36.
12. Ford Motor Company Publicity release, 23 October 1946, Ford Motor Company Industrial Archives, 65-71:22.
13. In a speech to Ford.Motor Company's top executives on 'Manufacturing Horizons', given 16 June 1956 at the Greenbrier Hotel, Harder said that he had coined the term, 'automation', in 1936 but that 'the term did not come into extensive use until about 1947.' Ford Industrial Archives, 71-20:34. Harder repeated this claim several times, including in an oral history with Frank E. Hill, Myra Wilkins, and Henry Edmunds, 12 Nov. 1959, the summary of which is in the Ford Archives, Henry Ford Museum, and Greenfield Village, Dearborn, Michgan, Accession 975, Box 1.
14. These data on Harder's tenure at Ford come from an undated note that had obviously been prepared for use at the time of Harder's retirement, located in Ford Industrial Archives, 71-72:5.
15. Oral History Interview with Charles H. Patterson, Ford Archives, Dearborn, Michigan, Accession 975, Box 1.
16. *Ford Motor Company Annual Report, 1947*, 18.
17. Patterson Oral History Interview.
18. For a compellingly written history of the struggle to organize Ford, see Nevins and Hill, *Ford: Decline and Rebirth*, 133–67.
19. Quoted in Nevins and Hill, *Ford: Decline and Rebirth*, 322.
20. This style of management did not change rapidly, however, in spite of Breech's and Henry Ford II's efforts. Donald E. Petersen, the CEO of Ford during much of the 1980s, writes that when he joined Ford in 1949, 'I was introduced into an environment that was largely run by fear.' Petersen and John Hillkirk, *A Better Idea: Redefining the Way Americans Work* (Boston, 1991), 95.

21. Quoted in Nevins and Hill, *Ford: Decline and Rebirth*, 305.
22. Nelson Lichtenstein, *Labor's War at Home: The CIO in World War II* (Cambridge, 1982), 227; Nevins and Hill, *Ford: Decline and Rebirth*, 305–6; Bruce R. Morris, 'Industrial Relations in the Automobile Industry', in Colston E. Warne *et al.*, eds., *Labor in Postwar America* (Brooklyn, 1949), 399–417.
23. For an excellent history of the FAA, see Nelson Lichtenstein, ' "The Man in the Middle": A Social History of Automobile Industry Foremen', in Nelson Lichtenstein and Stephen Meyer, eds., *On the Line: Essays in the History of Auto Work* (Urbana, 1989), 153–89. See also Nevins and Hill, *Ford*, 335–8.
24. *Ford Motor Company Annual Report, 1949*, 26.
25. Nelson Lichtenstein, 'La Vie aux usines Ford de River Rouge: Un cycle de pourvoir ouvrier (1941–60)', *Mouvement Social*, 139 (1987), 77–106.
26. On FMC's deliberate strategy of out-sourcing many of its components after 1933, see Hounshell, *From the American System to Mass Production*, 300. For quantitative estimates of FMC's and GM's degrees of integration, see Harold Katz, *The Decline of Competition in the Automobile Industry, 1920–40* (New York, 1977), especially ch. 6.
27. Oral History Interview with John Dykstra, Ford Archives, Accession 975, Box 1.
28. Nevins and Hill, *Ford*, 333. As Nevins and Hill note, by mid-Sept. 1946, FMC had abandoned the idea of a Light Car Division altogether. But by no means had a light car been scotched in the minds of many executives.
29. The committee consisted of the company's top executives (less the director of public relations) and included Ernest Breech, chairman, Mead Bricker, John S. Bugas, Lewis D. Crusoe, John R. Davis, Benson Ford (vice-president and director of the Lincoln-Mercury Division), Henry Ford II, William T. Gossett (whom Breech had brought in as general counsel), Delmar S. Harder, Graeme K. Howard (vice-president and director of the International Division), and Harold T. Youngren. Organization of Ford Motor Company, 1947.
30. 'Ford Puts on a Big Show to Introduce its New Models', *Business Week* (19 June 1948), 54.
31. 'Two-Year Wait Faces Average Buyer, Ford Warns', *Printer's Ink*, 223 (18 June 1948), 98. *Automotive Industries* reported that after the first three days of the model's public showing, dealers booked 465,646 new orders, which was on top of the 1.3 million order on FMC's books. Vol. 84, 1 Aug. 1948, 102.
32. *Ford Motor Company Annual Report, 1947*, 8. In the following year's annual report, Henry Ford II wrote, 'The task of designing, engineering and producing these new models represented an undertaking that could be compared with the three previous major developments in the Company's history—the introduction of the Model T, the Model A, and the V-8 engine.' *Ford Motor Company Annual Report, 1948*, 7.
33. The Light Car Division continued its existence until the third quarter of 1947 when the Policy Committee reversed itself within a 30-day period, going from a decision of 'full speed ahead' in mid Aug. 1947 to total abandonment of the Light Car Division in Sept. after the results of a Ford-commissioned survey showed that Americans wanted bigger, roomier, faster, more powerful cars. Nevertheless, the idea of a light car remained fixed in some

executives' minds and thus intruded into much of the company's long-term planning process. Nevins and Hill, *Ford*, 333.

34. The XA-1 engine (planned for the Ford economy car) was a 179 cubic inch, six-cylinder engine with a bore of 3.5 inches and stroke of 3.1 inches. The XB-1 engine (planned for the main Ford line) was a 207.8 cubic inch, six-cylinder engine with a bore of 3.5 inches and stroke of 3.6 inches. These two engines were designed to 'have common tooling for blocks, heads, and other major components'. Minutes of the Forward Planning Committee Meeting of 1 Feb. 1949, Ford Industrial Archives, 65-71:14. These minutes reveal the degree to which the idea of the economy car bedevilled forward engine planning in the company.

35. The manuscript records of the company—especially Breech's papers—suggest that the small car idea was brought up every six months or so between 1949 and 1952. Later, under the leadership of Robert S. McNamara, the company committed to the manufacture of an economy car: the Falcon (and its Mercury cousin, Comet).

36. For evidence of the Ford Division's bid to control its own engine production, see Robert S. McNamara to T. O. Yntema, 18 Nov. 1949, Ford Industrial Archives, 65–71:14.

37. On the Sales Department's desire to maintain the V-8 as an option for the new Ford, see Minutes of the Forward Product Planning Committee of 13 Dec. 1948, 28 Dec. 1948, Ford Industrial Archives, 65-71:21. See also the Minutes of the Forward Product Planning Committee of 8 Feb. 1949, Ford Industrial Archives, 65–71:14.

38. Ernest R. Breech, Executive Communication, 20 May 1949, Ford Industrial Archives, 65–71:15. Other members of this committee were Benson Ford, L. D. Crusoe, R. T. Hurley, R. S. McNamara, and S. W. Ostrander.

39. Ibid.

40. *Ford Motor Company Annual Report*, 14.

41. On the changeover from the Model T to the Model A, see Hounshell, *From the American System to Mass Production*, 263–301.

42. At the time of the plant's construction, outside suppliers manufactured more than 40% of the company's bodies. See the Project Appropriation Request, 23 June 1949, bound in 'Proposed New Pressed Steel Plant', Ford Industrial Archives, 65-71:41.

43. James M. Rubenstein, *The Changing U.S. Auto Industry: A Geographical Analysis* (London, 1992), 119.

44. FMC considered several other locations as well, including Pittsburgh, Pennsylvania; Youngstown, Ohio; and Moundsville, West Virginia.

45. Daniel Nelson, 'Origins of the Sit-Down Era: Worker Militancy and Innovation in the Rubber Industry, 1934–8', in Daniel J. Leab, ed., *The Labor History Reader* (Urbana, Ill., 1985), 335.

46. Project Appropriation Request, 23 June 1949.

47. Delmar S. Harder, 'Automation—Key to the Future', speech delivered before the Quad-City Conference on Automation at.Davenport, Iowa, 27 Aug. 1954, Ford Industrial Archives, 71-20:34, 1. Ford Motor Company's Speech Services office printed Harder's speech in pamphlet form. At the end of the pamphlet, the following text appeared: 'Exactly when Ford Motor Company first

visualized the concept of "automatic handling of parts between progressive production processes" has not been established. Certainly the idea was developed before the word "automation" was coined. The word, however, made its formal debut in April, 1947, when a group of Ford production engineers whose job was to design work handling devices was re-named the Automation Department.' In late 1948, an article in the *American Machinist* 92 (21 Oct. 1948), 107, noted the creation of this Automation Department, but there is an intriguing anomaly in that no organization chart of Harder's manufacturing organization—and I have seen many complete charts in the Ford Industrial Archives documenting FMC's organization between 1946 and 1950—shows any such 'Automation Department'. James R. Bright noted that the 'automation department was disbanded in a few years, partly because of Ford's decentralization program and partly because it was found that this organizational structure did not facilitate the integration efforts and functions required'. *Automation and Management* (Boston, 1958), 5.

48. *Automotive Industries*, 98 (Apr. 1948), 26.
49. Rupert Le Grand, 'Ford Handles by Automation', *American Machinist*, 92 (21 Oct. 1948), 107; Joseph Geschelin, 'Revolutionary Automation at Ford Operates with Iron Hand', *Automotive Industries*, 99 (1 Dec. 1948), 30–33.
50. The Project Appropriation Request, 23 June 1949, itemizes literally hundreds of automation devices that would be installed in the plant.
51. Le Grand, 'Ford Handles by Automation,' 109.
52. Douglas Reynolds, 'Engines of Struggle: Technology, Skill, and Unionization at General Motors, 1930–40', *Michigan Historical Review*, 15/1 (1989), 69–92.
53. Stephen Meyer, 'The Persistence of Fordism: Workers and Technology in the American Automobile Industry, 1900–60', in *On the Line: Essays in the History of Auto Work*, edited by Nelson Lichtenstein and Stephen Meyer (Urbana, Ill., 1989), 73–99, discusses GM's use of transfer machines in the immediate postwar period.
54. On Church's machinery, see Edward A. Marsh, *The Evolution of Automatic Machinery as Applied to the Manufacture of Watches at Waltham Mass., by the American Waltham Watch Company* (Chicago, 1896). One of these machines survives in the National Museum of American History, Smithsonian Institution, Washington, D.C.
55. Frank G. Woollard, *Principles of Mass and Flow Production* (London, 1954), 30. I am indebted to Wayne Lewchuk for this reference; see his *American Technology and the British Vehicle Industry* (Cambridge, 1987), 168–9.
56. *The Productivity Dilemma: Roadblock to Innovation in the Automobile Industry* (Baltimore, 1978), 204–5.
57. *Automation and Management*, 60–1.
58. Based on my assessment of articles cited in the *Industrial Arts Index* during the second half of the 1940s and the early 1950s, it is clear that transfer machinery was a 'hot topic' in the automobile industry, and an increasing number of firms were making major commitments to the acquisition of transfer machinery. See also Woollard, *Principles of Mass and Flow Production*, 111–59.
59. For a general overview of the Motor Building rearrangement, see 'Ford Streamlines Operations in Motor Building', *Mill and Factory*, 40 (Mar. 1947),

100–4. Installation of transfer machines in the Motor Building is discussed in 'Ford Retools Cylinder Block Line with Transfer Machines', *Automotive Industries*, 98 (1 Feb. 1), 38–9; 'First Transfer Machine for Drilling Crankshaft Oil Holes; Ford Motor Company', *Automotive Industries*, 98 (1 Apr. 1948), 36, 76; and 'Latest Transfer Machines at Ford Plant', *Automotive Industries*, 99 (1 Aug. 1948), 41, 90, 92.

60. Bright, *Automation and Management*, 59.

61. L. D. Crusoe to Henry Ford II, 'Preliminary Review of Proposed Engine Program', 13 Dec. 1948, Ford Industrial Archives, 65-71:14.

62. Ibid.

63. Minutes of Forward Product Planning Committee Meeting, 1 Feb. 1949, Ford Industrial Archives, 65-71:14.

64. This committee was formerly the Special Facilities Committee but had been renamed, made a permanent committee reporting to the Policy Committee, and charged with the responsibility of resolving production facilities issues. See Henry Ford II to E. R. Breech, 16 Sept. 1949 (with Breech's annotations on it from his conversation with Ford), Ford Industrial Archives, 65-71:15.

65. Minutes of the Executive Committee, Ford Motor Company, 4 Nov. 1949, Ford Industrial Archives, 84-63-1217:1. The full report—'Forward Engine Program Facilities Study', 4 November 1949—is also included in this same location.

66. Unfortunately, the records of the company do not reveal if this criterion was self-imposed by the Facilities Committee or established by the Executive Committee. Bits and pieces of information suggest that among executives there was a consensus of limiting Rouge production of most major components to 50% of total production, but I am unaware of any explicit policy statement to this effect. That FMC's actions became an issue with workers in Detroit, however, is clear. See Ernest Breech, 'Decentralization—What It Is and How It Works', n.d. (*c.* 1951), Ford Industrial Archives, 71-20:5.

67. Robert S. McNamara to T. O. Yntema (vice-president, Finance), 18 Nov. 1949, Ford Industrial Archives, 65-71:14.

68. Minutes of the Executive Committee, Ford Motor Company, 2 Dec. 1949, Ford Industrial Archives, 84-63-1217:1.

69. D. S. Harder to Henry Ford II *et al.*, 'Forward Engine Program—Facilities Study', 2 Dec. 1949, Ford Industrial Archives, 65–71:14. Harder noted in this document that 'the location of all three plants in Detroit provides the lowest cost point', but Breech annotated his copy of the locational cost comparison, pointing out that 'no effect on costs [is] given to labor efficiency by locations.'

70. Minutes of the Executive Committee, Ford Motor Company, 2 Dec. 1949, Ford Industrial Archives, 84-63-1217:1.

71. Ibid. That the organizational responsibility for the Buffalo Stamping Plant was problematic is indicated by a set of documents in Breech's papers in which Harder and Breech struggled to draft a document that laid out the issues surrounding organizational responsibility for the plant. The final draft was issued as an Executive Communication to all members of the Policy Committee, 15 Nov. 1949, Ford Industrial Archives, 65-71:7.

72. Henry Ford II to vice-presidents *et al.*, 27 Dec. 1950, Ford Industrial Archives, 65-71:39.

73. Ford Motor Company Press Release, 1 Jan. 1950, Ford Industrial Archives, 74-18056. See also *Business Week*, 21 Jan. 1950, 23.
74. Construction and start-up dates for the Cleveland Engine Plant and the Cleveland Foundry are contained in 'Review of Cleveland Operations', Engine and Foundry Division, Oct. 1954, Ford Industrial Archives, 65-71:8.
75. Charles H. Patterson to Facilities Committee, 9 Apr. 1953, Ford Industrial Archives, 65-71:15. This decision, of course, made FMC more vulnerable to localized strike power.
76. Joseph Geschelin, 'Engine Plant Operation by Automation', *Automotive Industries*, 106 (1 May 1952), 36.
77. *Automation and Management*, 63. As G. G. Murie, who had been one of the original employees of the Automation Department, explained to the Ohio Society of Professional Engineers in 1957, 'The entire department [i.e., plant] became a synchronized system and functioned as one large transfer machine.' 'Automation at Ford', 20 Mar. 1957, Ford Industrial Archives, Automation Notebook 1, 85-57043. Note that Murie's speech borrowed heavily from the texts of speeches that other Ford production managers, including Harder, had previously given.
78. See e.g. Minutes of the Facilities Committee, 4 May 1951, Ford Industrial Archives, 65-71:15.
79. *Automation and Management*, 64.
80. Minutes of the Facilities Committee, 18 May 1951, Ford Industrial Archives, 65-71:15.
81. *Automation and Management*, 63.
82. R. S. McNamara to D. S. Harder, 12 Jane 1952, Ford industrial Archives, 65-71:10. 'Standard volume' was a concept that the recruits from GM brought to FMC. It represented an optimal output for a given plant and was based on the productivity characteristics of machine tools and other indices of production. Standard volume and standard volume costs served as benchmarks for performance on output and costs. Annual volume was simply an annualized figure that was calculated on the basis of a given output for a given period (e.g. a week's production times 52 weeks).
83. 'Comparison of Monthly Costs with Planned Costs, 1952 Ford 6 Overhead Valve Engine', 24 Nov. 1952, Ford Industrial Archives, 65-17:10.
84. Ibid.
85. M. L. Katke to E. R. Breech and Henry Ford II, 8 July 1959, Ford Industrial Archives, 65-71:13. Sometime between 1953 and 1954, the Engine and Pressed Steel Group was realigned to become the Engine and Foundry Division, and the Metal Stamping Division was created. Both these divisions reported to a vice-president and group executive, who in turn reported to Delmar Harder, executive vice-president, basic manufacturing divisions. I have not pinpointed exactly when this reorganization was effected.
86. W. H. Guinn to L. P. Hourihan, 3 Feb. 1954, Ford Industrial Archives, 71-2:4.
87. Perhaps the enormous popular and technical interest in the plant served as the catalyst for the board's visit. Either way, such a group inspection tour of a Ford production facility by the board was highly unusual.
88. 'Review of Cleveland Operations', Engine and Foundry Division, Oct. 1954, Ford Industrial Archives, 65-71:8. The record is unclear on who actually

carried out the briefing of the board. No records survive to indicate the board's reaction to the briefing or whether it served to silence McNamara's carping on costs at Cleveland.

89. The horsepower race is treated in Lawrence J. White, *The Automobile Industry since 1945* (Cambridge, Mass., 1971), 216–20, and Abernathy, *Productivity Dilemma*, 97. See also the estimates of the social costs of the horsepower race (and other model changes) during the period in Franklin M. Fisher, Zvi Griliches, and Carl Kaysen, 'The Costs of Automobile Model Changes since 1949', *Journal of Political Economy*, 70 (1962), 433–51.

90. Ray H. Sullivan to the Administration and Executive Committee, 3 Oct. 1955, Ford Industrial Archives, 65-71:13. The Lima plant would also produce engines for the short-lived Edsel.

91. The improvements to automated engine production at the Lima plant are discussed in 'Taking the Worry Out of Automation', *Business Week* (23 Aug. 1958), 44–6. Abernathy, *Productivity Dilemma*, 240, discusses briefly the extension of automation to assembly operations at the Lima plant.

92. The Dearborn Engine Plant was the formal name given to the former Parts and Accessories Building at the Rouge that had been converted to an engine production facility at the same time that the Cleveland Engine Plant was built. Several articles in the technical and trade literature discuss the Dearborn Engine Plant and its machinery.

93. R. S. McNamara to C. H. Patterson, 1 Mar. 1955, Ford Industrial Archives, 71-2:4.

94. Sullivan to Administration and Executive Committee, 3 Oct. 1955. The press was well aware of the rapid obsolescence of Cleveland Engine Plant no.1's equipment, as indicated by 'Ford Retools "Last-Word" Plant', *Business Week* (9 Feb. 1957), 109–10.

95. 'Manufacturing Engineering's Point of View on UAW-CIO's Report on Automation', 30 Dec. 1954, attached to D. J. Davis to C. C. Donovan, 30 Dec. 1954, Ford Industrial Archives, 85-57043, Notebook no.1 on Automation.

96. Ford Motor Company Press Release, 20 Mar. 1956, Ford Industrial Archives 71-20:3. Davis's testimony will be discussed below. Harder had hired Davis away from Avco Corporation in 1949, but Harder had no doubt known Davis because of Davis's 21 years in production engineering at GM's Cadillac Division. Davis entered FMC as the director of the Manufacturing Engineering Office, which was responsible for production engineering development and integration.

97. D. J. Davis, 'New Manufacturing Horizons', 12 Mar. 1957, Ford Industrial Archives, 71-20:13. Ford had also pushed related concepts of standardization in production systems, including standardized, plug-in control panels for transfer machines and automation equipment. See D. J. Davis's Eli Whitney Memorial Lecture, 'Planning for Profit', given before the American Society of Tool Engineers, May 1958, Ford Industrial Archives, 71-20:13.

98. Ford Motor Company Press Release on Automation, 15 Apr. 1957, Ford Industrial Archives, 85-57043, Notebook no.1 on Automation.

99. John F. Randall to D. J. Davis, 'Automation Obsolescence as a Result of New Tooling Programs in Stamping Plants', 23 Jan. 1958, Ford Industrial Archives, 66-5:1.

100. See Patterson's various speeches: 'Manufacturing Flexibility in a Changing Market', 12 Oct. 1961; [untitled] speech before the American Society of Quality Control, 15 Nov. 1963; 'Engineering for Productivity', Cleveland Engineering Society, 16 May 1963; [untitled] speech before the Ohio Chamber of Commerce, 19 June 1967. All speeches in Ford Industrial Archives, 71-20:55.
101. M. L. Katke to C. E. Bosworth, 1 July 1960, Ford Industrial Archives, 73-13881:4.
102. 29 Mar. 1952, 146–8, 150.
103. Text of R. H. Sullivan Remarks, 15 Sept. 1953, Ford Industrial Archives, 71-20:70. Sullivan had been Harder's assistant at Bliss and before that had been a production engineer at Yale and Towne. See his biographical sketch in the Ford Industrial Archives.
104. 3 Mar. 1953, Ford Industrial Archives, 71-20:5.
105. 'The Promise of Automation', *Harper's*, 210 (Apr. 1955), 41.
106. Space constraints prevent me from discussing the 'automation panic' that spread throughout the United States in the 1950s. But for a rich contemporary view of this panic, see the 82 minute See It Now Series documentary, 'Automation', that Edward R. Murrow prepared for CBS television broadcast 9 June 1957. So loaded had the term 'automation' become by 1953 that GM banned its employees from using it to describe the company's manufacturing developments and methods.
107. For Davis's testimony see Ford Industrial Archives, 71-20:13, or the Special Subcommittee's Report, *Automation and Technological Change: Hearings before the Subcommittee on Economic Stabilization of the Joint Committee on the Economic Report* (Washington, 1955). His testimony was also reprinted in Howard B. Jacobson and Joseph S. Roncek, eds., *Automation and Society* (New York, 1959), 34–43. For an earlier statement arguing that automation embodied Henry Ford's ideas about mass production, see Ford Motor Company News Release, 13 May 1954, Ford Industrial Archives, 85-57043, Notebook no.1 on Automation. This document concludes: 'Regarded as an evolutionary development, rather than a revolutionary one, automation, Ford men believe, has started a trend which will continue and will expand as its benefits become known.'

4

Design, Manufacture, and Quality Control of Niche Products: *The British and Japanese Experiences*

TIMOTHY R. WHISLER

Nearly all studies of the British-owned motor industry investigate the decline of the sector during the postwar period by focusing upon the design, production, and distribution of volume cars. However, the success in terms of export sales, unit profitability, and product image of British niche products, such as Range/Land Rover, Jaguar, Austin-Healey, MG, and Triumph sports cars, is often ignored or cast aside as irrelevant. Clearly, attention should be given to products, offered by the major indigenous firms, that did not fit the general trend of the industry.[1]

Historians of the motor industry who study the national industry or individual enterprise frequently overlook the product mix, which influences design, manufacturing, and distribution methods as well as the overall cost structure. The examination of a specific product reveals that a firm usually adopts different techniques to produce different models. An analysis of the sports cars built by British Motor Corporation, Standard-Triumph, British Leyland, and Nissan provides insight into the methods that contributed to the success of niche products and, in the case of the British, the failure of volume cars.

Essentially custom-built to meet domestic demand for high performance cars during the 1920s and 1930s, MG sports cars created a niche market in the USA after the war on the basis of differentiated characteristics and product image. The growing demand for sports cars encouraged complementary Austin-Healey and rival Triumph models to enter the US market. To meet increasing overseas demand and lower manufacturing costs, the sports car companies adopted mass production methods and standardized components. Sports cars consistently returned the largest unit profits and achieved the largest export sales of the parent firms' model range. Nearly 80 per cent of MG and Triumph sports cars were sold overseas, particularly in the USA. In addition to the badly needed sales revenues, sports car component requirements increased capacity utilization rates at corporate manufacturing plants that low mass market car sales frequently left underutilized.

Ironically, the causes of the British difficulties in designing, engineering, and producing volume cars explain the success in sports cars. The sports car designers were adept at differentiating their product, thereby creating demand. They were aware not only of the market, but also that production methods had to be compatible with low annual output. Studies of production frequently overlook this vital connection of anticipated demand and the choice of manufacturing methods.

Lewchuk argues convincingly that rigid British socio-economic institutions prevented the motor firms from successfully replacing traditional British production methods with capital-intensive high-volume or Fordist techniques during the 1960s. However, traditional labour-intensive low-volume methods were ideally suited to the production of sports cars and niche products. British sports cars remained successful as long as other traditional institutions, such as the 'practical men' in management and engineering, small and simple corporate structure, and limited market competition, existed.

When, in the mid-1970s, British Leyland responded to Nissan's Datsun Z series by abandoning inherited design and production methods in favour of dedicated Fordist techniques, the results of sports and volume cars became analogous. The failures in design, production, and quality control indicated that the corporate crisis was acute in engineering. Early sports car design and production at Nissan was remarkably similar to British methods. However, Nissan was able to adapt the fundamental concepts of sports car design and production to meet ever-changing market and manufacturing conditions.

4.1. The Design Process

A motor vehicle is the result of a firm's engineering resources, manufacturing capabilities and methods, available investment funds, intended market target and anticipated sales volume, and rival product characteristics. The design process is influenced by and affects these factors. The British sports car makers compensated for low annual volumes by using robust designs, which lend themselves to incremental innovation (known as 'face-lifts') rather than lean designs, which are conceived as an entirety and require radical redesign of basic specifications to accommodate further development.[2] Robust sports car designs also spread large initial costs and reduced future design and retooling costs by incorporating standardized components and facilitating long production life.

As part of the British government's 'Export or Die' campaign, which regarded the motor industry as potentially a large earner of scarce hard currency, the Nuffield Organisation entered the US market in late 1947. The Nuffield Board assumed that its sedans would gain market share since Detroit could not meet domestic demand, but the British volume cars offered lower levels of quality and mechanical characteristics at prices equivalent to or higher than American models. Increasing orders for the MG TC sports car convinced the Board that 'unique appeal' was necessary for sales success in the USA.[3] The MG was differentiated from domestic cars by higher levels of braking, steering, road-holding, and acceleration characteristics, and a distinctive product image created by its styling, convertible top, and participation in motor sport. The differentiated nature of the cars permitted sales at slightly higher prices than most US-built and imported sedans. In 1950 the price of an MG TD Midget was $1,850, as compared to a Chevrolet that, depending upon optional extras, ranged from $1,329 to $1,994. MG sales in the US rose from 1,473 units in 1948 to 17,693 units ten years later.

Standard-Triumph, the smallest of Britain's six volume car producers, ended MG's monopoly in the USA during the mid-1950s. While Standard-Triumph had achieved considerable export sales with its newly developed farm tractor, the firm had been unable to sustain overseas car sales during the late 1940s. The British government's Korean War rearmament campaign, which rationed steel to the motor manufacturers on the basis of their export sales, threatened the viability of Standard-Triumph's car range. Given its competitive advantage in tractors and the disadvantage of competing with MG's product image, Standard's decision may appear curious. However, management hoped to gain sales in the growing US sports car market by filling the price gap between the MG and Jaguar ($3,945).

Standard-Triumph's designers understood the fundamental connection between the design of a product and its intended place in the market (a concept later forgotten by British Leyland [BL] engineers). To justify the price and establish a product image, the innovative model featured more modern body styling and greater engine performance relative to MG. Time and financial constraints compelled the designers to use modified versions of the transmission, suspension system, axle, and chassis from the company's sedans and the engine fitted to the farm tractor. This strategy allowed Standard-Triumph to utilize existing tooling and spread production costs over combined sedan and sports car output.

In the absence of direct substitutes, MG had developed the pre-war T series Midget design with minor facelifts.[4] The TR2, launched in March 1953 and priced $500 more than the MG TF Midget, failed to displace MG as the sales leader, but began a design competition between BMC and Standard-Triumph. Over the next fifteen years the rivals sought to expand market share and maintain product image through a chain reaction of innovative models and niche segmentation. Whereas in 1952 BMC offered one MG sports car, by 1967 the firm featured two badges (Austin-Healey was launched in 1953) and six models ranging in price from introductory to up-market with corresponding levels of mechanical characteristics and comfort, including a fixed head. Triumph's range expanded from one to five sports cars during the same period.

MG was the first mover by periodically releasing revolutionary designs that were updated during long production runs. Triumph responded with an evolutionary development strategy. The original design was frequently face-lifted, allowing Triumph to offer a 'new' model more often than BMC. Both strategies, however, revealed the cost constraints the designers faced. After all, the corporate parents' main business was volume cars, and investment capital was limited by low profits and the British tendency to distribute a high percentage of annual profits as dividends.[5]

The early sports car designers were frequently torn by the conflicting market demand to differentiate their product and the corporate requirement to reduce production costs. MG retained design and engineering responsibility, but BMC Chairman Leonard Lord forced the designers to use one of the three corporate engines and base new model development upon racing car designs to conserve funds. Lord also purchased the first Austin-Healey designs from the Healey consulting firm and later entrusted MG with updating them. The 'Zobo' design revealed Triumph's cost and production advantage of having a consolidated design office. The design was so robust that its floor pan (the most costly body pressing) served as the basis for the introductory priced Herald sedan, the moderately priced Vitesse sedan, as well as the introductory priced Spitfire and moderately priced GT6 sports cars. The four cars shared common engines, exhaust manifolds, front suspensions, radiator grilles, waist rails, and screen rails. Even the bodyshells of both sports cars were so similar that the company combined them for tooling amortization accounting.[6]

At both Triumph and MG, however, the designers were 'practical men',[7] who had mechanical aptitude and learned their trade on the job rather than through university training. The development of Britain's motor industry and the country's shortage of trained engineers[8] encouraged the firms to use practical men as substitutes for engineers. The early motor men were entrepreneurs and mechanics, not engineers. They generally trusted and promoted men with similar skills and outlook. Formally trained engineers were regarded with suspicion since, in the opinion of the practical men, they were outsiders who relied on theoretical rather than practical notions. In this climate, Britain's shortage of engineers mattered little.

The practical men had displayed an uncanny understanding of the market and basic mechanics. However, changing market demands, particularly US safety and pollution legislation enacted in the mid-1960s, required expertise rather than experience. In 1966 Triumph's Chief Engineer reported that his department was having 'very grave difficulties' meeting exhaust standards.[9] In the absence of a corporate research centre, MG relied upon a writer for its owners' magazine, who held a degree in chemistry, to reduce pollution emissions.[10] While the British firms struggled, Nissan devoted a section of its engineering department to conducting short- and long-term exhaust studies and participated in a cooperative research programme sponsored by the Ministry of International Trade and Industry.[11] The British companies simply lacked a sufficient quantity of adequately trained engineers to conduct long-term research and viewed collaboration more as an opportunity for industrial espionage than as teamwork.[12]

Old and overwhelmed, the practical men passed from the scene as stricter American legislation and an ambitious corporate new model development programme created an engineering crisis at BL. In a 1973 survey of world manufacturers, the US Environmental Protection Agency (EPA) concluded that BL was one of three firms whose engineering expertise appeared incapable of meeting US regulations. Privately the corporation agreed.[13] BL exacerbated the situation by mismanaging its limited engineering resources. The firm failed to consolidate the various engineering offices it inherited. Rather than developing a specialization of labour, the sports car design offices were duplicating research efforts. BL even fulfilled Triumph's request for similar collision and exhaust equipment held by MG.[14] The firm's limited engineering resources were also depleted by a constant loss of the few professional staff it possessed. As early as 1966 internal studies indicated competing motor and non-automotive firms were luring BL employees and recruits with higher compensation. BL management did not recognize this as a simple case of supply and demand. The nation's chronic shortage of engineers had manifested itself in the form of higher prices for a scarce commodity. Despite pleas for action from the line divisions, corporate management resisted pay increases, fearing parity demands.[15] The result was an absolute loss to BL of essential human capital.

The design competition between the Triumph TR7 and Datsun Z series sports cars illustrated the divergent paths of BL and Nissan. Cusumano argues that in the early 1950s the Japanese firms studied and selectively adopted design and production techniques from Western manufacturers to improve their competitive position. A formal arrangement allowed Nissan to build the Austin A50 sedan and provided access to BMC manufacturing technology and engineers from 1953 until 1959.[16] Nissan unveiled its first sports car the year the agreement expired and the British influence was unmistakable. The design of the S211 resembled the Austin-Healey 3,000 and the subsequent Fairlady series appeared similar to the MGB. The Japanese sports cars never posed a serious sales challenge as the imitational design and inferior product image limited Nissan annual sales in the USA.

In the 1970s, however, the tables were turned. Launched in 1969, the Datsun 240-Z became the sales leader two years later on the basis of innovative styling and higher levels of mechanical characteristics compared to British models. By the mid-1970s Nissan had transformed the Z series from a near substitute for the British sports cars into an up-market model by developing (far more quickly than Triumph's evolutionary strategy) the characteristics and comfort of a robust design.

BL sought to rapidly develop a corporate model that would simultaneously replace its existing range and reduce Nissan's market share. The lack of a corporate design and engineering office meant both MG and Triumph sought control of the project. The wrangling resulted in a sports car engineered by Triumph and styled by the volume car division Austin Morris. MG played no role whatsoever.[17] BL's Triumph TR7—just as Nissan's models had been twenty years earlier—was a near perfect example of an inferior market share firm imitating the dominant product. The inability of the designers to duplicate rival characteristic levels did not offset the lower price of the TR7. In contrast, the old practical men had understood the role of innovation in design.

4.1.1. Sports Car Production Scale

Analysis of car company cost structures centres upon the firm's installed capacity and annual output, since considerable cost savings can be achieved through economies of scale in the manufacture of power trains, castings, and bodyshells and in final assembly. Capacity utilization rates also indicate whether plant and labour is being effectively employed. In this aggregate approach individual models and components are combined without regard to different cost structures and production methods in an attempt to assess the scale and efficiency of a concern.

The study of niche products requires a more detailed approach. Rhys[18] and Pratten[19] argue that 'model-specific' economies of scale—number and type of models, installed capacity of specific models, number and output of basic components, length of production life and capacity of individual plants—were important cost considerations. Sub-optimal model-specific output can raise the cost structure of a firm that theoretically possesses sufficient installed capacity. Additionally, the type of manufacturing and assembly methods used will influence off-standard rates and consequently capacity utilization rates. Since power train production and final assembly accounted for about 70 per cent of a car firm's total internal costs during the mid-1970s,[20] sports car makers had to manage 'dimensions of scale' to pass the 'survivor test' of competing in a high cost niche market and by doing so contribute to the corporation's output and profit.[21]

According to former BL Vice-Chairman and Managing Director, John Barber, the main function of a sports car from a manufacturing perspective was to 'load' the factories and thereby earn incremental profits.[22] This strategy served the dual function of lowering sports car component costs (as argued earlier) as well as increasing total individual power train output. Each of the three BMC corporate power trains and two Standard-Triumph basic engines were fitted to sports cars. In contrast to the British, who would, over time, increasingly rely on sports cars engine requirements to increase annual output of its entire basic power train range, Nissan placed only one of its engines in a sports car. The firm's annual output of the 2.5-litre six-cylinder engine was consistently the lowest in Nissan's power train range.[23] Not surprisingly, this unit was placed in the Z series sports car.

During the 1950s and early 1960s, MG and Austin-Healey performed the ideal function of 'loading the factories'. BMC was achieving minimum efficient scale (calculated to be about 150,000 units per year) in the production of its A and B series corporate engines,[24] and sports car annual requirements ranged from about 6.5 per cent[25] for the A series small engine (under 1.2 litres) fitted to the Sprite/Midget to about 10 per cent for the B series engine (1.5–1.8 litres) installed in the MGA/MGB. Clearly, the high annual volume of the BMC A series (315,922 units in 1962), reduced power train costs to the point where a small sports car such as the Austin-Healey Sprite was feasible. The cost structure of the B series allowed production of the MGA even though annual output never surpassed 24,000 units.

In contrast to BMC, where sports cars benefited from relatively high-volume output, Standard-Triumph relied upon common components to reduce the cost structure of both low-volume mass-market and sports cars. Standard-Triumph's attempt to achieve volume production through vertical expansion in the late 1950s resulted in near bankruptcy. Leyland, a truck and bus manufacturer, purchased the company in 1961 and immediately abandoned the volume approach for a niche model, low-volume, high-capacity utilization strategy.[26] Sports cars became crucial in loading the factories. For example, between 1963 and 1967 Triumph produced an average of about 80,000 units of its 1.3 litre engine; Spitfire sports car requirements comprised an annual average of 23.6 per cent of total production. Sports car production kept total annual engine output high enough to allow the use of low-volume transfer machinery.[27] Despite low annual volume, higher prices based on the image and niche market target of Triumph models contributed to a low break-even point.

During the 1960s and early 1970s BMC and Nissan concentrated upon the output of sub-compact and compact models. Both firms used up-market sports cars to lower the cost of large capacity engines built in low annual volumes. In 1960 the Austin-Healey 3,000 accounted for 22.8 per cent or the 30,644 BMC C series 3-litre engines built. Six years later the output of the sports car declined to only 5,494 units, but this was nearly one-half of 11,335 3-litre units built. Although model-specific figures are unavailable, annual production of Nissan's early Fairlady sports car models, sold primarily in Japan and the USA, was about 7,000 units. The Fairlady shared engines with the Cedric sedan and the Z series utilized the same six-cylinder power train as the Skyline and Laurel sedans. These large sedans had the lowest annual output in Nissan's model range.

During the latter half of the 1960s and 1970s the relationship between volume and sports car requirements changed considerably at BMC and BL. An unintentional interdependency emerged as a result of the firms' decreasing sales in the growing mid- and full-sized sedan segments. Increased demand for BMC sports cars mitigated somewhat the cost effect of lower-volume car sales, but corporate profits and capacity utilization rates declined. Between 1963 and 1966, annual MG demand for B series engines rose from 17 per cent of 136,984 corporate units to 26.7 per cent of 123,122 units. BL failed to reverse the sales trend of BMC's volume cars and then compounded the problem of model-specific scale by adopting a half-hearted rationalization strategy even though it built the widest range of models and components in the world.[28] The firm planned to gradually withdraw inherited models and power trains as new cars and components, built in high volume capital-intensive plants, met sales targets. However, sales resistance to BL's new models forced the company to retain many of the old models to defend market share and contribute revenues to offset the recently incurred capital costs. When BL entered bankruptcy in 1974, it had sufficient installed capacity to achieve minimum efficient scale. However, the firm was prevented from achieving cost savings by sub-optimum model-specific output.[29]

As volume car sales declined, sports cars were regularly accounting for one-quarter to one-third of the individual power train output in the early 1970s and then nearly one-half and more in the latter half of the decade. Sports car requirements significantly increased model-specific engine output, spreading volume-car unit fixed costs and preventing diseconomies of scale. The British experience suggested that BL depended upon a delicate balance between volume and sports car components requirements. Yet the corporation, mysteriously convinced that its models could achieve high annual output, chose to fit unique, but similar capacity, engines in the TR7 and various mass market models. This decision would contribute to the subsequent TR7 unit losses.

The scope for sharing bodyshells was obviously limited and reducing the bodyshell production costs—about 15 per cent of total unit costs according to Pratten[30]—posed a challenge, given the low annual output of sports cars. Initially, Triumph purchased complete TR2 bodyshells from an independent body company based upon projected annual output of 2,400 units. As market demand steadily increased and accounting revealed that the bodyshell accounted for 45 per cent of total unit costs, management moved body production in-house. Thereafter the model returned a unit profit.[31] Purchasing bodyshells remained a viable option for up-market models produced in very low annual volumes. For example, annual output of the Austin-Healey 3,000 rarely exceeded 5,000 units and BMC purchased the bodyshell during the model's 13-year production life.

Long production life was the primary method used by the sports car makers to spread the cost of low annual in-house bodyshell production. Pratten estimated that the bodyshell costs of a model with an annual output of 25,000 units decreased by 30 per cent as its production life was extended from two to ten years.[32] Not surprisingly, the shortest sports car bodyshell life was 7 years and the longest was 21 years. Clearly robust designs and the development strategies of Triumph and MG extended bodyshell life. Triumph built 80,149 virtually identical TR4, TR4A, and TR5 units, but many of the major body panels were also incorporated in the production of 91,850 TR6 models. Between 1962 and 1980, BMC/BL body factories manufactured 521,843 MGB bodyshells even though annual average volume was about 29,000 units and annual production never surpassed 40,000 units. The use of larger but fewer body pressings and the maintenance of separate chassis/body construction, particularly at Triumph, also lowered body costs.

4.2. Production Methods

Lewchuk suggests that in the early years of the British motor industry management adopted a strategy of low-capital, labour-intensive production methods rather than Fordism. Management largely relied upon shop stewards and a piece-work payment system to coordinate and control production. Relatively high returns on investment perpetuated this method until the late 1950s and 1960s, when more capital-intensive techniques used by rivals threatened the British position. The firms responded by attempting to superimpose modern methods upon traditional British production institutions. However, management lacked the skills to supervise and plan highly mechanized mass production, while labour, based upon its strong position in the manufacturing process, successfully negotiated favourable employment levels. A productivity crisis emerged as the new system was characterized by unregulated worker effort and product flow.[33] Low capacity utilization caused by high off-standard rates—defined as the difference between the plan and actual production time (and consequently output) resulting from mechanical failure, material shortages, rectification, sub-optimal workpace, absenteeism, and labour disputes—was a prominent symptom of the British predicament.[34]

While Lewchuk's analysis explains the demise of British volume car operations, the maintenance of traditional British assembly methods was a factor in the continued success of sports car production. MG's unusual labour-intensive and Triumph's more traditional low mechanized techniques proved to be better managed and more cost effective compared to the more dedicated capital-intensive volume car methods instituted by the parent firms in the 1960s and 1970s. The sports car techniques, easily supervised by 'practical men', provided the flexibility and low off-standard rates necessary to pass the survivor test. In other words, production, much like design, was tailored to the market position. These processes, however, did have drawbacks. Annual output was limited at MG and control problems arose at Triumph, while both assemblers were affected by inefficiencies at corporate manufacturing plants. Nissan modified the British approach, thus reaping the benefits and eliminating the disadvantages.

4.2.1. *Labour-Intensive Methods: MG*

The plant and assembly techniques used by MG were developed in the 1920s and maintained until 1980. At no time was Abingdon equipped with the usual conveyor belts and mechanized assembly line. The most significant change occurred in the 1960s when air-powered tools replaced hand tools and monocoque bodyshells eliminated the jigs and trolleys needed to construct chassis and then attach body panels. Even the labour count remained steady throughout the postwar era at 800 in assembly, 200 in the racing department, and an additional 100 in the drawing office and management. In contrast to the stereotype of the British industry, Abingdon was rarely afflicted by industrial unrest. Most MG employees stayed at Abingdon for their entire careers.

The assembly track was composed of two sets of blocks, one grooved to guide tires and one flat. Chassis and later bodyshells were pushed manually to each station, which was supplied by runners with component wagons. The system permitted the simultaneous assembly of various low-volume models and the quick accommodation of newly introduced models. In addition, output could be adjusted according to market conditions or material shortages originating from upstream manufacturing plants, without regard to fixed cost considerations of dedicated machinery. The conversion of a track to accommodate new model production required two weeks.[35] From 1945 until 1980, 56 different sports cars and sedans were assembled at Abingdon and annual output at the factory ranged from less than 2,000 to nearly 60,000 units.

Abingdon's cost considerations centred upon variable costs, primarily component and labour charges. Component costs were determined by the corporate parent on a cost plus basis. Therefore attention was focused upon labour costs. The prospect of lower wage rates apparently influenced Lord Nuffield to locate assembly in the agricultural Thames Valley during the 1920s rather than in the West Midlands, the traditional motor region.[36] But, clearly, wage-to-effort ratios had to be managed carefully. A former MG manager noted that under piece-work, production and costs were interrelated almost by definition, even during slumps. In periods of high demand, bonus incentives and high effort levels boosted output without additional labour. After the advent of measured day work (MDW) in the early 1970s, wage-to-effort and labour count management became more crucial especially during swings in demand. In this case the good industrial relations at the plant facilitated adjustments in time cycles through mutuality. Nonetheless, it is probably not a coincidence that annual output tended to be lower after the introduction of MDW, resulting in higher wage-to-effort ratios.[37]

Comparisons of productivity reveal that, as expected, MG could not match the rates of capital-intensive firms, such as GM or even some British companies. In 1977 the average automated UK assembly track produced 50–52 cars per hour compared to 22 per hour at MG. Three years earlier GM's Lordstown plant in Ohio was assembling cars at a rate of 101 per hour.[38] On the other hand, compared to other semi-specialist firms and British Leyland, MG's performance was competitive. In 1968 Porsche's Zuffenhausen factory was assembling 7.5 cars per hour, which was about one-third the rate achieved at Abingdon.[39] The assembly of a complete MG sports car required about 15 man hours compared to 40 man hours for the assembly of the Mini volume car. The difference was reflected in off-standard rates at Abingdon of about 10 per cent, compared to 90–130 per cent for the Mini.[40] This reflected MG's harmonious labour relations, the absence of machinery down time, simpler product and assembly design, and the ability of the practical men to supervise rudimentary operations. In other words, MG was one of the most efficient plants in an otherwise very inefficient firm.

The obvious drawback of MG's labour-intensive method was limited annual output. Total annual production never exceeded 56,000 units per year, and it is doubtful that output could have been increased beyond this level without a change in methods or location. In 1964 BMC's cost accountants warned of increased costs when MG built 55,400 units.[41] This is consistent with the sudden and steep increase in marginal costs as output at a low-fixed-cost plant meets designated capacity. Conversely, during periods of low demand for sports cars, sedans were assembled at Abingdon to reduce labout costs, especially after the introduction of MDW. The plant's cost structure and the historic small fluctuation in annual output suggest that a cost 'window' existed whereby Abingdon was efficiently utilized. In a theoretical comparison to a mechanized factory, Abingdon had a lower break-even point and lower average total costs at low output levels, but higher marginal costs.[42]

Restricted annual output, antiquated plant, pre-Fordist assembly methods and chronically underutilized volume car factories seemed to make a case for the transfer of MG assembly to a mechanized plant. This option was bypassed in the 1950s and 1960s when BMC expanded capacity by simply duplicating existing machinery instead of instituting new production technology.[43] British Leyland moved to more capital-intensive methods in the 1970s, but the mechanized Gateline assembly system employed to build the mass market Marina, which shared engines with the MGB and Midget, was dedicated to producing only the volume car in high annual output.[44] Moreover, the coordination of assembly processes and basic product specifications necessary for compatible multi-model production was virtually impossible as long as sports and volume car design and engineering remained separate.

The labour-intensive methods used at Abingdon simultaneously reflected and corresponded with the industry's rigid production institutions. BMC was simply unable to produce a variety of low volume niche products using modern methods. The labour-intensive technique was the only option capable of achieving the flexibility and control needed to produce sports cars profitably. In 1974 each MGB returned a profit of £368 which compared favourably with the average unit profit of £38 earned by British Leyland in 1969, one of the corporation's most profitable years.[45]

4.2.2. Mechanized Sports Car Production: Triumph

After noting the 'relatively large' unit profits of sports cars during the 1960s, the Triumph Board argued in 1971 that the profitability of the division, which also offered mass market and executive models, depended upon sports car output.[46] However, a comparison of unit profitability reveals the developing productivity crisis within the indigenous firms since Triumph used more mechanized sports car production methods. The £57 unit profit of the GT6 sports car was the highest in Triumph's range during the mid-1960s, but Ford (UK) volume cars returned an average unit profit of £54.[47] The comparison with Ford highlights the low profitability of Triumph's volume cars and indicates that sports car manufacturing costs were relatively high in relation to market position.

Triumph's assembly methods theoretically should have lowered unit costs and raised unit profits compared to MG. This was especially true since Triumph utilized both flexible mechanized and traditional dedicated assembly techniques. The flexible assembly originated from the common Zobo design of the Spitfire and GT6 sports cars and the Herald and Vitesse volume cars. The flexible method offered the product-mix control inherent in labour-intensive assembly combined with the potential for higher annual output at lower cost. Whereas dedicated assembly tracks have to be operated at sub-optimal speeds or shut down in periods of low demand, the multi-model line could be balanced with little cost penalty.[48]

Nevertheless, the command and coordination problems that plagued volume car production were emerging at Triumph. Poor supervision prevented the firm from consistently balancing the product mix to meet demand, thus negating the main advantage of the flexible method. Only output of the GT6 and Vitesse sedan, both built in very low annual volumes, met or exceeded company production targets.[49] The Triumph Board also noted that the Zobo and dedicated TR sports car lines rarely operated at designated speeds, resulting in shortages of sports cars. Numerous problems reduced the track speeds and raised off-standard rates. Triumph was plagued with recurrent disputes over piece-work rates, manning levels, and work loads. Chronic absenteeism and late starts/early finishes were routine. Limited capital and a reluctance to invest in new technology kept old and often unreliable machinery in service. Work flow was disrupted by frequent shortages of bought-out components resulting from production problems and strikes at independent suppliers. The Board acknowledged that the only sports car that could be manufactured and assembled profitably with dedicated mechanized techniques was the high-priced TR series.[50]

Similar problems in corporate manufacturing plants exacerbated the situation downstream. During the 1960s the Triumph Board attributed the inability of the Speke no.1 facility to consistently meet TR sports car bodyshell targets to a 'lack of management at all levels', 'rigid practices of labour', and the duplication of functions at the assembly and manufacturing plants.[51] The Speke Works Manager replied that the problems were rooted in bottlenecks in the production process, inefficient machinery, and insufficient capacity.[52] At one point management actually welcomed low bodyshell output because design and casting difficulties in its foundry subsidiary had reduced production of engines. Low-quality components often overwhelmed rectification staff. Management, frustrated by high scrap rates, increased part supplies simply by lowering acceptance standards.[53]

Various measures by BL failed to reduce off-standard rates and in some cases actually increased them. The firm agreed to union demands for generous manning levels in return for the institution of MDW as part of the move to capital-intensive production methods. Labour also used mutuality to receive wage or effort concessions in exchange for the introduction of cost accounting or industrial engineering exercises. The habitual use of additional labour and overtime to increase output raised variable costs further and revealed management's misunderstanding of highly mechanized production. Capital replacement programmes were undertaken without regard to work organization and updated technology, resulting in bottlenecks and low product quality. Management resorted to holding large stocks of bought-out components to avoid production disruptions caused by outside sources. The failure to consolidate basic functions and responsibilities generated interdivisional and division–central staff conflicts within BL's multidivisional structure, which in turn prevented the development of a coherent corporate strategy and managerial hierarchy.

4.2.3. Capital-Intensive Production Methods: British Leyland and Nissan

Consistent demand in North America, based on the cars' differentiated design and the absence of non-British competition, combined with shared components and flexible traditional assembly techniques, allowed Triumph and MG to maintain profitability. In the 1970s, however, the sales and unit profit results of the Triumph TR7 became analogous to that of British volume cars when Datsun ended the British monopoly and BL discarded the historic sports car production formula. The TR7 provided a clear example of BL's inability to integrate resources, design and engineer a car according to market demand, and then plan and execute capital-intensive production.

Triumph anticipated that first-year (1974–5) sales of the TR7 would reach 52, 844 units and rise to 64, 417 units the following year. Although this estimate was roughly equivalent to yearly Datsun and combined MG and Triumph annual sales in the USA, the TR7 sales forecast was based more on corporate production requirements than realistic market prospects. No British model had ever achieved Triumph's projection. Annual output of the TR7 had to reach these targets for BL to achieve standard utilization rates at the proposed capital-intensive high-volume engine and assembly sites.[54]

Triumph's strategy dissolved with serious cost consequences for the sports car. Output of the volume cars that were to share the TR7 engine and assembly facilities declined markedly. The vastly underutilized plant, declining corporate sales, and Nissan's growing US market share engulfed the TR7 project with urgency. Much to the dismay of the Specialist

Car Division's Chief Engineer and Director of Manufacturing, the development period was abbreviated to accelerate pilot production.[55] The design had been conceived to simplify production and ensure product quality.[56] It did neither. During the bench and road testing period, 300 quality problems were identified by engineering and the engine failed the EPA exhaust test.[57] Each of the pilot cars required at least 20 hours' rectification, forcing the suspension of production for three weeks while 40 major and 1300 minor design and manufacturing modifications were made.[58] The Chief Engineer reported that some significant flaws could be resolved only by re-engineering basic specifications, but management refused to delay the January 1975 launch. Instead the rectification area was enlarged.[59] Unlike the Japanese who emphasized quality throughout the design and manufacturing stages,[60] BL viewed quality control as the last phase of the production process. Creative quick-fixes could not overcome fundamental engineering defects and the emphasis upon the production schedule. The TR7 was one of the most unreliable cars available in the US market during the 1970s, just as the consumer awareness movement was reaching its height.[61]

By the time sports car production came on-stream in late 1974, it was clear that the TR7 would carry the costs of a unique engine and a highly mechanized dedicated assembly facility. During the first month of full production, the Specialist Car Division Board learned that TR7 assembly off-standard rates were very high, and consequently the labour count was out of proportion with lower than anticipated output. The efforts of corporate and division industrial engineers failed to improve volume and efficiency. In April 1975 the Chief Engineer reported to a concerned Board that his engineers were considering new manufacturing techniques to raise weekly output from about 300 units to the reduced plan of 800 units.[62]

The Chief Engineer was caught in a cost catch-22. Even if he had managed to achieve the model-specific target, the assembly and engine plants would have remained vastly underutilized since the volume cars had been withdrawn. Weekly TR7 output briefly attained 695 units (70 per cent capacity utilization rate) in 1977, but model-specific engine and assembly capacity utilization rates rarely surpassed 50 per cent when extrapolated on a yearly basis during three years of TR7 production at the Speke no.2 facility.[63] Management's unilateral imposition of lower manning levels and work reorganization in autumn 1977 suggests that off-standard rates had remained excessive. Moreover, sales resistance in the USA, resulting from low characteristic and quality levels compared to the Datsun Z series, was so strong that BL was forced to lower, not raise, prices and hold a substantial number of cars as inventory. A Speke union official's claim that each TR7 built had lost £850[64] was a reasonable estimate.

BL tried to preserve the TR7 by transferring assembly in 1978 and again in 1979 to lower-capacity sites, instituting more flexible production methods, correcting the engineering defects, and spinning-off innovative convertible and high-performance versions. It was not a coincidence that steady improvement in TR7 capacity utilization and off-standard rates corresponded with the gradual return to traditional methods. However, the initial failure to integrate design with production techniques and market target created an enduring negative product image that prevented sales from responding. In 1981 BL, struggling to survive, withdrew from the niche the British had created.

As in design, Nissan appeared to modify the basic principles developed by the British to produce sports cars. Nissan centred sports car production at its Shin-Nikkoku (Nissan Auto Body) subsidiary, which also manufactured bodies and assembled all the firm's niche products such as light trucks, buses, and all-terrain vehicles. The plant specialization and lower labour wage rates at NAB resembled BMC's decision to locate sports car assembly at Abingdon. Nissan's integration of body production and sharing of plant fixed costs among models was similar to Triumph's production strategy.

Unlike the British, however, NAB benefited from the high effort-to-wage levels of labour, increasing use of high speed machinery and automated processes and computerized control and coordination of production. The modifications to manufacturing techniques resulted in dramatic increases in productivity levels. Beginning in the 1960s Nissan and its subsidiaries increasingly relied upon automation and computerization to reduce costs by controlling production. By the 1970s, sophisticated computers monitored component supply, assembly track speed, and product mix at Nissan and NAB.[65] Computerization provided Nissan with the supervision and flexibility missing in BL's mechanized TR7 build system.

Interestingly, Hutton claims that the average monthly capacity utilization rates of the Z series assembly tracks was about 60 per cent, [66] which was slightly higher than the TR7 line, but below the standard rate of 80 per cent and Nissan's corporate rate of between 82 and 97 per cent.[67] How could Nissan continue to reap considerable unit profits while TR7 losses forced BL to withdraw its model? Two factors provided Nissan with a comparative advantage. TR7 output had to achieve high-capacity utilization rates to offset the high fixed costs of dedicated plant and machinery and high variable costs of the large labour count. The inability of BL to manage its chosen production system resulted in extraordinary off-standard rates and inflated the cost structure. In contrast, Nissan's production technique virtually eliminated off-standard rates thus raising the productivity of labour and machinery at relatively low annual output levels.[68] The innovative design and higher quality of the Z Series allowed

Nissan to increase annual sales while substantially raising product prices. The combination of high productivity and considerable unit profit margins factored into prices provided the Z series with a break-even point far lower than the TR7 or even Nissan's compact mass market sedans.

4.3. Conclusion

The study of semi-specialist niche products must be conducted by examining model-specific dimensions with the understanding that product differentiation by characteristics and image must be strong enough to overcome higher production costs compared to mass market models. The sports car producers had to pass the survivor test of continuously offering sufficiently distinguished products and lowering production costs. Therefore the connection between design and production was even more important than in the mass market where price is a major factor in competition. The British successfully designed models with characteristics in demand until the development of the TR7. Production rather than market considerations influenced the development of the model resulting in dismally low sales which exacerbated existing production problems.

The British could only achieve success in sports car production by adhering to processes essentially formulated in the late 1940s. BL could not combine the fundamental concepts of flexible assembly, common components, innovative design, and niche market demand with Fordist production techniques and sophisticated engineering. This failure—the deficiencies were especially pronounced in management and engineering—starkly revealed the rigid institutions that existed in the industry. In contrast, Nissan proved capable of learning from and then modifying the basic principles developed by the British to meet the market requirements of the 1970s. The Japanese firm acquired its competitive advantage by retaining innovative design and innovating basic production methods.

NOTES

1. MG was part of Morris Motors/Nuffield Organisation since the sports car company was founded in 1923. Austin briefly offered Austin-Healey before the merger with Nuffield in 1952 created the British Motor Corporation. In 1954 Triumph sports cars were introduced by Standard-Triumph, which, after nearing bankruptcy in 1961, was purchased by Leyland Motor Corporation. In 1968 Austin-Healey, MG, and Triumph sports cars came together when Leyland-Triumph merged with BMC to form British Leyland Motor Corpor-

ation. British Leyland was nationalized in 1974 after the firm experienced a severe cash-flow crisis. Although Rootes built a sports car range, this paper will concentrate upon Austin-Healey, MG, and Triumph. For a complete history see Timothy Whisler, *At the End of the Road: The Rise and Fall of Austin-Healey, MG and Triumph Sports Cars, 1945–81* (Greenwich, CN, 1995).

2. Richard Whipp and Peter Clark, *Innovation and the Auto Industry: Product, Process and Work Organization* (London, 1986), 54–7.

3. British Motor Industry Heritage Trust (hereafter BMIHT), minutes of the Morris Motors Board of Directors Meeting, 15 Oct. 1947.

4. e.g. the 1946 MG TC Midget was so similar to the 1939 TA Midget that components ordered prior to the war and delivered in January 1946 were sent directly to the assembly line. BMIHT, minutes of the Morris Motors Board of Directors Meeting, 11 Jan. 1946.

5. George Maxcy and Aubrey Silberston, *The Motor Industry* (London, 1959), 176–8.

6. Modern Records Centre (hereafter MRC), MSS 226/ST/1/1/14, minutes of the Standard-Triumph Board of Directors Meeting, 10 Oct. 1963 and MSS 226/ST/2/3/2, GT Spitfire Project, 4 May 1964.

7. Robert R. Locke, *The End of the Practical Man: Entrepreneurship and Higher Education in Germany, France and Great Britain, 1880–1940* (Greenwich, CT, 1984), 101.

8. David C. Mowerey, 'Industrial Research', in Bernard Elbaum and William Lazonick (eds.), *The Decline of the British Economy* (Oxford, 1986), 189–222.

9. MRC, MSS, 226/ST/1/1/15, minutes of the Standard-Triumph Board of Directors Meeting, 10 Feb. 1966.

10. Tony Felmingham, former MG manager, interview with author, May 1988.

11. John B. Rae, *Nissan/Datsun: A History of Nissan Motor Corporation in USA, 1960–80* (New York, 1982), 112.

12. Merton Peck, 'Science and Technology', in R. E. Caves (ed.), *Britain's Economic Prospects* (Washington, 1968), 473–4. Peck's general observation applied to the motor industry. For a discussion of the British motor industry's reluctance to cooperate in engineering research, see *The Times*, 1 Dec. 1966, 11.

13. MRC, MSS 226/ST/1/11/1, minutes of the Advisory Board Meeting, Specialist Car Division, 30 May 1973.

14. MRC, MSS 226/ST/1/7/1, minutes of the Triumph Board of Directors Meetings, 12 Feb. 1971, 12 Mar., 1971, and 9 July 1971.

15. Ibid., 27 Apr. 1973 and 29 June 1973.

16. Michael Cusumano, *The Japanese Automobile Industry: Technology and Management at Nissan and Toyota* (Cambridge, Mass., 1991), 90, 241, 375.

17. Ibid., MSS 266/ST/1/7/1, minutes of the Triumph Board of Directors Meeting, 9 July 1971.

18. D. G. Rhys, 'European Mass-Producing Car Makers and Minimum Efficient Scale: A Note', *Journal of Industrial Economics*, 25 (1977), 313–20.

19. C. F. Pratten, *Economies of Scale in Manufacturing Industry* (Cambridge, 1971).

20. Rhys, 'European Mass-Producing Car Makers', 315–16.

21. George J. Stigler, 'The Economies of Scale', *Journal of Law and Economics* 1 (Oct., 1958), 56.

22. John Barber, former Managing Director and Vice-Chairman of British Leyland Motor Corporation, interview with author, May 1988.

23. Cusumano, *Japanese Automobile Industry*, 108.

24. *Motor Business*, 28 (Oct. 1961), 44.

25. The source for all British production figures is BMIHT, Austin Rover Group Production Records.

26. MRC, MSS 226/ST/1/1/14, minutes of the Standard-Triumph Board of Directors Meetings, 16 May 1961, 19 Dec. 1961, and 2 Aug. 1962.

27. MRC, MSS 226/ST/1/1/15, minutes of the Standard-Triumph Board of Directors Meeting, 6 May 1965.

28. *Autocar* (3 May 1975), 16; (10 May 1975), 16.

29. British Parliamentary Papers 1974–5, XXV, The Fourteenth Report from the Expenditure Committee (Trade and Industry Sub-Committee), *The Motor Vehicle Industry*, para. 99, Q 2287 and Q 2289.

30. Pratten, *Economies of Scale*, 135–6.

31. Ibid. and MSS 226/ST/3/A/PR/1, Triumph Sports Car Costs, 12 Nov. 1952.

32. Pratten, *Economies of Scale*, 136.

33. Wayne Lewchuk, *American Technology and the British Motor Vehicle Industry* (Cambridge, 1987) and 'The Motor Vehicle Industry' in B. Elbaum and W. Lazonick (eds.), *The Decline of the British Economy* (Oxford, 1986), 135–61.

34. Paul Willman and Graham Winch, *Innovation and Management Control: Labour Relations at BL Cars* (Cambridge, 1985), 58–60.

35. Ibid.; *Motor* (8 Apr. 1959), 355–6; and Tony Felmingham, former MG manager, interview with author, May 1988.

36. Political and Economic Planning Group, *Motor Vehicles* (London, 1950), 6.

37. Tony Felmingham, interview with author, May 1988.

38. D. G. Rhys, 'Employment Efficiency and Labour Relations in the British Motor Industry', *Industrial Relations*, 5/2 (summer 1974), 6.

39. *Motor* (16 Nov. 1972), 33.

40. Tony Felmingham, interview with author, May 1988.

41. BMIHT, note to Sir George Harriman from Thornton Baker and Co. Chartered Accountants, 15 Nov. 1965.

42. D. G. Rhys, *The Motor Industry: An Economic Survey* (London, 1972), 277–8.

43. Karel Williams, John Williams, and Dennis Thomas, *Why are the British Bad at Manufacturing?* (London, 1983), 220–1.

44. Willman and Winch, *Innovation and Management control*, 49.

45. MRC, MSS 226/ST/1/11/1, minutes of the Advisory Board Meeting, Specialist Car Division, 27 Nov. 1974 and Rhys, *The Motor Industry*, 329.

46. MRC, MSS 226/ST/1/10/1, minutes of the Triumph Executive Management Meeting, 9 Sept. 1971.

47. MRC, MSS 226/ST/2/3/2, GT Spitfire Project, 4 May 1964 and British Parliamentary Papers 1974–5, XXV, The Fourteenth Report from the Expenditure Committee (Trade and Industry Sub-Committee), *The Motor Vehicle Industry*, para. 31.

48. Pratten, *Economies of Scale*, 144.

49. MRC, MSS 226/ST/1/1/5, minutes of the Standard-Triumph Board of Directors Meeting, 7 July 1966; MSS 226/ST/1/10/1, minutes of the Triumph

Executive Management Meeting, 1 July 1971, and MSS 226/ST/1/7/1, minutes of the Triumph Board of Directors Meeting, 11 Dec. 1970.

50. MRC, MSS 226/ST/1/10/1, minutes of the Triumph Executive Management Meeting, 9 Mar. 1973, and MSS 226/ST/1/1/14, minutes of the Standard-Triumph Board of Directors Meetings, 4 June 1964, 16 Jan. 1964, 2 June 1963, and 6 June 1963.

51. MRC, MSS 226/ST/1/1/14, minutes of the Standard-Triumph Board of Directors Meetings, 6 Sept. 1962, 8 Nov. 1962, 16 Jan. 1964, and 5 Dec. 1963.

52. Ibid. 4 June 1964.

53. MRC, MSS 226/ST/1/1/15, minutes of the Standard-Triumph Board of Directors Meeting, 10 Sept. 1964; and MSS 226/ST/1/1/14, minutes of the Standard-Triumph Board of Directors Meeting, 2 Apr. 1964.

54. MRC, MSS 226/ST/1/10/1, minutes of the Triumph Executive Management Meeting, 11 June 1971.

55. MRC, MSS 226/ST/1/11/1, minutes of the Advisory Board Meetings, Specialist Car Division, 11 Mar. 1974, 4 Apr. 1974, 2 May 1974, and 29 May 1974.

56. *Motor* (25 Jan. 1975), 42.

57. MRC, MSS 226/ST/1/11/1, minutes of the Advisory Board Meetings, Specialist Car Division, 31 July 1974 and 28 Aug. 1974.

58. Ibid. 30 Oct. 1974 and 27 Nov. 1974 and Whipp and Clark, *Innovation and Auto Industry*, 161.

59. MRC, MSS 226/ST/1/11/1, minutes of the Advisory Board Meetings, Specialist Car Division, 25 Sept. 1974 and 30 Oct. 1974.

60. Cusumano, *Japanese Automobile Industry*, see ch. 6.

61. *Consumer Reports*, Aug. 1980, 499.

62. MRC, MSS 226/ST/1/11/1, minutes of the Advisory Board Meetings, Specialist Car Division, 29 Jan. 1975 and 30 Apr. 1975.

63. BMIHT, Austin Rover Group Production Records.

64. *The Times*, 9 May 1978, 3.

65. Cusumano, *Japanese Automobile Industry*, 196, 198, 200, 215, 316–18.

66. Ray Hutton, *The Z-Series Datsuns* (London, 1982), 126.

67. Cusumano, *Japanese Automobile Industry*, 198.

68. Ibid. 200 and Martin Kenney and Richard Florida, *Beyond Mass Production: The Japanese System and its Transfer to the US* (New York, 1993), 8, 14–17. Martin and Kenney argue that computerized control of production and continuous process innovation, including worker empowerment, are elements of inno-vation-mediated production. The authors contend that this Japanese devised production organization is a supercession to Fordism.

5

The Road from Dreams of Mass Production to Flexible Specialization: *American Influences on the Development of the Swedish Automobile Industry, 1920–39*

NILS KINCH

5.1. The Logic of Mass Production and the Swedish Automobile Industry

The Ford way of organizing production had a great impact on the Swedish debate in the 1920s on the possibility of establishing a domestic industry. Irrespective of the conclusions drawn about the prospects for Swedish manufacturers to match their American counterparts, references were made by the debaters to the mass production of standardized cars. Those who believed in the possibility of establishing a competitive automobile industry on a fairly large scale proposed that potential entrants should look to the mass market and should enter those segments where the American manufacturers were already selling large quantities. It was assumed that a competitive advantage could be developed for cars and light trucks of Swedish origin, with higher quality and better durability than their American counterparts. It was thus suggested that Swedish manufacturers would have the potential more or less to mimic the American manufacturers and to compete head-on in these mass markets for standardized products.

The Ford system of mass production was based on a few basic principles: the standardization of products, subdivision of the work, and the extensive use of machinery to replace manual labour (see Hounshell 1984). Contrary to the common belief, a precondition for the mass production system is not the continuous assembly line but 'the complete and consistent interchangeability of parts and the simplicity of attaching them to each other' (Womack *et al.* 1990: 27). The principle of not catering to any customer's individual demands in production planning is one of the keys to the mass production system. The focus on the production of a single model, or a very limited range of models, allows the full use of a mechanized work process and specialized machinery. However, an automobile manufacturer organized along these lines becomes rather inflexible and vulnerable to variations in demand. Because of high fixed costs, a minimum proportion of the planned production capacity has to be reached for the system to be effective. Thus a precondition for the effectiveness of mass production logic is that a large and stable demand can be guaranteed. A company operating on a standardized mass market has to be able to offer an attractive model at a price that is on a par with that of its competitors. This means that it is very important in the development of new models to follow the general trend very closely, and to put on the market a car that appeals to the customers and reaches volumes that enable the use of the most efficient production methods.

Against this background it must have been questioned whether it would be possible to design a competitive car in Sweden (a country with a population of slightly more that 6 million in 1925), to organize a rational production system, and to build up a sales organization that could match the imported mass-produced American car.

The position of the Swedish automobile industry seems to be unique in a global perspective.[1] Despite a very limited domestic market for automobiles, Volvo and Scania-Vabis have become two viable actors in the global automobile industry. The way they developed in the early years in order to reach this position will be examined below. The description will primarily be limited to the period up to 1939 and will focus mainly on Volvo.

As will become evident below, an analysis only centred on the introduction of Fordism would in the Swedish case be too limited. Thus, the following discussion will examine the impact that influences from the American automobile industry in general have had on Swedish manufacturers' choice of product/market strategy and production methods. In this analysis attention will focus on the conceptualization of a Swedish car industry that emerged from the debate at the time Volvo was founded, and on the direct implications of American competition on the development of Swedish efforts to start a domestic car industry. The natural point of departure for an overview of influences from the United States on the development of the Swedish automobile industry is an account of the situation prevailing in the 1920s. By that time motoring had achieved its first breakthrough, which resulted in rapidly increasing imports of American cars. This stimulated a debate on the possibility of establishing a domestic industry to replace a portion of the imports, which were deemed to have reached 'alarming' proportions. The question of custom duties on automobiles was also taken up in the Riksdag (the Swedish Parliament) in the mid-1920s, and a series of investigations were launched to discover the preconditions for an increase in the production of cars. Besides the call for the establishment of a national industry, efforts were also made to encourage the American manufacturers to establish assembly units in Sweden. With the exception of Saab, which was not founded until 1949, all the companies that have ever made any serious attempts to establish large-scale production of automobiles in Sweden were already active on the scene in the 1920s. The years between the end of World War I and the beginning of World War II can thus be described as a dramatic period in the history of the Swedish automobile industry.

The paper will be organized as follows. As a start, an overview of the automobile industry in Sweden at the time Volvo was established in 1926 will be given. In the next two sections the discussion in the mid-1920s on the possibility of establishing a domestic car industry and the debate in the Riksdag on the issue of duties on cars is summarized. This is followed by a description of the development of AB Volvo up to 1939. Next, different ways in which influences from the American automobile industry affected the Swedish manufacturers will be presented. Further, the unsuccessful attempts to compete on a mass production market and the niche strategy that emerged instead will be discussed. Finally, a very brief comparison between the Volvo and Toyota production systems will be made, followed by a conclusion.

5.2. An Overview of the Swedish Automobile Industry, Except for Volvo, up to 1939

A distinctive feature in the history of the Swedish automobile industry is that the number of companies that have tried to enter the industry is very small. Apart from the manufacturers presently on the scene, only two other major ventures have been started.[2] This might tempt us to conclude that the industry has developed in a fairly unproblematic way. However, this is not the case at all. Scania-Vabis, and on certain occasions Volvo, too, have been in serious trouble and a discontinuation of the business has been considered. Saab has over the years also faced problems.[3]

When plans for a major car industry first came up in Sweden around 1925, two companies were already operating in the automobile industry. In order to provide some background to the debate and the general situation prevailing at the time of the establishment of the Volvo company, an overview of the development of the automobile companies operating in Sweden will be given below. As only one company survived—Scania-Vabis—most of the description will be related to this company, and a brief account of its development up to 1939 will be given. The history of AB Thulinverken, Tidaholms Bruk, and the assembly units established by foreign automobile manufacturers in Sweden can provide our point of departure.

5.2.1. The Failures

The first major attempt to establish an automobile industry that ended in failure was at AB Thulinverken.[4] During World War I the company had manufactured aircraft for the Swedish air force. In an attempt to change from wartime to peacetime production, Thulinverken started a major project to produce a car of German design; a volume of 1,000 units was planned. This project was undercapitalized and the timing was unlucky, as it coincided with the start of the depression in 1920. Only 300 cars were produced before the company went into the hands of the receivers. Later efforts to restart the project coincided with the establishment of AB Volvo, which may have deterred any potential investors from supporting the risky venture.

The second company to go out of business was Tidaholms Bruk.[5] This company produced its first automobile in 1903, and went on to become a manufacturer of trucks, buses, and special vehicles designed to customer specifications. Though it produced a few cars, it soon dropped this side of the business. The company had an almost entirely integrated manufacturing plant and an annual capacity of about 150 units. In the late 1920s and early 1930s it introduced some successful new designs for heavy trucks and buses, and in order to meet increasing demand production capacity was expanded. The greater financial burden resulting from this expansion caused the company serious difficulty when demand slumped in 1932 at the onset of the depression, and in 1933 the company went into the hands of the receivers.

Common to both these unsuccessful companies was the fact that they lacked the support of a financially strong owner and thus were not given a second chance.

5.2.2. Foreign Assembly Units

In 1928 General Motors established a production unit in Stockholm, and the same year 10,000 cars were assembled from imported components. Later, in the 1930s, they were followed by other American automobile manufacturers and a number of assembly units operated by the Swedish distributors of American cars. Until World War II these assembly units and car imports totally dominated the market for automobiles in Sweden.

The economic rationale for these operations was that transportation costs and customs duty were lower for parts than for complete cars, and Swedish wage rates were also lower. After the war the assembly of American cars in Sweden gradually diminished in importance. This may be explained partly by import restrictions on cars and car components after 1947, and partly by a switch in demand away from large, heavy, petrol-consuming cars in favour of small European cars. As a result General Motors and the Ford Motor Company closed their Swedish assembly units in 1957 (see Giertz 1991; Glimstedt 1993).

5.2.3. Scania-Vabis

In 1925 Scania-Vabis had a workforce of 329 and produced 190 trucks and buses a year in more or less wholly integrated factories, using basically manual production methods.[6] The company can be described as one of the pioneers in the automobile industry, as its roots go back to 1896. Scania-Vabis is the result of a merger between two companies in 1911: Scania (founded in 1900) and Vabis (established in 1891). Scania had started with the production of bicycles but extended its operations to automobiles in 1903. Vabis started by supplying the railway companies with carriages, but as a result of severe competition in this business was looking for other uses for its production capacity. It produced its first automobile in 1897. Scania-Vabis produced cars, buses, trucks, and special vehicles such as fire engines.

In 1919 plans for a considerable extension of operations were drawn up. According to the new strategy the company was to concentrate on the production of standardized trucks. The idea was to transform Scania-Vabis into a large-scale export industry, ready to face the competition of cheaper foreign trucks when the expected increase in the demand for transportation by truck materialized. By then the truck version of the Model T-Ford dominated the Swedish market for trucks.

Scania-Vabis was building up its resources for rapid expansion in the years immediately preceding the recession following World War I. This year proved to be a disaster for the company. The automobiles produced by Scania-Vabis were expensive and outdated, and in 1920 only 81 trucks and 80 cars were delivered. The situation for Scania-Vabis was aggravated by a lengthy labour dispute in 1920. This, together with soaring demand and heavy debts, put a great strain on the company's liquidity and in 1922 it went into the hands of the receivers. Scania-Vabis's biggest creditor was the Stockholms Enskilda Bank, which took over control of the operations with a view to securing their claims. The company was given a period of grace, so long as it did not make any demands on its owner.

The attempt to establish a position on the large market for standardized trucks had thus failed. The new company was to develop along a totally different road. It was to concentrate on the demand from institutional customers for heavy vehicles with special characteristics. In close cooperation with its customers Scania-Vabis acquired experience and succeeded in designing engines and chassis of excellent quality in performance. These institutional automobile users had their own repair and maintenance shops. Thus, without any investment in a sales or service organization, customer-driven product development and a process of consolidation of the company were initiated. By 1927 the operations were profitable again, and by the following year the accumulated losses had been made up. However, investment in the operation was almost non-existent. According to Giertz (1991), Marcus Wallenberg at the Stockholms Enskilda Bank was considering the possibility in the early 1930s of selling Scania-Vabis to the newly established but rapidly expanding Volvo company.

In 1938, 205 buses and 89 trucks were sold, or 294 units altogether. That year 38 vehicles were exported, representing 20 per cent of the total turnover. During this period the sales of trucks and buses in Sweden had multiplied, but Scania-Vabis's development did not match this growth in the market.

The workshop at Scania-Vabis was manned by a highly skilled team of craftsmen and was very flexible, which suited the strategy of producing to the customer's orders. However, the incentive to standardize the products and to organize production along rational lines was low. Although the new managing director, Gunnar Lindmark, was one of the pioneers in using time studies in Sweden, he applied his experience to the organization of work at Scania-Vabis only to a very limited extent. The engines and chassis were produced directly from blueprints, and the design of the working methods was left to the workers. Available sources indicate that the way production was organized left much to be desired as regards discipline, order, and the principles for calculating wages. It was not until 1937 that a specialist on rational production methods was appointed as head of the workshop at Scania-Vabis.

5.3. Calls for a National Automobile Industry

Around 1925 suggestions appeared in trade journals and elsewhere about the necessity of establishing a Swedish car manufacturing industry to reduce dependence on the rapidly increasing importation of foreign cars.[7]

5.3.1. A Quest for a Swedish Car Manufacturing Industry

The point of departure for this discussion was the American dominance in the car manufacturing industry, and the measures that European companies could take to counter it. The American strategy was based on the standardization of products and the production in long runs of cars designed to appeal to the needs of large segments of the population. Up to that time the European manufacturers had been producing quality cars in short runs for a limited group of customers. The new situation, it was explicitly stated, called for a changeover to American production methods, and for the focus on small groups of customers to be abandoned. The primary reason why the question of a Swedish automobile industry was taken up just at this time was the large, and increasing, amount of imports, but another reason was idle capacity in the Swedish mechanical engineering industry. A few Swedish companies had already started to manufacture parts for the automotive industry, and some of them had even started exporting successfully to the American automobile industry. One of the companies referred to was probably SKF, which produced ball bearings and which by then could already boast of large exports and had a number of subsidiaries abroad.

It was concluded that if it was possible to make a profit from delivering parts manufactured in Sweden to the American manufacturers, it would also be feasible to produce a complete car for the Swedish market at competitive prices. This car should be positioned as regards quality and price above the Ford Model-T, and, as the trend was set by American cars, it should have a sturdy and American-influenced design. In the situation of the automobile industry at the time it was not considered too difficult to choose a suitable design. The car had to be adapted to Swedish conditions and to fulfil the requirements expected of a modern car abroad.

In this debate developments in the UK were held up as an example for Sweden to follow. It was claimed that, without the protection of a customs barrier, the automobile industry in the UK had managed to capture 80 per cent of the market. From this it was concluded that a Swedish automobile industry would be able to capture 50 per cent of the demand, provided the model and the price were suitable. Swedish imports amounted to 12,000 cars in 1924, and the figure was expected to increase considerably over the next few years. It was claimed that at a volume of 6,000–8,000 units rational production at competitive prices would be possible.

This very optimistic prediction was based on incorrect assumptions about conditions in the UK automobile industry. Since 1915, with the exception of a period between 1 August 1924 and 30 June 1925, the industry had benefited from protection at 33.33 per cent custom *ad valorem* on imports of automobiles and automotive parts (cf. Sloan 1964: 318). Added to this, the taxation of vehicles and fuel discriminated against

American cars with their larger engines. This protection is not mentioned at all in any of the articles reviewed, in spite of its importance in explaining the success of the automobile industry in the UK *vis-à-vis* the American manufacturers.

The reason why no production had yet been established in Sweden was that until then demand had been insufficient to allow for rational production. However, a lack of entrepreneurial spirit was also mentioned. The way a Swedish automobile industry should be organized was more or less taken for granted. All that was needed was the establishment of an independent plant for assembling the parts produced by different companies in the Swedish mechanical engineering industry. It was assumed that several Swedish companies already possessed the necessary competence and machinery. This meant that the required investment in machinery would be limited, and a Swedish automobile industry could be made possible simply by coordinating the resources already available in the country. In 1925 the establishment of a car manufacturing industry in Sweden was thus regarded as primarily a question of organization. One of the journals concluded that all the conditions and requirements that were needed for the production of automobiles on a larger scale than hitherto were now to hand. 'What we still lack is the person who will take on the task, and who has the ability to gather all the good forces around himself' (*Teknisk Tidskrift*, 1 May 1926, 164).

5.3.2. *Focus on Trucks and Buses at First, Instead of Cars*

During the debate the situation regarding the production of trucks was also touched upon. This part of the business differed in that a wide range of models was needed to satisfy the variations in demand. Sweden already had two companies producing large trucks and buses; in quality and price they could match the foreign competition. However, there was no production of small models with a loading capacity of $1/2$ to $1^1/2$ tons. This was by far the type in most demand, and the demand was mainly covered by the Ford 1-ton model, which was described as cheap but of limited durability. It was suggested that a Swedish quality product would have a chance of finding a market in this segment. Whether this production of small trucks should be combined with a potential new production unit for cars, or whether it should be taken up by one of the existing truck manufacturers, depended on what model of car would be produced. There was an opportunity to use the front part of the car for the truck, and thus to lower the production costs. One drawback of the approach, it was pointed out, was that such a division of resources might make both lines of production unprofitable.

Not everyone shared the optimism about the possibilities for establishing a competitive automobile industry at that time. It was argued that Sweden had been hopelessly left behind, and that the American advantage was too great unless it were possible 'to count on strong support from the state and the workers and injection of sufficient capital' (*Svensk Motortidning*, 22 November 1925, 551). This was not considered a realistic expectation, and instead of a bold large-scale car venture, a long and gradual development whereby conditions were improved step by step was suggested. The main difficulty lay not in organizing the available resources, but in raising the substantial amount of capital that was needed to develop an effective sales organization. The prospect of developing Swedish car production on the basis of the existing production of buses was thought to be far more promising. It was assumed that the venture's chances of success would increase, if the experience of the Swedish bus manufacturers and the resources they could make available were utilized. 'However, a development along these lines might take several years, but the success would without any doubt be considerably more reliable' (ibid.).

5.4. The Riksdag on Car Import Duties

The question of the establishment of a Swedish automobile industry also came up when the Riksdag discussed the question of duties on cars during the period 1924–7.[8] The subject was first brought up in 1924 when a motion was put forward that the customs duty on automobiles shoud be increased from 15 to 30 per cent on assembled cars. This demand for increased protection was initiated by one of the companies in the industry, Tidaholms Bruk. In the following year a member of the Riksdag, on behalf of the Ford Motor Company, moved a resolution to reduce the custom tariff on parts and components used in the assembly of cars.

5.4.1. Arguments For and Against a National Industry

One way to sum up the debate that followed until the issue was finally determined in 1927 is to say that it focused above all on what would primarily benefit industrial development in Sweden: a Swedish-owned automobile industry, or assembly units controlled by foreign manufacturers (Nordlund 1989: 155). In 1927 the plans for AB Volvo were widely known, although no car had yet been produced. With this in mind many members of the Riksdag felt uncomfortable about reducing the customs duty on imported parts for automobiles, in order to promote the establishing of foreign-owned assembly units. The argument in favour of a national industry was based on the need for automobiles adapted to Swedish conditions, and the necessity of guaranteeing the requirements of the military as regards vehicles, should foreign supplies be blocked. A crucial argument in this context also concerned the assumed high qualifications of the Swedish mechanical engineering industry and its need for a large project to promote employment. At the same time it was emphasized that historical developments had clearly shown that during its build-up period the Swedish industry would need protection, if it was to cope with competition from the North American manufacturers. Due to the exceptional position of the automobile industry and in the light of the foreign examples (cf. the customs duties on automobiles in the UK), many members of the Riksdag, although generally in favour of free trade, were prepared to abandon such a policy in this case, or to disregard any principled objections to the use of government subsidies to stimulate a particular industry or company. In the parliamentary proceedings it was quite clear that the plans for the new car manufacturer, AB Volvo, as presented by its founder, Assar Gabrielsson, in a parliamentary committee, had made a strong impression on those present. Gabrielsson argued for a temporary increase in customs duty to 30 per cent on imported cars, the reason being that it would be possible in this way to price the Volvo cars at a level that would generate the capital required for increasing production, from a planned series of 1,000 units to a volume of 8,000 units. The project needed a further infusion of SEK 2 million to be viable, and this was his last resort. Even if the members of the committee were not prepared to support a temporary increase in customs tariff on imported cars, many of them were convinced by his presentation that the Riksdag ought to look for other ways of facilitating the financing of the company.

The advocates of the development of a national automobile industry, to be protected by an increase in the customs duty on imported cars, were countered by the following arguments. On grounds of principle some people rejected the idea of an increase in customs tariffs or of selective government subsidies for the automobile industry. These measures would impose costs on the users and would prevent the expansion of motoring. A national venture had to be justified on its own merits, and must have sufficient inherent strength to attract capital on the conditions prevailing in the market. Others questioned the need for a car adapted to Swedish conditions and were dubious about the possibility of starting large-scale car production in Sweden at that time. It would have been possible ten years earlier, but now the American manufacturers had gained an advantage that would be impossible to overcome.

5.4.2. Arguments For and Against Foreign Assembly Units

The Ford Motor Company had plans to start an assembly unit in Stockholm. However, the imported parts for cars were subject to a much higher customs duty than the 15 per cent prevailing on an imported assembled car. The tariff that would hit an unassembled Modle-T Ford was estimated at 26 per cent. Under these circumstances it was not considered profitable to start an assembly unit in Sweden and a change was urged.

Those who argued in favour of the establishment of foreign assembly units in Sweden demanded reduced customs duty on imported parts for automobiles on the one hand and simplified customs clearance for such imports on the other. The primary reason for the promotion of these ventures was that it would create employment in the assembly units, but another reason was to provide Swedish ironworks and the Swedish mechanical engineering industry with an opportunity to become subcontractors in this industry. One argument was that after a time, and on a basis of the experience gained at these assembly units, a competence would develop that could eventually be used to establish a national car industry.

The counter-argument here was that the customs tariffs were already low, and that the further favouring of foreign automobile manufacturers would make it more difficult for the national industry to hold its position. If the idea of increasing the difference between the customs tariff on assembled automobiles and that on parts was to stimulate the establishment of an assembly industry, it would be better to do this by raising the tariff on assembled cars and keeping the tariff on parts at its present level. This would favour both the established Swedish companies and the development of a subcontracting industry. Others maintained that the foreign assembly units would find it in their interest to make more use of Swedish subcontractors to a very limited extent only. The experience of other countries supported this supposition.

5.4.3. The Decisions Taken by the Riksdag

A first step in the resolution of this issue was that the tariff on imported automobile parts was reduced to 10 per cent. When the question was finally decided in 1927 this tariff was increased to 12 per cent and the custom clearance process was simplified. The suggestion that the customs tariff should be increased from 15 to 30 per cent was rejected. Thus, the Volvo company was not given any protection from the tariff system; nor is there any evidence that it was blessed with any kind of subsidies, either. Maybe this was a necessary if not a sufficient condition for the development of a competitive Swedish automobile industry.

5.5. The Development of Volvo, 1926–39

This section will first describe the general background and the plans drawn up for the Volvo company at its start.[9] Next, the way production was organized is presented. This is followed by a description of the actual course of events and covers the development of the car and truck operations respectively.

5.5.1. *General Background to the Establishment of Volvo*

The venture that would develop into the present Volvo Corporation started in August 1924, when Assar Gabrielsson, the sales manager of the Swedish producer of ball bearings, SKF, on his own initiative and with his own money had a car designed. In 1920–2 he had worked in SKF's French subsidiary and had the opportunity to follow closely the automobile industry in that country. The task of designing a car adapted to Swedish conditions was given to Gustaf Larson. He was a qualified engineer who had been employed at SKF some years before. He had also worked for an English motor manufacturer in Coventry, UK, in 1911–14, and had been in contact with William Morris when he was designing his first car. The blueprints were completed in June 1925. However, no one was prepared to invest in the project on the evidence of the blueprints alone, and Gabrielsson had ten test cars produced, using his own capital. The parts for these cars were manufactured by different Swedish companies in the mechanical engineering industry, and these companies later became Volvo's main suppliers.

In a confidential memorandum written in June 1926 the draft plan of the car venture was presented in detail (see Ellegård 1983). According to this plan, 1,000 cars were to be produced in the first year, 4,000 in the second, and 8,000 in the third. Of the first year's production 400 were to be exported, and once production had reached 8,000 it was planned that around 60 per cent should be sold abroad. Cars were mainly sold during the spring and summer, and sales to countries in the southern hemisphere were suggested as a way of evening out production over the year. Argentina was mentioned specifically. This would also limit the dependence on the Swedish market.

The model chosen had to be a utility car selling at a price that could help to increase the use of automobiles in the country. The scale advantages of the American industry would be offset by the lower wages in Sweden. It had been calculated that an annual production of 8,000 cars would result in a profitable venture. Production should be based on the extensive use of Swedish subcontractors, and apart from design it was planned that only the assembly work would be carried out by the company. It was for this concept that Gabrielsson managed to obtain the support of SKF. The whole project was taken over by that company in September 1926; it was organized as a subsidiary with Gabrielsson as managing director and Larsson as technical director. Gabrielsson occupied this post until 1956, when he became chairman of the board, a position he held until his death in 1962.

In an interview Gabrielsson declared that 'the most meticulous work has been put into preparing for the production of cars. The problems have been examined down to the last detail, and the programme drawn up is felt to be feasible' (*Stockholms Dagblad*, 31 Mar. 1927, 4). In 1935 shares in Volvo were issued to the shareholders of SKF, and ever since Volvo has had no owner with a dominating controlling interest in the company.

5.5.2. 'Producing the Volvo Way'

The way in which Volvo organized its production system differed considerably from that of the other Swedish companies in the industry.[10] Apart from the final assembly of the car, Volvo only dealt with the upholstery and the assembly of the bodies in its factory in Gothenburg during the start-up period. It may be noted that the operations performed by Volvo did not require any substantial investments in machinery or tools. To begin with, 88 per cent of the materials in the car came from external suppliers (Olsson 1993). Special equipment for cars, such as carburettors and electrical components, were bought abroad. In the late 1930s approximately 15 per cent of the components were imported. However, most of the components were produced to Volvo's specifications by five main Swedish subcontractors and a large number of smaller suppliers. The necessary working capital could be kept to a minimum by letting the suppliers allow Volvo credit of such long duration that it was possible to sell the car before the materials were due for payment. This system was adopted because of a lack of capital, but it also meant that Volvo could benefit from the production experience possessed by various renowned Swedish companies.

Components for the chassis and the forged goods for the engine and gearbox were supplied by AB Bofors, an arms manufacturer. The engine was made by AB Pentaverken, which made, for example, motors for lifeboats. The gearbox was supplied by AB Köpings Mekaniska verkstad, which specialized in the production of machines for the engineering industry. Pressed metal sheets for the bodies and other pressed parts for the chassis and engines were supplied by Svenska Stålpressnings AB, Olofström, a subsidiary of AB Separator. In 1930 the assembly of car bodies was transferred to Olofström. Wooden parts for the bodies and wheels were supplied by AB Åtvidabergs Industrier, a furniture manufacturer. Ball bearings, and later on castings for the engines and brakes, were provided by SKF. These companies were the main suppliers in terms of volume. Alongside their relationship with Volvo, these five companies were also linked to each other as suppliers of components and semi-finished parts. In addition Volvo used many other Swedish firms, which supplied direct to Volvo or acted as subcontractors to the above mentioned firms. Although the production directly controlled by right of ownership was very limited, Volvo nevertheless influenced production in the different subcontractors in many ways. The companies became very dependent on each other, linked together in a network in which materials and parts were processed in several stages, as they were transferred between various companies and finally assembled by Volvo.

No other Swedish company needed automotive parts of the kind Volvo required. The other companies in the industry, Scania-Vabis and Tidaholm, were each producing just 200 trucks and buses annually in more or less wholly integrated factories, using basically manual production methods. A prerequisite for a decentralized production system is the interchangeability of parts. This was facilitated by the pioneering work of the Swede C. E. Johansson, who in 1901 had invented a set of gauge blocks that made accurate measurements in a decentralized production system possible. In 1923 he was employed by the Ford Motor Company in the United States. During the years when he worked in Sweden he had helped to spread the art of precision measurement in Swedish industry (*Ratten* (1939)3: 2). Generally speaking, the level of the mechanical engineering industry in Sweden by the standards prevailing at that time was good. However, only SKF, supplying large quantities of ball bearings produced in long series to the automobile industry abroad, ranked as a specialist. For most of the parts Volvo could not follow a more traditional purchasing strategy in the sense that potential suppliers were set against one another with the one offering the lowest price being chosen (see Gadde and Håkansson 1993). With the exception of the wooden parts of the body and the wheels, there was generally only one possible subcontractor that had the right qualifications, machinery, and capacity.

Thus a necessary condition for the launching of Volvo's venture was to convince a number of potential suppliers to make long-lasting commitments and to take the steps required to achieve the quality and productivity necessary for making the venture a success. The support of SKF was certainly very important in making the realization of the project possible. Besides providing the required capital and laboratory resources, for the suppliers it was a guarantee in itself that the famous SKF company supported this risky venture. However, in order to get the needed support Volvo in many cases had to make commitments that later on limited its freedom of action. This way of organizing the production was known as 'producing the Volvo way'.

It had been doubted whether, because of its limited size, Volvo would be able to survive the competition from mass-produced cars. When the size of the company was mentioned, Gabrielsson made a distinction between 'the smaller Volvo', which constituted the company as such, and 'the larger Volvo', which also included the dealers and subcontractors. He argued forcefully that Volvo should be regarded as a large industry.

Our vehicles have been exported all over the world and shown to be competitive as regards both price and quality. We do not fear the giant American manufacturers with their series of millions of low-priced cars.

Volvo's production system, by means of which a fruitful and capital-saving collaboration with a large proportion of the Swedish quality industry has come into existence, makes Volvo into a larger industry with the resources of such an industry. That our own plants cannot meet all the demands that are made on Volvo, is—in view of our special system of production—only desirable.

(Gabrielsson 1936: 46)

The skill possessed by the companies in the initial stage should not be overestimated. As events moved on, we can see that the adaptation and development of these resources to fit Volvo's requirements were probably of greater significance. Volvo had to introduce to the Swedish supplier firms the methods and mentality that had given the American industry its exceptional position. In the words of Gabrielsson:

One of the big difficulties has been to obtain the necessary change at our subcontractors, who in many places were almost all used to what could be called more or less manual production methods and who, when it came to automobile parts, had to be adapted to production in large runs with all that meant not only in the way of changed work methods but also in the way of a changed mentality. We have been working hard on this, and we have partly succeeded in introducing an atmosphere of Americanism in several places.   (Gabrielsson 1937: 22)

Volvo did not possess the required know-how on its own, but had to organize the transfer of various competences from the United States. This was achieved in a variety of ways. Swedish-Americans with experience of the American automobile industry were recruited by Volvo, licence agreements were made with American companies, and Swedish engineers were sent to the United States on study trips.

The decentralized production system that was initially introduced, which made extensive use of subcontractors producing parts to Volvo's specifications, was gradually transformed into a more integrated system as Volvo acquired the most important suppliers. Thus, because of their inablility or unwillingness to increase their capacity at the required pace, the supplier of engines was taken over in 1930, and the company producing gearboxes for trucks in 1941.

5.5.3. The Volvo Cars, 1927–39

Volvo's first models were equipped with a four-cylinder engine and either an open or closed body. The first car was assembled on 14 April 1927. This model proved to be a failure; a misjudgement had been made regarding the proportion of open and closed-body cars. In 1927 only 297 cars of a planned volume of 1,000 were sold. In September of that year a decision was taken to speed up the introduction of a light truck based on components for the car. The Volvo management realized that it was necessary to try another line of products if the company were to survive. The first car had been designed in such a way that it could easily be modified as a light truck, and the first model introduced in February 1928 was immediately successful on the market. In the second year only 983 cars and trucks were sold; 200 of the open cars had to be scrapped, and in order to sell all 500 closed-body cars the price had to be reduced. This involved Volvo in considerable financial difficulties, and SKF had to cover big deficits in Volvo. The enormous problems faced by the new company almost resulted in its sale to the American car manufacturer Nash in 1928. This was avoided only by Gabrielsson's personal intervention (*AB Volvo Annual Report* 1976: 45).

The Volvo car, designed to the standards prevailing in 1924, had become obsolete by the time it was introduced in April 1927. The new trend was for 8DC-cylinder cars with closed metal bodies. Thus it was imperative to introduce a new model as quickly as possible, and the production of the first six-cylinder closed metal-body car started in 1929. Until 1936 Volvo continued to experiment with a number of variants on this car model, but annual production remained in the range of 600–900 units. Most of these were designed as taxis. In 1935 a new streamlined model, the PV36, was introduced, but it took four years to sell the series of 500 cars. It was not until the PV51 was introduced late the following year that Volvo managed to design a car that was received favourably by the market. For the first time the demand for Volvo cars exceeded the available production capacity, and almost 1,700 units were produced in 1937. In 1939 the production of cars amounted to slightly over 2,800 units, of which almost 500 were taxis. Volvo's share of the market amounted to 5 per cent, and hardly any cars were sold abroad. Thus the volumes projected in the 1926 plan were still out of reach after thirteen years in operation.

In the period up to World War II Volvo launched a number of car models. At least five different families of engines can be distinguished and as many as ten different chassis and bodies. In addition Volvo also experimented with convertible models in small runs. Even if some of the components were used in more than one model, this meant that the conditions for using mass-production techniques did not obtain. Volvo had great difficulty in designing a car that was accepted by the market, and the company experimented with a large number of models produced in series that often did not exceed 500 units.

5.5.4. *The Volvo Trucks, 1927–39*

The development of the truck and bus operations was quite different from that of the problematic car venture. From 1928 to 1957 these vehicles represented the major part of Volvo's business. In the first two models of the trucks and cars, the same types of engine and of some other important components were used. However, as Volvo introduced heavier models in the 1930s, the production of trucks and buses gradually developed to become a separate business with very little direct connection with the troubled car operation. In the period up to 1939 Volvo developed a large number of truck models with a loading capacity ranging from 2.5 to 13 tons. The rate of innovation was high; it was driven by close cooperation with the customers and a willingness to satisfy the needs of different segments of the market. The niches developed by Volvo grew satisfactorily in the 1930s and the total market for trucks expanded. Although the volume of trucks produced annually by Volvo gradually increased over the years, this was not the result of any attempt to follow a mass market strategy. The fact that some of the models introduced could be sold in large numbers did not slow down the rapid rate of innovation. Volvo stuck to the idea of continuing to improve the design of the truck concept and to produce many variants, rather than cutting down the range of models in order to obtain a standardization of its products. Had Volvo wanted to mimic the prevailing American product/market strategy, it would have concentrated on a few models once it had designed one or a small number that were well received by the market. This was not what happened. The developments in truck design were later to be the foundation for the truck models introduced after the war, when what had once been models for niche markets proved to belong to the mainstream. Although it may be difficult to establish what represents a new model and what is only a variant of an existing one, it can be said that Volvo introduced approximately twenty truck models in the period up to 1939. The total numbers of each model, sometimes produced over a period of five years or more, ranged from 30 to almost 7,000. However, the typical average was well below 2,000 units (see Glimstedt, 1993, p. 122). To this can be added a number of bus models produced from 1932 onwards. In 1938 Volvo produced about 4,600 trucks and buses. Its market share for buses was now 43 per cent and for its much larger sales of trucks about 30 per cent. Exports amounted to 22 per cent of sales and consisted mainly of trucks sold to twelve different countries, including Argentina and Brazil.

Even though some components were used in more than one truck model, this indicates that Volvo's strategy for truck and bus operation was by no means geared to mass marketing.

5.6. American Influences on the Swedish Automobile Industry up to 1939

Generally speaking, knowledge of developments in the American automobile industry was spread by technical journals and by people visiting the United States. Thus it can be assumed that the general trends were widely known among interested parties in Sweden at the time, and certainly also among the technicians at Scania-Vabis and Tidaholms Bruk. However, there is no evidence of any more direct influence, for instance through cooperation with American manufacturers or by employing technicians with experience from the American automobile industry.[11] In this respect the establishment of Volvo represented a significant break with the earlier tradition. This section will describe the American influence on Volvo in some detail.

5.6.1. American Influences on Volvo

Although people with experience of the large-scale production of automobiles were lacking in Sweden in the mid-1920s, it was still possible to recruit Swedish-speaking individuals with up-to-date knowledge of the design and production of cars from the other side of the Atlantic.

An example of a Swede making a career in the American automobile industry is John Björn. He left Sweden in 1891 and got a job in Jeffery's Bicycle Company in Chicago. By the time that company started the production of automobiles in 1901 he had advanced to being a partner in the firm. The name given to the first car was Rambler, but after some years it was changed to Jeffery, after the founder of the company. In 1917 Nash acquired the majority of the shares in the company and the Nash car was introduced. Björn had been responsible for engineering and production in the Jeffery company. He worked for ten years as General Superintendent in the growing Nash factories and organized the production along modern principles. Björn recruited many Swedes and other Scandinavians; this resulted in a large Scandinavian population in Kenosha and Racine, the places where the Nash factories were located (*Svensk Motortidning*, 8 Aug. 1927).

After a long strike in 1909 that affected more or less the whole of

Sweden's industry, a great many engineers and workers emigrated to the United States to seek a living, and many of them had careers for themselves in the automobile industry. This flow of people to the United States was to continue for several years. The Volvo management was well aware of this potential resource, and over the years many people were recruited to fill various positions in Volvo itself or at its subcontractors.

In the development work that began in the autumn of 1924 J. G. Smith played an important part.[12] In 1910 he had emigrated to the United States, where he worked for a number of automobile manufacturers and gathered a good deal of information on automobile design. When he returned to Sweden in 1924 he came into contact in some way with the project for the Swedish car. His 'hands-on' experience of American car production left its mark on the design of the test cars and the first serially produced models. After that, however, he disappeared from the scene. The body of the first car was designed by A. O. Ström, who between May 1923 and November 1926 had worked for various coachworks and car manufacturers in the United States.[13]

Another person who was to influence the production of cars at Volvo for many years was Ivan Örnberg. On a temporary visit to Sweden in the autumn of 1926 he had examined the test cars, before orders for parts and components for the first series were given to the subcontractors. He also had emigrated to the United States in 1910, and had worked for various automobile manufacturers there. On the occasion of his visit to Sweden he was chief engineer at Hupmobile.

One of the points made by Örnberg was that the four-cylinder engine would have to be replaced by a six-cylinder version, as the trend in the States was in that direction. He suggested that the Volvo management should study—or to put it bluntly, copy—the engine that General Motors had developed for its new Pontiac model. The four-cylinder engine was retained, however, as it was felt that there was insufficient time to make the change. But Örnberg did help to solve the problem of the vibrations in the first Volvo engine, and he supplied the Volvo people with information about American engine designs. Shortly afterwards Carl Einar Abrahamsson became linked with Volvo. Between 1923 and 1926 he worked in the United States, and since 1925 had been a designer at General Motors and had taken part in developing the new six-cylinder engine for the Pontiac. He was made head of the drawing office at Volvo, a position that he held until 1955 (Lind 1977: 10).

In the course of developing the new six-cylinder engine for the car launched in 1929, Volvo had consulted Continental Motor Corporation in Detroit. Another change made at this time was that, instead of producing a gearbox of Swedish design, it was decided to buy one from Warner Gear Corporation. In the early 1930s Gabrielsson and Larsson made several study trips to American component producers and automobile manufacturers, and also visited Continental Motors and Warner Gear Corporation.

In the 1930s many Swedes with experience of the American automobile industry were brought into the company. In 1933 Ivan Örnberg was called from Hupmobile by Gabrielsson to a position as head of Volvo's car business. Until his death in 1936 Örnberg contributed substantially to the renewal of the design of the Volvo cars. The results of his efforts were the two new models released in 1935 and 1936. Lind (1977) mentions that he also brought figures with him that made it possible to compare production costs in the United States and Sweden.

Together with Örnberg two other designers were recruited from America. One was Edward Lindberg, with nine years experience at Studebaker, who was employed as a designer of car bodies at Volvo. This meant that in 1933 four out of eight technicians in the leading technical group dealing with Volvo car operations had experience in the American automobile industry. To this group should be added at least three more engineers with this kind of experience at lower levels.

After Örnberg's sudden death in 1936 there was a delay of two years until in 1938 Gabrielsson once again went to the United States to recruit a leader for the car operations. This time his choice fell on Olle Schjolin, who, after nine years with Yellow Coach in Chicago (a subsidiary of General Motors), had been called to GM's headquarters in 1930 to take up a leading position in the design department. Among other things he had been in charge of developing the new Opel car (*Ratten* (1938)4: 6). At the same time Carl Lindblom was recruited from GM. Together they brought with them a project for a small car that they had patented, but the development of this car had to be abandoned when they were recalled to the United States in May 1940, as they had been drafted for military service. Schjolin was responsible for the design of the last Volvo to be built to an American-influenced design. The PV 60 had been planned for launching in 1940, but because of the war its introduction onto the market was postponed until 1946. Not until 1950 was the originally planned series of 3,500 units sold out. Lindblom returned after the war and became responsible for the design of cars at Volvo. Shortly afterwards, however, he re-emigrated to the United States.

It was not only to positions directly connected with the design of cars that Volvo recruited people with experience from the United States. The head of the purchasing department at Volvo, Anders Johnson, had worked for Sandvik Steel Inc. in New York (a subsidiary of AB Sandvik) between 1921 and 1925, before joining Volvo in 1928 (Harnesk 1965). Later, one more person from this company was attached to the purchasing department. Thure G. Gehre, who was recruited as head of the assembly factory in Gothenburg in 1929, had also worked in the United States between 1920 and 1929. Among other things he worked as a designer at Yellow Coach and Truck Co., Chicago (*Ratten*, (1939)10: 8).

Other kinds of American influence affected Volvo's subcontractors. However, with few exceptions the data on this subject are meagre, although there is evidence that American influence played an important part in developing an efficient production system (see Gabrielsson 1956). In 1933, Bernard Johansson, a Swede with nineteen years in the American automobile industry behind him, was appointed as head of the tool department at Volvo's subcontractor, Olofström (*Separatorblader*, (1960)4: 27). Olofström supplied Volvo with car bodies. When the expansion of the car market in 1937 made heavier demands on Olofström's production capacity, Volvo sent for ten experts from the American coach work manufacturers, The Budd Co., who between 1937 and 1939, when the war broke out, drew up plans according to American standards. New plants were built and a big new American stamping machine was installed (*Luftrenaren* (1944)3: 15). Volvo had a licensee agreement with The Budd Co. American consultants were also brought in to introduce new production techniques at AB Bofors, the armament manufacturer, which undertook forging operations for Volvo and which for many years was Volvo's single largest supplier (Steckzén 1946).

From all this it can be seen that during the period up to 1939 the American impact on Volvo's car operation was massive. No information is available regarding any corresponding influence on the truck or bus operations. However, it may be assumed that Volvo followed developments in the design of engines and certain components, but that the ideas for the design of chassis were very much the result of experience-based learning.

5.7. Discussion

In this section the reason why mass production of automobiles according to the logic of Fordism never succeeded in Sweden will be elaborated. To begin with, the rationale for the adoption of this strategy in the Swedish context is touched upon. This is followed by an analysis of some factors that may explain the unique position of the Swedish automobile industry and the product/market strategy on which this rests. Most work done in automobile history focuses on cars. In contrast to this trend, the analysis to follow will emphasize the development of the truck and bus venture and draw attention to some conditions specific to Sweden in this area. In a separate subsection the unexpected evolution of the car business is presented. This and the subsection following it, where a comparison of the production systems of Volvo and Toyota is made, extends the time period that is the main focus of this paper.

5.7.1. The Unsuccessful Search for Mass Markets

Why was a mass-market route along the logic of Fordism ever considered in Sweden? In retrospect, a strategy of avoiding the mass-market approach in the automobile industry does not seem to be particularly controversial. Arguments in favour of such a strategy were also put forward in the debate around 1925. However, it is no exaggeration to claim that in the 1920s the mass-market and large-scale logic was institutionalized (see DiMaggio and Powell 1991) in Swedish society and more or less taken for granted as the only proper method to start the national manufacture of cars. In hindsight, conditions in Sweden would make it seem more or less self-evident that a domestic automobile industry, if it was to have any chance of developing, should try to avoid direct competition with the major American companies, and instead should explore the advantages rendered by a flexible production system and proximity to the customers.

There are many grounds for supporting such an approach. Sweden lagged far behind the United States in taking up motoring, and no Swedish entrant into the industry could enjoy any kind of first-mover advantage. Further, the limited size of the home market made it unlikely that a sufficiently large market for cars specially suited to the domestic taste—if such became available—would ever develop. Nor is there any evidence that distinctive competitive advantages for the production of automobiles would appear in Sweden.

Against this it could be asked: if Swedish industry had managed to establish several companies that had become successful exporters in other areas, why could a similar development not be expected for a large-scale automobile industry as well? The general standard of the Swedish mechanical engineering industry could be considered as high. Among Swedish companies that succeeded in becoming competitive at an early stage the following can be mentioned: SKF (ball bearings), LM Ericson (telecommunications), AGA (equipment for lighthouses), and AB Separator (separators and dairly equipment). However, it must be remembered that these 'genius companies' based much of their success on technical innovations that put them ahead of their foreign competitors. In contrast to these companies, a Swedish automobile industry could not base its future development on any competitive advantage connected with Swedish inventions in engine technology or the design of chassis, for example.

The Swedish automobile industry received no protection, and the comparatively low customs tariff of 15 per cent was retained. Paradoxically, this may have been one of the most important reasons why, despite Sweden's limited population, it has been possible to establish three automobile manufacturers, of which two have at times been very profitable. The efforts of the Swedish manufacturers to compete head-on with imported American cars and trucks repeatedly failed. In the 1920s and 1930s it was not possible for them to establish a position within a market disposed to the mass production of standardized products. The Swedish manufacturers could not match the quality or the price of the more or less indistinguishable American automobiles, which were aimed at the larger segments of the market. In order to survive they were forced to adopt a niche strategy, with specialized designs produced in short runs.

5.7.2. *The Development of Niche Markets for Trucks*

The idea of the interchangeability of parts, which is central to mass production and is a necessary condition for a decentralized production system, was to be fully adopted first by Volvo and later also by Scania-Vabis. However, it was necessary for Volvo to abandon one of the other foundation stones of the mass production logic, namely, far-reaching product standardization. Thus the Swedish automobile industry eventually achieved a logic that, in the terminology of Piore and Sabel (1984), could be called 'flexible specialization'. As a consequence of the fierce competition on the market for cars Volvo was forced to switch to the production of trucks and later on buses. Another effect of the intense competition was that a large share of the cars produced in the 1930s were cabs. Common to these vehicles, often used by institutions, are very precise demands on performance data. By way of close cooperation with their sometimes very demanding users, the Swedish manufacturers learned to cater for their customers' special needs. Instead of supplying an anonymous market, they worked with identifiable users of their product and developed long-lasting relationships that promoted the exchange of information and technical development (see Håkansson 1982; Gadde and Håkansson 1993).

It is possible to distinguish a few clusters of users, or what Dahmén (1950) would call 'development groups', which promoted the evolution of the design of trucks and buses in Sweden. We can see an example of this in the 1930s, when developments in the forest industry and a restructuring of the dairy industry contributed to an increase in demand for heavy trucks with exceptional properties and high reliability (see Glimstedt 1993). The bad roads in the Swedish countryside, and the special conditions that faced forest transport, put a strain on the material and the durability of vehicles. Later, the floating of timber by river was gradually discontinued, to be replaced by transportation by trucks. This further increased the demand for vehicles with high loading capacity that were well adapted to rough conditions. In this market Volvo and Scania-Vabis were able to benefit from their closeness to their customers and from their very flexible production system, which allowed them to produce in small series while still covering their costs. It is also probable that the hard competition between Volvo and Scania-Vabis on these niche markets also helped to encourage a high technical standard in the trucks and buses. This process of experiential learning clearly helped to renew and improve the truck concept, and models were designed that could be successfully marketed abroad. Swedish trucks became renowned for their loading capacity and durability.

Thus, in order to survive, the Swedish manufacturers were forced to concentrate their production within segments in which proximity to the market and a flexible production system were advantageous. The limited size of the domestic market, which was further accentuated by the focus on niche markets, made exporting necessary at an early state in the companies' expansion programmes. In this way the Swedish automobile industry soon became internationalized, which proved to be a competitive advantage at a later stage, when the market opened up more generally.

5.7.3. From American Cars in Sweden to Swedish Cars in America

For many years Volvo tried to develop a car designed according to American ideals that could be sold in the Swedish market. As has been shown above, during the 1930s Volvo recruited most of the team of designers attached to its car operations from the United States. It took this team, with their American experience and outlook, almost ten years before they eventually managed to design a car achieving an annual volume exceeding 1,000 units. In the original plan this had been the target set for the first year. In 1938 less than 3,000 cars were sold by Volvo in Sweden. The target for the second year had been set at 4,000 units. It was not until after World War II, when Volvo gave up the idea of copying an American car and introduced a Swedish-style medium-sized car instead, that the volumes aspired to back in 1927 were finally achieved. Between 1947 and 1965 more than 400,000 units of the PV 444–544 were produced. Paradoxically, the breakthrough for Volvo as a manufacturer of cars coincided with the accidental but successful introduction of this model in the United States in 1955 (Kinch 1992). In less than three years every fourth car produced was sold on this market. Cars have dominated Volvo's business ever since, and North America has remained its most important market. Thus the success of the cars was based on a concept quite unlike the one that was originally conceived. Instead of developing an American-influenced model for marketing in Sweden as originally intended, the success of Volvo as a manufacturer of cars has been related to a Swedish-style car having its largest market in the United States.

5.7.4. 'Producing the Volvo Way': Fordism or a Pioneering Toyotaism?

The way Volvo organized its operations differed considerably from the Ford model. Volvo did not succeed in the efforts to standardize its products and reach out to the mass markets, and the American production methods introduced were not specific to Fordism. They were part of a more general tradition of applying systematic time and work flow studies advocated, for example, by Frederic Taylor. Taylor's book *Scientific Management*, published in 1911, was translated into Swedish as early as 1913, and his ideas were taught at the Royal Institute of Technology in Stockholm. This indicates that his methods were well known in Sweden.

The interesting aspect of the early Volvo history is the decentralized production system so vividly described by Gabrielsson. This initial set-up had some characteristics that resembled the system later to be introduced by Toyota (see Wada 1991). Right from the start Volvo made something of a 'permanent deal' with its main subcontractors and organized a network in which materials and parts were processed in several 'tiers'. In a sales handbook, Gabrielsson described the advantages of 'producing the Volvo way' and presented it as a deliberate strategy peculiar to Volvo.[14] The suspicion that his description was a way of making a virtue of necessity is confirmed in his later writings, where the system adopted is described as 'the poor man's wisdom' (Gabrielsson 1937). This was the only way to launch the project and not something that he had desired from the start. However, this system was quite successful for many years and Volvo managed to develop its full potential.

The decentralized production layout initially introduced was gradually transformed into a more integrated system as some of the most important suppliers were acquired by Volvo and the relationship to some others lost its 'permanent deal' status. In the 1930s AB Bofors had been the largest subcontractor by volume. However, after World War II it lost its position and was succeeded by Olofström. The start-up of production after the war made it possible for Volvo to reconsider its production design, and the purchasing strategy that had been more or less forced upon it was now to be abandoned. This was further accentuated in 1958 when plans for a new plant for the car assembly operation were considered. This stimulated thinking about the way production was organized and encouraged efforts to handle relationships with suppliers on strict business lines. As a result of its success with the car introduced on the American market in 1955, Volvo was negotiating from a position of strength. Earlier agreements made when Volvo's situation had been less favourable, or based on smaller volumes than the ones now projected, were reviewed. This process was retarded by the commitments Volvo had made over the years to some of its suppliers. In the early 1930s Olofström had been given an exclusive dealing right whereby Volvo agreed to buy all its requirements of bodies and pressed parts from Olofström. When in 1958 Volvo wanted to buy the company, Olofström turned down the bid Volvo offered. As the bids were so far apart no agreement could be reached but the relationship continued as Olofström had a contract valid until 1965. It was not until 1969 that this company was eventually acquired by Volvo.

When Gabrielsson in the 1930s described how the production system and the competitiveness of the Volvo company were related he treated it as an organization problem in 'the larger Volvo'. The important issue was *who* should do *what* and *how*. He stressed the importance of complementary resources and the development of the competence of the suppliers. However, by the end of the 1950s, it was described as a purchasing problem. Now it was a question of acting on a market and making a choice from a given supply. A description of the purchasing function of Volvo from this period reveals a new production policy in which the idea of single sourcing and long-term commitments was abandoned. It was clearly stated that the possibility of maintaining competition was a primary condition for achieving the right quality at the lowest cost (*Luftrenaren* (1957)3: 10). This statement is in line with the prevailing ideas of what constituted an effective purchasing strategy in the management literature of American origin. However, it is interesting to notice that this development at Volvo is quite opposite to the policy adopted by Toyota at the same time. In contrast to Toyota, Volvo did not have any serious problems with the quality of its automobiles. The reason for the failure with the cars in the 1930s was that Volvo did not manage to design a model that was accepted by the market. Toyota experienced some quality problems when the cars were introduced in the USA in 1958. This made Toyota management reconsider the way the production was organized, and they deliberately entered and developed a system that very much resembled the one Volvo had been forced into at the start but then gradually abandoned.

5.8. Conclusion

The ideas of a mass market for standardized cars dominated the Swedish debate in the mid-1920s and also had an impact on what the manufacturers conceived as the road to follow in order to be successful. However, in the period we have considered the adoption of Fordism cannot be claimed to be a central factor if we are interested in explaining how the Swedish automobile industry reached its competitive edge. The way in which the industry eventually came about was not in accordance with this logic. Swedish companies did not establish a position on the market based on the production of standardized vehicles in long runs at low cost. In brief, it can be said that fierce competition, primarily from the mass-produced American automobiles, forced the Swedish companies to adopt the product/market strategy that has characterized its success up to the present. Instead of standardized products for an anonymous market, they developed specialized products adapted to identifiable customers' needs and exploited the advantages of a flexible production system. This development was neither intended from the outset or based on the adoption of American ideas of mass production, nor contingent upon any conditions particular to Sweden. As the short histories presented above have clearly shown, the model was forced on the companies more or less against their will and not the logical way of organizing the business. The observed development of Volvo and Scania-Vabis is the result of a patient process of trial and error and the adoption, modification, and elaboration of ideas borrowed from abroad. In the Volvo case a large number of Swedes with experience in the American automobile industry were brought in, together with American consultants. All this resulted in a gradual development of competence in the design and production of specialized automobiles adapted to a very flexible production system that had very little resemblance to Fordism.

NOTES

1. Three different automobile manufacturers, Volvo, Scania, and Saab-GM, are operating in Sweden today. Volvo is a public company, while Scania and 50% of Saab-GM have since 1991 been owned by the Wallenberg-dominated holding company Investor. Among international manufacturers of heavy trucks Volvo-GM ranks second and Scania-Vabis fifth. For many years Scania-Vabis has been the most profitable company in the truck business. Volvo has been a profitable actor in the segment of high-priced quality cars, with a particularly strong position in the United States. The plans to extend the alliance between Volvo and Renault in a complete merger were turned down in Dec. 1993. Saab is in the not-too-enviable position of being the smallest producer of family cars in the world and is suffering from losses. Since 1989 General Motors has a 50% share in the Saab car business.
2. A few other attempts to assemble cars from imported parts on a very small scale were made in the early years of the twentieth century. Some minor efforts were also made by two companies in the 1930s to produce buses (see Sahlgren 1989: 38).
3. It may be questioned whether Saab has ever been profitable since its start in 1949, with the possible exception of a few years with a high dollar rate and the benefits of the early introduction of the turbo aggregate. Its losses have been covered by other companies in the 'Wallenberg group': first by the Saab

134 *Nils Kinch*

Aeroplane business, then later, when this company merged with Scania-Vabis in 1969, by Scanis trucks.

4. See *Svensk Motortidning*, 31 Aug. 1920; 1927:15.
5. See *Bofors Pilen* 1960:4; 1961:2; Giertz 1991.
6. This passage is based on Sahlgren 1989 and Giertz 1991.
7. *Svensk Motortidning*, 29 Mar. 1925, 115 and 12 Nov. 1925; *Teknisk Tidskrift*, 25 July 1925 and 1 May 1926.
8. This section is based on the records of the Riksdag and the minutes and working material of the parliamentary committee Beviljningsutskottet. A more detailed description of this debate is presented in Kinch 1993c.
9. This description is based on Lind (1997) and annual reports and magazines of Volvo Corporation. Different aspects of Volvo Corporation have been reported in Kinch (1987, 1992, 1993a, 1993b).
10. A more extensive description of Volvo's production system is given in Kinch (1987).
11. The attitude to outside influence at Scania-Vabis seems to be the exact opposite of Volvo's. This remained characteristic of the company right up to 1946. However, there is one exception to this closed outlook. In 1929 Gunnar Lindmark, the managing director of Scania-Vabis, went to the USA with the chief engineer August Nilsson to study the production of buses. Among other companies they visited the Twin-Coach Company in Ohio. Later a licence for this 'Bulldog' bus was acquired, and the first one was delivered in 1932 (Giertz 1991).
12. This section is mainly based on material collected in *Ratten*, Volvo's customer magazine, and on Lind (1977, 1984).
13. *Svensk Smidestidning* 1 (1927), 3–4, 7.
14. This may be questioned, as already in 1924 William Morris, in 'Policies that have built the Morris business', describes a similar system (first printed in *System*, Feb. and Mar., then reprinted in *The Journal of Industrial Economics*, (1924), 11: 193–206).

REFERENCES

AB Volvo Annual Report 1926–93.
Beviljningsutskottets arbetsmaterial 1924–27 (Working material of the parliamentary committee Beviljningsutskottet).
Bofors Pilen 1960–1 (the company magazine of AB Bofors).
Dahmén, E. (1950), *Svensk industriell företagsamhet: Kausalanalys av den industriella utvecklingen 1891–1939* [Swedish industrial enterprise: A causal analysis of the industrial development 1891–1939], Stockholm.
DiMaggio, P. and Powell, W. (1991), 'The Iron Cage Revisited: Institutional Isomorphism and Collective Rationality' in Powell, W. and DiMaggio, P., (eds.), *The New Institutionalism in Organizational Analysis* Chicago: The University of Chicago, 63–82.
Ellegård, K. (1983), *Människa-Produktion: Tidsbilder av ett produktionssystem* [Man-

Production: Pictures of a Production System], Meddelanden från Göteborgs Universitets Geografiska institutioner, series B, No. 72; Göteborg.

Gabrielsson, A. (1936), Försäljningshandbok utgiven av AB Volvo i samband med introduktionen av PV51 [A sales handbook published in connection with the introduction of PV 51], Göteborg.

——(1937) 'Volvo—Föredrag hållet inför Svenska Ekonomföreningen' [Volvo: A speech delivered to the Association of Economists at the Stockholm School of Economics], *Ratten* 10–12, 4–5, 20–7.

——(1956) 'Volvo under trettio år' [Thirty years of Volvo], *Transbladet* 19/4: 4–19.

Gadde, L.-E. and Håkansson, H. (1993), *Professional Purchasing*, London.

Giertz, E. (1991), *Människor i Scania under 100 år* [One hundred years of people in Scania], Stockholm, 1991.

Glimstedt, H. (1993), *Mellan teknik och samhälle: stat, marknad och produktion i svensk bilindustri 1930–60* [Between technology and society: State, market and production in the Swedish automobile industry 1930–60], Avhandlingar från Historiska institutionen i Göteborg 5; Göteborg.

Harnesk, P. (1965), *Vem är Vem? Götalandsdelen* [Who is Who?], Stockholm.

Hounshell, D. H. (1984), *From the American System to Mass Production 1800–1932*, Baltimore.

Håkansson, H. (ed.) (1982), *International Marketing and Purchasing of Industrial Goods: An Interaction Approach*, Chichester.

Kinch, N. (1987), 'Emerging Strategies in a Network Context: The Volvo Case', *Scandinavian Journal of Management Studies*, May, 167–84.

——(1992), 'Entering a Tightly Structured Network: Strategic Visions or Network Realities', in: Forsgren, M. and Johanson, J. (eds.), *Managing Networks in International Business*, Philadelphia, 194–214.

——(1993a), 'La Vision Strategique à L'Epreuve des Contingences: L'histoire de Volvo', *Décisions Marketing*, May, 9–17.

——(1993b), 'The Long-Term Development of Supplier-Buyer Relationship: The Case of Olofström and Volvo', Paper presented at the 9th IMP Conference, Bath, 23–25 Sept.

——(1993c), 'Riksdagens behandling av biltullar 1924–7' [The Riksdag on car import duties 1924–7], Working Paper 1993/13, Department of Business Studies, Uppsala University.

Lind, B-E. (1977), De tidiga åren [The early years], *Autohistorica* 1–2: 1–81.

——(1984) *Volvo Personvagnar från 20-tal till 80-tal* [The Volvo cars from the 1920s to the 1980s], Stockholm.

Luftrenaren 1944:3, 1957:3 (The company magazine of AB Volvo).

Olsson, K. (1993), Undated and unsigned draft. Department of Economic History, University of Gothenburg, Gothenburg.

Nordlund, S. (1989), *Upptäckten av Sverige: Utländska direktinvesteringar i Sverige 1895–1945* [The discovery of Sweden: Foreign direct investments in Sweden 1895–1945], Umeå Studies in Economic History; Umeå.

Piore, M. J. and Sabel, C. H. (1984), *Second Industrial Divide: Possibilities for Prosperity*, New York.

Ratten (The customer magazine of AB Volvo).

Riksdagstrycket 1924–7 (Records of the Swedish Parliament).

Sahlgren, U. (1989), *Från mekanisk verkstad till internationell industrikoncern: Scania-*

Vabis 1939–60 [From mechanical workshop to an international industrial firm: Scania-Vabis 1939–60], Uppsala Studies in Economic History 31; Uppsala.

Separatorbladet, 1960:4 (The company magazine of AB Separator).

Sloan, A. P. (1964), *My Years with General Motors*, Garden City.

Steckzén, B. (1946), *Bofors Industrier*, Stockholm.

Stockholms Dagblad, 31 Mar. 1927.

Svensk Motortidning, 1920–7 (A magazine published by the Royal Automobile Club of Sweden).

Svensk Smidestidning, 1927:1.

Teknisk Tidskrift (Published by the Swedish Association of Technology), 25 July 1925 and 1 May 1926.

Wada, K. (1991), 'The Development of Tiered Inter-Firm Relationships in the Automobile Industry: A Case Study of Toyota Motor Corporation', *Japanese Yearbook on Business History*.

Womack, J. P., Jones, D. I., and Roos, D. (1990), *The Machine that Changed the World*, New York.

PART II

THE FORD SYSTEMS AND PRODUCTION MANAGEMENT

6

The Development of Company-Wide Quality Control and Quality Circles at Toyota Motor Corporation and Nissan Motor Co. Ltd.

IZUMI NONAKA

6.1. Introduction

Although quality control systems were originally introduced into Japan from America, subsequent modification led to the appearance of what is now known as 'Japanese quality control'. The purpose of this paper is to examine some of the history of this metamorphosis.

Traditionally, American quality control has put great emphasis on final inspection at the end of the assembly line by professional quality control inspectors; mistakes are eventually corrected in a rework area.[1] In contrast, the approach in Japan has evolved into a system of 'building in' quality throughout the entire process, relying on 'company-wide quality control'. This in turn depends on quality control awareness all the way down from top management to the shop floor, and in every functional area, including planning and design, production, administration, and sales. Ideally, all employees in every department should be cooperatively involved in activities to improve quality. The Japanese approach is sometimes known as total quality control (TQC), but it will be standardized for the purposes of the present paper as company-wide quality control or CWQC.

The Japanese quality control system can be characterized by the following six elements:

1. The existence of a number of organizations dedicated to the promotion of quality control activities (QC organizations), which are rep-

This paper was originally translated by Brendon R. Hanna for presentation at the 21st Fuji Conference. It has been greatly modified for the present volume.

resented by JUSE (Japanese Union of Scientists and Engineers) and the Japanese Standards Association, to which virtually every enterprise seeking to introduce quality control sends employees for training.

2. The use of professors of engineering and retired employees as quality control consultants (in the United States of America specialist consultants are generally used).

3. The existence of the Deming Prize (America now has the Malcolm Baldridge National Quality Award, established in 1987 and given for the first time in 1988).

4. The responsibility of all employees, including top and middle management as well as lower-level workers, from planning and design, to production, marketing and sales, for quality control. This is called company-wide quality control (this contrasts with the American reliance on specialist quality control inspectors).

5. The building in of quality throughout the whole process of production, (unlike the typical American emphasis on final testing).

6. The presence of quality circles.

How were these features of Japanese quality control shaped? This paper will try to answer this question by surveying the process by which it took shape between 1949 and 1966. Surprisingly enough, QC organizations, along with the use of consultants, emerged early in Japan's postwar drive to reindustrialization, and it was they who introduced process-oriented company-wide quality control to Japanese enterprises, including Toyota Motor Corporation and Nissan Motor Co., Ltd. (hereafter Toyota and Nissan, respectively). The Deming Prize was initiated to commemorate Dr W. Edwards Deming's now famous series of lectures on statistical quality control, and quality circles were introduced by JUSE in 1962.

The next section of this paper will function as a survey of the general history of Japanese quality control, outlining the introduction of statistical quality control from America, the initiation of the Deming Prize, the use of consultants, and the formation of QC organizations. The third section will provide a comparison of Toyota and Nissan at the time that Nissan won the Deming Prize, forcing Toyota to play catch-up in the implementation of CWQC. The fourth section examines the creation of quality circles at both Toyota and Nissan.

6.2. Introduction of American Quality Control

6.2.1. Statistical Quality Control

Although statistical quality control was introduced into Japan from England and America before World War II, it was tried by only a handful of companies, such as Toshiba and NEC; organizations for the promotion of such ideas had yet to develop.

In 1949 and 1950, however, the Civil Communications Section (CCS) of the Occupation force's General Headquarters convened the CCS Management Seminar,[2] and in 1950 Dr Deming presented an eight-day series of lectures, thus introducing the quality control techniques of American engineers and statisticians to the Japanese.

Also of great significance is the formation of organizations dedicated to the promotion of quality control (QC organizations), such as the JUSE (Japanese Union of Scientists and Engineers) founded in 1946, and the Japanese Standards Association (JSA) founded in 1945. These two bodies, along with the Japan Management Association (JMA), began quality control education activities even before the CCS Management Seminar and Dr Deming's lectures.

In May of 1949, Dr Eizaburo Nishibori started a quality control training programme under the auspices of JMA. This was followed just one month later by a programme organized by JSA taught by Toshio Kitagawa, a statistician, and Yasushi Ishida, prewar quality expert at Toshiba. And in September of that same year, a group of instructors that included Dr Kaoru Ishikawa, who had studied the subject after the war, began a basic course in quality control for JUSE. Dr Nishibori, however, did not remain active in the field of quality control training, and the statistical bent of Kitagawa and Ishida proved too specialized for their techniques to become widely used and accepted. It was the approach of the group with Kaoru Ishikawa, then, that was to evolve into Japanese company-wide quality control, and that formed the basis for the development of quality circles.

6.2.2. *Two Early Influences: The Deming Prize and the Use of Consultants*

The lectures that Dr Edwards Deming gave 10–18 July 1950 on statistical quality control by way of control charts and the performance of sampling inspections, addressed to an audience of 220, were aimed at transmitting to the Japanese a rudimentary framework for statistical quality control. The lectures were enthusiastically received and were assembled into a book.[3] Because of occupation restrictions, Deming could not repatriate royalties from the book, so he chose to donate his share of the proceeds to JUSE. This was used to establish, from 1951, the Deming Prize for an Individual (*Deming-sho hon-sho*) and the Deming Application Prize (*Deming-sho jishi-sho*).[4]

The Application Prize, awarded annually to a Japanese company, has become a hallmark of the implementation of quality control. Winning the prize is an immediately recognized symbol of the superiority of that company's products, and what typically happens is that other competitors in an industry (automobiles, electronics, or steel, for example) make strenuous efforts to garner the Prize over subsequent years once one company in that industry has won it. This has the effect of spreading improved quality control over an entire industry following the award of the Prize to a particular firm. Nissan's winning of the Prize in its tenth anniversary year, 1960, sparked efforts at Toyota that led to the latter's capture of the Prize in 1965.

Japanese companies began to make use of their own particular brand of specialist consultants from about the same time that JUSE originated its basic course in quality control in September 1949, when course instructors began visiting factories in order to take surveys and to assist in the introduction and application of quality control. These instructors were mainly professors of engineering whose visits provided opportunities both for guidance to others and the furtherance of their own research activities. These individuals will be referred to as the QC consultants.

While these QC consultants were instrumental in spreading American-style statistical quality control in Japan, they also felt that, no matter how wonderful these principles were, it would be difficult to introduce them widely without modification. Unless statistical quality control was reformulated to better fit Japanese companies and attitudes, it could not be expected to take root. Dr Kaoru Ishikawa noted in a critique that,

What we learned when we offered the first basic course was that, while sciences like physics, chemistry, and mathematics are universal, principles of management like quality control have human, social elements strongly at work within them; no matter how excellent the American or British approaches, they will definitely not work if imported to Japan just as they are . . . we had to develop a Japanese way.[5]

Ishikawa and the other QC consultants had not as yet come up with a clear notion of what that Japanese method might be, but they struggled to synthesize one from foreign texts and their actual experiences in Japanese factories. In 1950, Ishikawa and Eizaburo Nishibori, along with three other university professors, undertook their first factory assignment, the results of which were less than successful. This was because they were trying to cram in too many statistical quality control techniques all at once. Ishikawa later remarked that, 'If you try to teach too much, they don't learn anything at all'.[6]

In response, the consultants boiled down the American techniques into seven tools for quality control, accompanied by cause and effect diagram (or Ishikawa Diagram, named for its originator). It is my belief that the reason the statisticians who knew about quality control even before the war were unable to lead the postwar quality control drive, was that they employed a mathematical approach in their explanations of quality control and were difficult to understand. The QC consultants, on the other hand, strove to use simple terms and few formulas so that quality control would be easy to understand. Furthermore, they did not hesitate to rearrange what they had learned of American quality control to suit Japanese tastes.

The role played by the QC consultants was huge; they eventually spread the concepts of company-wide quality control, building in quality throughout the entire manufacturing process, and quality circles. Their efforts at modifying the system of quality control to fit Japan were indispensable in spreading the idea to Japanese companies, including, of course, Nissan and Toyota.

Finally, there is a close relationship between the QC consultants and the Deming Prize. Most Japanese companies, upon making the decision to introduce quality control, utilize the services of groups of consultants, who visit the company/factory several times a month to provide guidance. After several years of steady improvement, when the company's level of quality control is judged to warrant a Deming Application Prize, these consultants suggest that the company might apply. Of course, the decision as to whether to apply or not is made by the company. Applying for the prize results in a series of submissions of documentation and inspections of the company's facilities. In addition, interviews with top management are carried out to determine the extent of their understanding of quality control. To ensure fairness, consultants who have assisted the firm are of course excluded from the panel of judges.

6.3. The Contest for the Deming Application Prize: Leader Nissan and Follower Toyota

6.3.1. Overview of the Deming Prize

The Deming Prize for an Individual has nearly always been won by consultants[7] (typically, university professors in departments of science and engineers) or by key employees of companies that have implemented quality control.[8] While Motosaburo Masuyama, friend of Dr Deming and renowned statistician, became the first winner of this prize in 1951, statisticians subsequently won the prize only twice, in 1953 and 1957. The prize in 1952 was shared by Shigeru Mizuno, who later became a consultant for Toyota, and Kaoru Ishikawa, who went on to consult for Nissan, as well as six others.

While the Deming Application Prize throughout the 1950s was won by companies in the process industries of chemicals, pharmaceuticals, and steel and by electrical equipment manufacturers, the 1960s saw the sudden rise to prominence of firms in the automobile industry: Nissan in

1960; Nippon Denso Co., Ltd. in 1961; Toyota in 1965; Kanto Auto Works, Ltd. in 1966; and Bridgestone Corp. in 1968, a pattern that continued into the early 1970s with Toyota Auto Body Co., Ltd. in 1970 and Hino Motors Ltd. in 1971. But then it was the turn of several office and equipment makers such as Ricoh Co., Ltd. and Pentel, to be followed by a number of subsidiaries and affiliates of previous winners. Construction companies were most obvious in the decade starting in 1979, although Toyoda Gosei Co., Ltd. won the prize in 1985, as did Toyoda Automatic Loom Works, Ltd. in 1986.[9]

Toyota was the first winner of the ultra-prestigious Japan Quality Control Medal (see note 4), and its affiliates, Aishin Industry Co., Ltd. and Toyota Auto Body Co., Ltd., were winners in 1977 and 1980 respectively. As recently as 1992, Nissan's Oppama Plant captured the Deming Application Prize for a Division. Finally, the Deming Application Prize for Small Enterprise, while not taken up explicitly here, has been won consistently since 1960 by affiliates of Nissan and Toyota.

6.3.2. Nissan

For both Nissan and Toyota, 1949 was the year that statistical quality control was introduced.[10] Nissan's programme began when, at the urging of their supervisor, three members of the Yokohama Plant Inspection Division attended JUSE and JSA quality control lectures.[11] One of these original three, Akira Iwata, studied control charts and general statistical quality control techniques, while the other two concentrated on sampling inspection and experimental design. Iwata's impression was that 'Statistical quality control is difficult but interesting.' He recalled, 'For a period of two years, I wound up meeting several times a month after work with [JUSE and JSA] QC organizations.'[12]

The three men were quickly given the opportunity to try out the new concepts at what was then Nissan's Atsugi Plant. The facility, which has since been spun off into a separate company, manufactured gears and screws, and it was an excellent setting for sampling experiments, so Iwata and the others began sampling screw production. Lots that failed were returned to the workers who had made them; this showed them that they were the ones responsible for quality control in manufacturing. The use of statistical quality control (SQC) at the Atsugi Plant both reduced the percentage of defective parts and proved the value of SQC techniques, such that Nissan was soon able to substitute sampling inspection for its former method of inspecting 100 per cent of production.

Following the successful introduction of experimental SQC at Atsugi, a Quality Control Department was established in 1954 within the Testing Division at Nissan's main Yokohama Plant, becoming an independent department in 1957.[13] It was originally the job of the people in this Quality Control Department, the QC staff, to train inspectors in the use of sampling inspection and correct testing techniques. The QC staff were also responsible for the improvement of quality through better manufacturing and testing, and it was the inspectors themselves who first made analysis and adjustments, and who drew up control charts. Only after the supervisors on the shop floor and the production workers under them had been trained in quality control could they be expected to make control charts on their own.

From 1954, an internal QC training programme was begun by the QC staff for supervisors, foremen, and technicians. Simultaneously, section and division chiefs were actively encouraged to participate in the activities of external QC organizations, such as the Juran Lectures sponsored by JUSE in 1954 and the Basic Course in 1955.[14] In addition, from 1954 to 1955 inclusive, one of the aforementioned consultants, Dr Nishibori, together with Nissan's QC staff, transmitted the basics of SQC to the company's subcontractors.

In the early 1950s, most Japanese companies, Nissan included, equated the implementation of quality control with the use of statistical methods and control charts. From the middle of the decade, however, quality control and testing began to be perceived very differently at Nissan, with a new focus on the production process and company-wide quality control. In early 1956, Nissan's in-house publication, *Nissan Technology*, noted:

From its founding in 1954, the Quality Control Department was primarily concerned with its own organizational establishment and with training, particularly long training sessions put on by the famous Eizaburo Nishibori and other outsiders, and the result was that the production people thought that all they had to do was produce. [Subsequently] *the old way of thinking, that the inspectors would decide what was OK and what was not, has been dropped in favor of the idea that [quality] is made by the production people themselves, which is the real essence of quality control.*[15] (italics mine)

The same edition of *Nissan Technology* also stated:

[Quality control] has to be a company-wide phenomenon, starting with the president and reaching through middle management to technicians and production workers. That is, the desired result cannot be hoped for without the cooperation and effective implementation on the part of headquarters, manufacturing, technical support, research, administration, materials, warehousing, sales, general affairs, personnel, labor relations, etc.[16]

In the mid-1950s the idea of company-wide quality control slowly spread to a few Japanese companies. The people who introduced this idea were the QC consultants. Attention to the importance of process and company-wide quality control really began in 1955, five years before its appearance in Toyota, although there is obviously a difference between the introduction of ideas and their correct application on the shop floor.

In order to better understand Nissan's shift to CWQC, let us first review quality control in general in Japan from 1954 to 1960. The QC consultants and QC Organization that had dealt with the problem of information overload by distilling the essence of statistical quality control methods into 'seven tools for quality control', had another problem to address. Lack of understanding of quality control on the part of management was preventing its spread within Japanese industry. Ishikawa, still in his early thirties (rather young by Japanese standards), faced a constant battle to be taken seriously by older managers, but he knew that in the mid-1950s Japanese listened to foreigners.[17] He invited J. M. Juran, a widely respected American consultant, to Japan in order to teach two courses on quality control for top managers and for division and department heads; the latter lecture series was attended by Nissan managers in 1954. JUSE had him back the next year for similar courses, and again in 1957 for a special course for top managers. The concept, first accepted by top and middle management, was thus to spread, from 1960 on, in the directions of design and sales, 'bearing the fruit of company-wide quality control between 1963 and 1965'.[18] Consequently, the period from 1955 to 1960 was one of groping in the dark for a Japanese approach to CWQC.

On 3 February 1959 Nissan arrived at a formal decision to commit the company to the symbolic goal of applying for the Deming Application Prize.[19] One of the primary motivations for this decision was the issue of trade liberalization, an obligation stemming from Japan's admission to the IMF in 1952 and to GATT in 1955. The significance of this issue for the passenger automobile industry was not lost on Nissan, and the company pushed forward with improving the quality of its cars.

A secondary reason for the decision to apply for the Deming Application Prize was the growing importance of quality as export sales increased. Nissan had created a separate Export Department in 1958 and, it was planning to set up Nissan America in Los Angeles in 1960; if the company were to be successful in its export strategy, quality was critical.

In light of these two factors, Nissan's president, Katsuji Kawamata, announced his policy of 'more cars, made better' along with the decision to apply for the Deming Application Prize and to introduce CWQC. This left only one year to do the huge amount of work necessary to become the first automaker to win.

Following the February decision, in June of 1959, Ishikawa and several other consultants began instruction at the headquarters, the Yokohama Plant, and the Yoshiwara Plant. Two months earlier, management had begun quality control training. As a result, by 1960 some 900 division heads, section heads, supervisors, foremen, technicians, and clerks (out of a total of roughly 8,000 Nissan employees) had undergone such training. At the Yokohama and Yoshiwara Plants, assembly processes were rationalized and standardized, and defects were reduced. Over the fourteen months from February 1959 to March 1960, the Yoshiwara Plant saw internal losses from defects down by 20 per cent and losses from customer claims down by 40 per cent from Nissan's average 1954 loss rates.[20] This was achieved despite the fact that there had been no increase in the number of inspectors from 1954, the number of models produced had doubled, and the number of units produced had quadrupled. Process-oriented quality control was beginning to pay off.

From the late 1950s, design and sales, areas that were once thought of as unconnected with quality control, began putting it into practice. In 1959 designers were given quality control training, and a report was made indicating how early in the design process defects could be eliminated by employing quality control methods. In the area of sales, a series of pre-delivery checks was instituted to ensure that quality was maintained at each stage as the finished car passed from factory to dealer to customer. Additionally, customer comments and complaints were collected from dealers by headquarters sales staff, who in turn passed on problems to inspection staff. This feedback was redirected by inspection staff to the appropriate areas so that problems could be rooted out and eliminated. Quality control training was given to administrative staff as well, so that work could be standardized, and a set of administrative standards was established.

After the requisite investigation, Nissan's application was approved, and it received the Deming Application Prize. The prize was handed to Nissan president Kawamata by Dr Deming himself in May 1960, in commemoration of the tenth anniversary of the award's foundation.

6.3.3. Toyota

Just six months after Nissan's achievement, in November 1960, Toyota executive vice-president Eiji Toyoda made an important announcement to all employees. 'Quality is something that is made during the process [of our work]. I want you to help rationalize the control of quality by making products so well that testing will not be necessary. And it is a mistake to imagine that administration is unrelated to the improvement of quality'.[21]

The announcement was preceded in June of that year by distribution within the company of a pamphlet entitled 'An Appeal Concerning Testing', which explained that 'the best testing is no testing'. Toyota had also modified its organization in 1959 by converting its Testing Department into a Quality Control Department.

In conjunction with the tenth anniversary of the Deming Prize, 'Quality Month (Hinshitsu Kanri Gekkan)'[22] was established in 1960 to propagate quality control throughout Japanese companies. The publicity connected with this anniversary year was at least partially responsible for Toyota's 180-degree turn away from its previous reliance on testing; the company introduced CWQC in 1961.

Toyota's pre-prize efforts in the area of quality control can be divided into three distinct periods: 1949–55; 1956–9; and 1960–5. Like Nissan, Toyota introduced statistical quality control in 1949. When a JMA-sponsored QC lecture course was held at Nippon Denso, the head of Toyota's testing department, Hanji Umehara, had several subordinates attend, and they were quick to try out control charts and sampling inspection at a machining plant;[23] the similarity with Nissan is unmistakable.

In 1950, with the start of American military procurement in Japan for the Korean War, Toyota began to study American testing methods. 'At each step, from materials all the way to the finished products, American military inspectors performed strict tests according to American military standards, and they weren't easy to pass'.[24] In that same year, Eiji Toyoda and Shoichi Saito were dispatched to America on separate three-month observation missions to a number of sites that included Ford's River Rouge Plant. Because these trips had been decided upon when Kiichiro Toyoda was president, the new president, Taizo Ishida, decided to have the two executives go as planned even though the company had become busy as a result of Korean War procurement contracts. Upon his return, Eiji Toyoda stated:

The scale [of the River Rouge Plant] is certainly impressive, but once I got used to [their] system, I found some faults in it. There is not a great difference [between Ford and Toyota] in terms of manufacturing technology and production methods. In fact, Toyota, with methods like just-in-time, may be ahead. If we modernize our factories and invest in the newest equipment, we should be able to surpass Ford.[25]

After the two executives came back, they proposed a 5-Year Production Equipment Plan, and the modification of Ford's suggestion system to become to Toyota Creative Ideas and Suggestion System.[26] One of the things that had most impressed the two Toyota observers about the River Rouge Plant was the motto 'Safety and Quality'. After an in-house contest in 1953, Toyota's own motto, 'Good Thinking, Good Product', was adopted.

From 1953 to 1954, quality control training sessions were attended by a total of 700 production workers, QC staff, and team leaders from the manufacturing and inspection divisions. In 1955, Toyota released the Crown, Toyota's first completely domestically made passenger automobile, and this was followed by the Corona (ST10) two years later. From 1958 to 1962, when Toyota was again the beneficiary of American military procurement, it produced a total of more than 50,000 military trucks, accounting for about 6 per cent of sales during this period.[27] It is likely that Toyota's reliance on testing was confirmed by this experience. 'The achievements, with regard to quality and in other areas, brought about by the wide-ranging influence of stringent APA [American Procurement Agency] testing, cannot be overlooked'.[28]

The 1960s were a decade filled with important events for the Japanese car industry. Mazda (then Toyo Kogyo) and Honda Motor both entered the market for four-wheeled vehicles, further intensifying domestic competition; with the lifting of certain restrictions on truck and bus imports in 1961 and on passenger automobile imports in 1965, international competitiveness in both quality and quantity was a pressing issue. The first round of capital liberalization in 1967, along with complete liberalization scheduled for 1973, made the achievement of competitiveness all the more urgent. In response, Toyota built new factories one after another and undertook extensive capital investment. The Motomachi, Kamigo, Takaoka, Miyoshi, and Tsutsumi plants were built in quick succession. Meanwhile, Nissan was building its own Oppama, Zama, and Tochigi plants. All serious participants in the industry were planning increases in quality and quantity, accompanied by decreases in prices, to meet the perceived threat of a foreign invasion.

Besides the problems faced by the industry in general, Toyota had a special headache of its own. From August 1959 Nissan's Bluebird had topped the domestic sales charts for 64 straight months, while Toyota's second-generation Corona (RT20), released in 1960, received an unexpectedly lukewarm reception.

As mentioned, it was in June of 1961 that Toyota decided to initiate company-wide quality control in earnest.[29]

Company-wide quality control was introduced [at Toyota] in response to the trade liberalization measures of 1961, and, in order to effectively deal with and even prosper under expected harsh conditions, both domestic and international, the decision [was made] to initiate new, epoch-making management that would allow the development and manufacture of high-quality, low-priced products.[30]

The then president of Toyota, Fukio Nakagawa, was, like Kawamata of Nissan, a former banker. In the case of Nissan, the internal promotion of CWQC revolved around seven middle managers, including Yutaka Kume, a future president, and the two who had initially suggested a prize attempt to Kawamata. At Toyota, however, the Internal QC Promotion Headquarters, established in 1964, consisted primarily of elite managers led by executive vice-president Eiji Toyoda, managing director Shoichiro Toyoda (both future presidents), and managing director Hanji Umehara. And, like Nissan, Toyota relied on consultants and initiated a multifaceted programme of quality control education.

As a first step in this education programme, 80 technical managers and 35 key employees in 29 subcontractors were given a 132-hour course in quality control at Toyota from September to December of 1961.³¹ Subsequently, all 19 directors attended outside lectures, and the scope of quality control training was broadened to include division and department heads, administrators, foremen, and production workers.

Quality control all the way from design to sales was instituted with the third-generation Corona (RT40). The first-generation ST10, released in 1957, had used essentially the same chassis as the Crown, and was therefore not really the small car that it was claimed to be. As noted, market reaction was not entirely favourable.³² The second-generation Corona (PT20) appeared in 1960 with the marketing slogan, 'It still has four [tyres], but everything else is new!', yet it was beset by problems and was eclipsed by the Nissan Bluebird. Following the awarding of the Deming Application Prize, Umehara noted that the relative success of the RT40 was brought about by the complete implementation of quality control from planning to after-sales service, and by the enthusiasm and determination of all Toyota employees.³³

Sales of Toyota cars were conducted by a separate entity, Toyota Motor Sales Company, spun off from Toyota in 1950. This structure required that the two companies cooperate closely in demand forecasting and market research; consequently, the latter introduced CWQC in 1962.³⁴ Toyota was also active in guiding its subcontractors towards improved quality.

Still, in Toyota's manufacturing section, 'the traditional idea, that strict testing would result in good quality, was firmly planted'.³⁵ The QC education programme was expanded to included foremen and line workers in an effort to root out this idea, but it was not an easy task. However, with education and accompanying progress towards company-wide quality control,

The awareness that quality is built into each process was instilled in the mind of every employee; in addition, the system whereby defective parts were simply discarded was changed in favor of one in which the parts were not allowed to reach subsequent process [*sic*].³⁶

Still, as long as the words 'quality control' were used, the impression remained that this was the job of the Quality Control Department or the Testing Department, and that other parts of the company need not be concerned. Accordingly, a new phrase was adopted: 'Quality Assurance—systematic activity undertaken by the producer to ensure that the quality demanded by the customer is completely supplied'.

Toyota production methods such as just-in-time, *kanban*, and *jidoka* (automation) are well known, but it should be stressed that, in relation to quality control, if 100 per cent of the parts reaching a given process are not defect-free, Toyota methods will not work smoothly. In other words, quality is the foundation of Toyota production methods. From about 1963, just-in-time and *jidoka* were adopted in all Toyota factories, and a close relationship between these methods and quality was immediately established.

By 1962–63, devices had been installed that automatically stopped the equipment or operation when there was equipment malfunction or quality deficiency. On the [final] assembly line, if anything prevented standard practice from being followed, the production workers were empowered to stop the line, [after which] they were to locate the cause and make improvements.[37]

Under this system, production workers themselves had to think about what to do in case of a problem, how to go about locating the problem, and how to make sure that it did not recur. These issues were addressed in QC meetings, which were to become QC circles.

The fruition of Toyota's company-wide quality control, the New Corona (RT40), was taken for its driving test on the Meishin Expressway, Japan's first limited-access highway, in September 1964. The running condition of the first three New Coronas was publicized on radio and television with the message, 'Engulfed in a whirlpool of rumor, [the Coronas] have completed 100,000 km in 58 days!'[38] Assisted by the poor reception given Nissan's 1963 model change of its Bluebird, the New Corona surged ahead to become the best-selling automobile for 33 straight months. Toyota employees had learned; 'QC [*sic*] is useful. Let's continue with QC'.[39]

6.4. Quality Circles and Their Formation

6.4.1. The Origin of Quality Circles

The concept of quality circles was first expounded in 1962 in the first issue of a JUSE publication, *Quality Control for the Foreman* (Japanese title, *Genba to QC*; now called *Quality Control*) the front cover of which featured the Nissan Oppama Plant. Let us examine the beginning of quality circles. As mentioned earlier, courses for both top management and for department and section heads were held in the mid-1950s in order to spread the ideas behind quality control. Even so, the key to manufacturing good products was further down, at the level of the foremen and line workers. Because these individuals were so numerous and geographically widespread, however, their education presented a real challenge. JUSE and JMA solved this problem by producing QC radio seminars. From 1956 to 1962, seventeen quality-control seminars were broadcast over the Japan Short-Wave Broadcast Network and NHK (Nihon Hoso Kyokai, Japan's public broadcasting organization). This radio series inspired comments at a discussion sponsored by the editors of the monthly magazine named *Total Quality Control* in support of a magazine that would be easier to understand for workers and foremen, and of a forum for expression of their ideas, such as a foremen's conference.[40]

Quality Control for the Foreman was the immediate response of JUSE, which took careful steps to ensure that it would be read by its intended audience of workers and foremen. Main points were highlighted in easy-to-read large fonts, the cover price was made equivalent to a pack of cigarettes, and the B5 size allowed it to be easily folded to fit into the pocket of workers' overalls.[41]

Another idea was also published in the magazine. When visiting factories throughout the country to gather information for *Total Quality Control*, JUSE representatives and consultants discovered small groups of foremen and workers discussing quality control and solving problems using SQC methods. The idea, then, was to institutionalize this in the form of a quality circle. A magazine aimed at foremen would serve no purpose unless it was actually put to use, and Ishikawa pointed out that, instead of individuals studying quality control alone, it would be better to have small groups which would provide continuous mutual stimulus and reinforcement, meaning that each member would tend to continue his (or her) studies. When the magazine first appeared, many people were not sure just what a quality circle was, and many circles with more than ten members were registered (circles generally have under ten members). Many companies, their employees intrigued by the first issue, decided to register existing groups with similar functions as quality circles; Toyota was one of these.

6.4.2. Toyota's Lead Over Nissan in the Formation of Quality Circles

Toyota, then in the midst of implementing company-wide quality control, organized QC meetings in November 1962 in order to promote quality consciousness in its foremen and line workers in all its production facilities, and to form a direct line of communication from top management to the shop floor. These QC meetings took the form of an extension of the morning greetings and exercises, and they provided a forum for the dissemination of information such as quality control solutions. When necessary, groups got together to work out particularly difficult problems. Virtually corresponding to work groups on the shop floor, they were registered by Toyota as quality circles after JUSE publicized the idea. Toyota was not particularly quick to do this, however, as 3,000 circles from production facilities around the nation had already been identified by this time. Toyota was cited in the March 1965 issue of *Quality Control for the Foreman* as registering, all at one time, 80 circles in its casting, machining, press, die, and body operations.[42] Subsequently, the company steadily expanded the number of its quality circles and continued to refine their organization.

It should be pointed out that Toyota gained much of its quality circle know-how from closely related Nippon Denso, which won the Deming Application Prize in 1961, four years ahead of Toyota. Nippon Denso had selected individuals at each of its plants to conduct quality circle forerunners, known as 'pilot circles'. Toyota began its own programme after observing these.

In any case, as quality circles had just come into being (as such), a great many Japanese factories simply made their foremen quality circle leaders with subordinates as members, and registered them with *Quality Control for the Foreman*. In this atmosphere, it was not unusual in the 1960s for large companies simultaneously to register dozens of quality circles.

It was also in the March 1965 issue of *Quality Control for the Foreman* that Nissan's Yohohama Plant engine assembly 'ENG Circle' (ENG for engine) was registered. With 178 members,[43] this particular circle could hardly be called a 'small group' according to the standard definition of a quality circle. Nissan called this its first circle registration, but its first real quality circles were formed in March 1966, when 38 of them were registered for the Zama Plant. The plant manager, Shiro Yamazaki, was one of the original proponents of Nissan's bid for the Deming Application Prize and a key supporter of company-wide quality control. Kaoru Ishikawa advised Yamazaki that, even though Nissan had won the Deming Application Prize, progress in QC would not take root without quality circles. Yamazaki also sent a charismatic subsection chief, Sanosuke Tanaka, who had the abiding trust and respect of the production workers, to JUSE quality circle training sessions, and Nissan's first real quality circle revolved around Tanaka. He said: 'From 8 o'clock in the morning, the conveyor just kept on moving. Some days we couldn't even talk to our fellow workers. But after [we started] quality circles, a group of six or so would get together after work to talk for about ten minutes. The circle leader would start talking to one of the line workers, and then all sorts of problems would come to light'.[44]

Talk would always lead to work-related problems, and the circle leader would use the seven tools for quality control (for example, cause and effect diagrams), or a copy of *Quality Control for the Foreman* to discuss in a practical manner how to improve quality. According to Tanaka, quality circles were effective in fostering better communication and work relationships. Rather than simply serving to alleviate through communication the fractionalization and monotony of the assembly line, quality circles were instrumental in addressing new problems between people and machines brought about by the sudden increase in automobile production. I think this is one of the reasons why the number of quality circles in Japan increased in step with mechanization.

One example was the relationship between workers and an automatic metal plating machine installed in the assembly division of Toyota's Motomachi Plant. After the new machine was installed, breakdowns occurred frequently in the initial period of use. First, a graph of the breakdown times was made. Next, cause and effect diagrams were used to consider possible causes. All possible causes were explored in a quality circle, and each was tested, until the actual cause(s) was (were) determined.

The supervisor may understand the design of the machine and how to run it, but is probably unaware of its detailed tendencies or weaknesses. The people who know best about the condition of the machine are the workers, and quality circles provide an opportunity to get important information from them.[45]

We may conclude with another example from the body assembly division of the Toyota Motomachi Plant. The section responsible for producing the body of the New Corona (RT40) found that most unevenness occurred in the rear fender. This problem was taken up by a quality circle, which succeeded in reducing the unevenness by changing the order of pressures exerted on the metal fender blank by the press.[46]

In order to build in quality during each process, the foreman and line workers use the quality circle in their efforts to eliminate defects before they go on to the next process, making the system function. Although Toyota borrowed the individual suggestion system from Ford, it was subsequently modified to allow suggestions from groups. This shift from individual to group suggestion was also made in Nissan.

Let us trace here the historical paths that Toyota and Nissan followed in pursuit of quality. From 1949 to the mid-1950s, the two companies were travelling essentially the same road. While from 1955, Nissan introduced the idea of process-oriented quality and the beginnings of CWQC, Toyota was still wedded to the concepts enshrined in SQC. This can in large part be attributed to the fact that, in order to renew its contract for vehicles with the US Army Procurement Agency in Japan, Toyota had to rely on American testing-focused quality control.

While Nissan required only a one-year preparatory period in advance of its bid for the Deming Application Prize, it took Toyota nearly five years, from 1961 to 1965, to prepare. Although there were a number of differences between the two companies' approaches, the process leading up to the award of the prize was basically similar; it was after the awards that significant divergence between the companies became clear. It is often noted that Japanese companies tend to perceive the Deming Application Prize as a kind of examination which, once 'passed', deserves a bit of rest, and Nissan was one of these firms. Toyota set out to avoid this tendency and took the time to consider its post-prize policies even before it applied for the prize. The post-prize differences between the firms were reflected in Toyota's being the first winner of the Quality Control Medal and, in the creation of quality circles, Toyota preceded Nissan.

In 1966, the year that Japanese passenger automobile production overtook that of trucks and buses, Toyota undertook to strengthen its quality control, and Nissan introduced quality circles. It was largely because of the company-wide efforts, aimed at achieving greater quality and quantity, by firms in the automobile industry in the 1960s, that Japanese enterprise was able to recover from the oil shocks of the 1970s.

6.5. Conclusion

Let us return to the six elements that I listed in the introductory section. There we saw that Japan's QC organizations consistently responded to changes in education and training needs. They introduced quality control to top and middle management, and then spread the concept of company-wide quality control. It was JUSE that identified the beginnings of quality circles and initiated a system for their registration. The Japanese QC consultants also played a significant role, digesting and modifying Western theories and methods, and trying out the results at enterprises and their factories. The experience that they accumulated was not only useful for their own research purposes, but was also transmitted back to the QC organizations. Throughout his lifetime, Kaoru Ishikawa was to repeat over and over that 'Quality must be built in through the process'. The legacy left by him and the other consultants, along with the QC organizations, is the diffusion throughout Japanese enterprise of 'building in quality throughout the process'.

The basic contribution of the Deming Prize to Japanese enterprise has been the establishment of a competitive ethic. Dr Deming taught the Japanese statistical quality control and encouraged their efforts to apply it, but he did not teach them about quality circles or CWQC. I think one of the great contributions of Dr Deming was that he became the symbol of superior quality products throught his Deming Prize for Applications.

The processes of introducing CWQC and of forming quality circles have been illustrated in the cases of Nissan and Toyota. It is clear from these cases that the concept of CWQC was introduced from the QC organizations and/or QC consultants, although the particular processes were different at the two companies. From the viewpoint of the introduction, modification, and adaptation of American statistical quality control to Japan, it is also clear that the primary introducers of these ideas were the consultants and QC organizations. Japanese consultants immediately recognized the need for modification of statistical quality control, as noted in the cases of Ishikawa and Nishibori. It was they who learned and adopted American statistical quality control with enthusiasm and then selected those ideas or methods that would fit Japanese conditions. Their transformation of American quality control bore fruit in such Japanese developments as CWQC and quality circles. From that Toyota learned to develop quality circles and the Ford suggestion system took the form of group suggestion. This was one example of modification.

Thirty years have elapsed since the early 1960s. Nishibori, Ishikawa, and Mizuno (a Toyota consultant), the three best-known QC consultants, have all passed away in recent years, and on 20 December 1993 Deming joined them, to mark the passing of an era. Lately there has been criticism in Japan of the Deming Prize and of the second- and third-generation consultants.[47] Some young people complain about quality circles, and there is an increasing number of companies claiming that, with reductions in working hours, there is no time available for quality circles. It seems that the Deming Prize, QC consultants, and quality circles have come to a turning-point in Japan. It remains to be seen whether or not their applicability will extend into the future.

NOTES

1. J. P. Womack, D. T. Jones, and D. Roos, *The Machine That Changed the World* (New York: Rawson Associates, 1990), 55.
2. See I. Nonaka 'Origin of Quality Control in Japan (3): The Introduction of Statistical Quality Control', *Hinshitsu kanri* [*Total Quality Control*], 41/3 (March 1990), 55–62.
3. K. Koyanagi (ed.), *Dr Deming's Lectures on Statistical Control of Quality* (Tokyo: JUSE, 1950). A revised edition was published two years later: *Elementary Principles of the Statistical Control of Quality: A Series of Lectures* (Tokyo: JUSE, 1952).
4. The prizes are now financed entirely by JUSE. In addition to the original two prizes, the Deming Application Prize for Small Enterprise and the Deming Application Prize for a Division have been added. Also, in commemoration of the world's first International Conference on Quality Control in 1969, the Japan Quality Control Medal (*Nihon Hinshitsu Kanri Sho*) was established. This prize is reserved for companies that have won the Deming Application Prize or the Deming Application Prize for Small Enterprise at least five years previously, and it outranks the original prize. The first winner of the Japan Quality Control Medal was Toyota, in 1970. In response to requests from overseas for non-Japanese companies to be allowed to compete for Deming Prizes, the Deming Prize Committee revised its by-laws from 1984 to permit

foreign firms to apply. The first foreign recipient was Florida Power & Light Co. in 1989, and the second was Taiwan Philips in 1991.

5. K. Ishikawa, *Nihonteki hinshitsu kanri: TQC towa nanika* [*Japanese quality control: What is TQC?*] (Tokyo: JUSE, 1984), 22. Eizaburo Nishibori had the same idea of modification of quality control. He said, 'It is important to study American quality control. But not only imitate foreign quality control; we should devise a Japanese one. I went to RCA and GE before World War II, and saw the separation between manufactures and inspectors. They seemed not to trust one another. I strongly feel the necessity for a Japanese way of quality control.' H. Karatsu *et al.* (eds.), *Gijutsu no sozoryoku to hinshitsu kanri* [Creativity in technology and quality control] (Tokyo: Yuyusha, 1991), 171.

6. K. Ishikawa, 'Nihonteki hinshitsu kanri no tenkai' [The development of Japanese quality control] (interview), *Keiei to rekishi* [Management and history], (Tokyo: Nihon Keieishi Kenkyujo [Japan Business History Research Institute], 1986), No. 9, 10–12.

7. According to JUSE, *Deming Prize 35* (JUSE, 1985), 13 and 15, and Deming Prize Committee (ed.), *Deming-sho iinkai komon/iin meibo narabi ni jushosha ichiran* [Index of Deming Prize Committee advisors, committee members, and prize winners], (1992), 8–11, consultants have won the prize in 1954 (Eisaburo Nishibori), 1962, 1965, 1966, 1973, 1977, 1978, 1979, 1984, 1985, 1989, and 1991.

8. According to the same sources, individual winners from companies, some of whom went on to university positions and consultant status, include Yasushi Ishida of Toshiba Corp., in 1956; Takeo Kato, renowned industrial engineer with Mitsubishi Electric Corp., in 1961; Masumasa Imaizumi, then of Nippon Kokan K. K., in 1965; he later assumed a university post; and Shoichiro Toyoda, son of Toyota founder, Kiichiro Toyoda, in 1980.

9. Toyoda Automatic Loom Works, Ltd. was the parent of Toyota Motor Corporation. The origninal company continues to undertake work subcontracted from Toyota.

10. For more on the history of QC at these two companies, see M. A. Cusumano, *The Japanese Automobile Industry* (Cambridge: Harvard University Press, 1985), and M. Udagawa, 'Nihon jidonsha sangyo ni okeru hinshitsu kanri katsudo: Nissan to Toyota' [Quality control activities in the Japanese automobile industry: Nissan and Toyota], (Tokyo: Hosei University, Center for Business and Industrial Research, Working Paper Series, No. 36, 1993).

11. I. Nonaka, 'Nissan Jidosha ni okeru zenshateki hinshitsu kanri no tenkai katei: 1950–60' [Company-wide quality control at Nissan Motor Co., Ltd.: 1950–60], (*Aoyama shakai kagaku kiyo* [Aoyama social science register], 15/2, 1987).

12. Interview with Akira Iwata. At the time of the interview, he was president of the Nissan Institute of Automotive Mechanics, located near the Nissan Tochigi Plant.

13. *Nissan Jidosha no ayumi to hinshitsu kanri oyobi QC circle no rekishi* [The history of Nissan Motor, quality control, and quality circles] (Tokyo: Nissan Motor Co. Ltd., 1981), 1–2.

14. Y. Yamashita, 'Hinshitsu kanri do'nyu kara konnichi made' [From the introduction of quality control up until today], *Nissan gijutsu* [*Nissan technology*], No. 26 (Tokyo: Feb. 1956), 10–11.

15. Ibid. 9.
16. Ozaki Tatsuo, 'Hinshitsu kanri no tekio ni tsuite' [Adaptation of quality control], *Nissan gijutsu* No. 26 (Tokyo: Feb. 1956), 41.
17. Interview with Kaoru Ishikawa, and 'Nihonteki hinshitsu kanri no tenkai', 24.
18. Y. Iizuka, 'Deming-sho senko riyu ni miru TQC no ayumi' [The progress of TQC as seen from Deming Prize Awards], *Hinshitsu kanri [Total Quality Control]*, 33/4 (JUSE: Apr. 1982), 48.
19. Nonaka, 'Nissan Jidosha ni okeru zenshateki hinshitsu kanri no tenkai katei: 1950–60', 36.
20. Nissan Motor Co., Ltd., *Deming-sho jisishi sho ni tai suru Jjtsujo setsumeisho [Explanatory report on the present conditions of the company]*, (1959), 21.
21. *Toyota Shimbun [Toyo News]* (house organ of Toyota) No. 381, (Dec. 1960).
22. Sponsored jointly by JUSE, JSA, and the Japan Chamber of Commerce, and dedicated to the national promotion of quality control, Quality Month was founded in 1960. Each November is now known as Quality Month.
23. Toyota Motor Company, *Toyota Jidosha 20 nenshi* [A 20-year history of Toyota Motor], (Aichi: 1957), 343.
24. Toyota Motor Company, *Sozo Kagirinaku: Toyota Jidosha 50 nenshi* [Creation without limit: A 50-year history of Toyota Motor] (Aichi: 1987) 248.
25. I. Masajiro, *Oukoku no rirekisho* [All of Toyota], (Tokyo: Keiei Shorin, 1993), 159–60.
26. Toyota Motor Company, *Toyota Jidosha 30 nenshi* [A 30-year history of Toyota Motor], (Aichi: 1967), 329, 331–2.
27. For details, see Wada Kazuo, 'The Development of Tiered Inter-Firm Relationships in the Automobile Industry: A Case of Toyota Motor Corporation', *Japanese Yearbook on Business History*, Vol. 8, (Tokyo: Japan Business History Institute, 1991).
28. *Toyota Jidosha 30 nenshi*, 464.
29. *Sozo kagirinaku*, 380.
30. Deming Prize Committee, '1965 nendo Deming-sho iinkai hokoku narabi ni Deming-sho Jisshi Sho, Nikkei Hinshitsu Kanri Sho Bunkensho senko riyu oyobi Deming-sho Jisshi Sho kijusho kaisha chosa hokoku' [1965 Report of the Deming Prize Committee on the Deming Prize for Individuals, the Deming Application Prize, the Nikkei Quality Control Literature Prize, and on companies that have previously received the Deming Application Prize] (Tokyo: December 1965), 16.
31. *Toyota shimbun (Toyota news)*, no. 423 (Sept. 1961).
32. Toyota Motor Corporation, *Toyota no ayumi* [History of Toyota] (Aichi: 1978), 215.
33. '1965 Nendo Deming-sho Jisshi Sho wo jusho shite' [After we got the Deming Prize], (discussion), *Hinshitsu kanri (Total Quality Control)*, 17/1, (Jan. 1966), 19.
34. Deming Prize Committee, '1965 Nendo Deming-sho Jisshi Sho jushosha hokoku koen yoshi: Toyota' [Summary of the acceptance report of the winner of the 1965 Deming Application Prize: Toyota) (Dec. 1965), 19.
35. *Sozo kagirinaku*, 381.
36. Toyota Motor Corporation, *Toyota: A History of the First 50 years* (Aichi: 1988), 157.

37. Udagawa, 26.
38. *Toyota no ayumi*, 262.
39. Ibid. 264.
40. 'Genbacho wo meguru iroiro na mondai' [Various problems faced by supervisors], *Hinshitsu kanri* [*Total Quality Control*], 12/9, (Sept. 1961), 9–10.
41. See I. Nonaka, 'Origin of Quality Control in Japan (8): Birth of QC Circles in Japan', *Hinshitsu kanri*, 41/9 (Sept. 1990), 77–84.
42. Toyota's first circle to appear in *Quality Control for the Foreman* was listed in May 1965 as 'Circle 25231', consisting of 17 members (plant and operation unknown).
43. Registration was accompanied by an order for 179 copies of the magazine; presumably the assembly section chief ordered these copies.
44. Interview with Sanosuke Tanaka. Tanaka's quality circles met during workers' break time or after work for a total of two hours a month, which was paid.
45. Y. Shibata, 'Jido mekki souchi no kosho to torikunde' [Dealing with breakdowns of an automatic plating device], *Genba to QC* [*Quality Control for the Foreman*], No. 17 (1965), 14–19.
46. Y. Suzuki, 'Corona no ria fenda ototsu kaiseki jirei' [Case of analysis of unevenness in the rear fender of the Corona], *Genba to QC*, No. 28 (1966), 10–15.
47. The August 1992 edition of *Total Quality Control*, published by JUSE, was a special issue devoted to criticism of company-wide quality control, the Deming Prize, QC consultants, and quality circles.

7

Fordism and Quality: *The French Case, 1919–93*

PATRICK FRIDENSON

In the 1970s automobile manufacturers in Europe and North America became aware of the Japanese success in quality management, itself associated with the emergence of a 'neo-Fordism'. But quality had been a concern for car makers since their beginnings. So this paper will undertake a long-term history of quality in the French car industry in its Fordist era, in order to try to understand how the implementation of Ford production methods led French manufacturers first to a complex set of initiatives aimed at preserving the quality of their products, then—after World War II—both to difficulties and to the institution of specific quality departments. Yet it is obvious that only the pressure of consumer organizations worldwide and the competition of Japanese automobile companies raised quality to the level of a top priority among French car makers in the 1980s—a symbol of which was the power Renault's CEO gave to his Director for Quality in 1988, the power to delay new models coming into production as long as their quality is not satisfactory.

However, a caveat is necessary here. Over the seventy-four years under review there have been considerable changes in the ways in which the quality of cars is measured. Moreover, the results of car quality measurement have often prompted the making of new quality tools, all the more so as benchmarking for quality was repeatedly modified. So the reader should be aware that we shall deal with changing concepts and measures of quality (which hence will be surveyed chronologically), and should not overestimate the continuity from one period to the other, even within the same firm.

By focusing on quality, we hope to illuminate several aspects of the adaptation of Fordism to the culture of French car firms and to their markets: the development of new managerial structures; the incentives offered by the composition of the labour force and by the influence of its trade unions; the position of engineers between management and labour; the changing status of the customer (from a major reference to a subordinate place in management thinking, then to a more central one); the differences between the three major car makers: Renault, Peugeot, Citroën; the reception of foreign production models (first American, then Japanese) in an old industrial nation like France. Simultaneously, we would like to show why the French, who were among the first to devise statistical methods applicable to quality control of manufactured products, were much slower to introduce them in industry, although quality had been the traditional motto of French industry since the early nineteenth century.

However, before an analysis of the successive approaches to quality in the French car industry since 1919, it is necessary to give a brief outline of the ways in which French car manufacturers have implemented and modified the Ford system.

7.1. Three Stages in the Adoption of Fordism

In a recent paper, Fujimoto and Tidd have argued that, in contrast to the foreign subsidiaries of the Ford Motor Company, which carried out a direct transplantation of Fordism based on technology push, both British and Japanese domestic mass producers were much more successful by adapting Fordism selectively, i.e. following a demand pull strategy.[1] In the light of these two models, where do we place the pattern of the largest French car companies? Given the size of the French domestic market and the decline of French automobile exports between 1926 and 1952, it is tempting to suppose that the French firms' approach was similar to that of British and Japanese mass producers. A closer look at their policies shows, however, significant differences.[2] They can be explained by four features. The availability of existing or new factories was a key issue in the introduction or later development of Fordism. The persistent shortage of labour (due to France's low levels of fertility and to the numerous casualties of World War I) was a powerful incentive to transfer as many of the original Ford production methods as possible. The dominance of technology-oriented engineers in the management of French car makers made it all the more attractive. Of the leaders of the French 'Big Three' in the interwar period, two had personally toured Henry Ford's plants shortly before the introduction of the assembly line: Louis Renault in 1911 and André Citroën in 1912, and evidently both had been immensely impressed by the quality of the Model T, by the interchangeability of parts, and by the detailed costs accounting methods at each stage of the production process. They had met Henry Ford, James Couzens, and Charles Sorensen.[3]

In keeping with the conclusion of Fujimoto and Tidd, those car makers who limited themselves purely and simply to transplanting Ford's strategy and methods performed poorly. This was obviously true of Ford's subsidiary in France, which although founded in 1913, achieved only a small market share and showed mediocre results up to the mid-1930s.[4] It therefore exerted no direct influence on the domestic producers. The same was true of GM's subsidiary, which limited itself to producing components. But this also applied to one domestic producer, Berliet. On paper, its potential was good. During the World War I, Berliet had achieved mass production of trucks for the army, moving 'further toward standardization of trucks than the British, Germans, or Americans' (J. M. Laux, p. 63), powering each machine with its own electric motor, introducing motionless assembly lines for trucks and tanks. He had also been able to build an immense new factory. So, immediately after the war, he embarked on a Fordist policy, characterized by the production of one model car and one model truck. Such single-mindedness did not pay. By April 1921 Marius Berliet had to call in the receivers. His car was based on the American Dodge, and it did not attract the interest of many French customers. Moreover, the steel he made from French iron had less strength than American steel, so the car suffered many breakdowns. Thus the first example of real Fordism in France was associated with poor quality in the eyes of the French public.[5] Standardization was clearly not enough 'when not supported by thorough engineering and a correct analysis of the market' (J. M. Laux, 81–2).

Also consistent with the conclusion of the study by Fujimoto and Tidd is our analysis of the product policy chosen by the Big Three in the interwar period. Generally sensitive to demand, it was targeted at the French middle classes and did not aim at low-priced cars for workers. So, contrary to the Fordist model, it emphasized product variety, quality of design, and change. But this overall presentation needs to be qualified. The other prime mover in Fordism, André Citroën, had initially decided upon a low-priced people's car but could neither hold the advertised price nor the concentration on one model. Soon he turned to 'producing two or three models available in various body styles, including six-cylinder types from 1928 on' (J. M. Laux, 77). In contrast, Peugeot kept its prewar variety initially, but had to gradually reduce the number of models. In between, Renault maintained a rather varied range and short production runs. And Berliet recovered by making a variety of cars and trucks. Consequently, changes in the production system were slow, difficult to implement, and adapted to conditions in France. They were all the more gradual as they generally took place in existing plants (and with engineers and foremen accustomed to small lots). Like the British, the French manufacturers rejected day-rates, but, unlike them, they abolished craft demarcations. The reorganization of factory space and of product layout was carried out actively; later, the specialization of certain factories followed. But the numerous machines that were then bought were mostly not single-purpose ones. A number remained multifunctional, as would be expected when a certain degree of variety was retained. So foremen's tasks were not as narrowly subdivided as in America. Consequently, the introduction of the assembly line was quite gradual. Citroën introduced the stationary assembly line in 1919, the mechanically operated one in 1921, the continuously mobile one in 1926; Renault did the same in 1922, 1924, and 1926, respectively, and Peugeot had to wait till 1928 for such lines.[6] In 1929, as a result, French automobile production was five times higher than in 1919.

Yet once again there were striking differences. At Citroën most of the equipment was American, American engineers stayed in the plants for several years, and André Citroën himself paid a visit to Henry Ford in 1923. At the other companies things were more patchy: no American engineers were hired, Renault and Berliet themselves produced a significant part of the required machine-tools, and so on. In a sense, it might be concluded that the demand pull product policy did not lead to the flexible organization of production because of the interest of leading French engineers (such as Ernest Mattern, first at Citroën, then at Peugeot) in the major technical aspects of Fordism.

Thus this first stage of Fordism was characterized by a long learning process, which usually lasted seven to eight years. It entailed the recruitment of a throng of enthusiastic young engineers attracted by the automobile who gradually became experts in the production process. It at last brought about the interchangeability of the main mechanical parts by the late 1920s.[7] This in turn led to a massive influx of unskilled and semi-skilled labour, but the necessary growth of investment in capital goods damaged the cost structure, and this finally resulted in financial problems in the three major companies.[8]

A second stage of Fordism emerged at the end of the 1920s. Both Peugeot and Renault established brand new plants that reduced process flexibility, increased work standardization, and improved the flow of production (particularly for chassis and body assembly) in order to obtain greater economies of scale and lower production costs. This meant in fact moving toward a more faithful adoption of original Fordism. Contrary to most of their European competitors of the 1930s, who usually dissipated the benefits of such production methods by an inadequate product policy, Peugeot and especially Citroën (when the latter company had no choice but to rebuild and re-equip its main factory, which it did in haste) were able to reap the profits arising from a more focused strategy.[9] In the late 1930s, competition by foreign subsidiaries increased as they, too, built large-scale plants with high technical standards and a coherent production process in order to reach the mass market. This was the case first at Fiat's new subsidiary, Simca, then at Ford-France, whose brand-new factory had at last fully mechanized lines and more specialized machine-tools. But the war interrupted this mounting confrontation between various styles of mass production.[10]

The economic and cultural shock that World War II inflicted on France allowed the transition to a third stage in the development of Fordism in France. It helped to lift the reluctance to move into the bottom segment of the market, i.e. popular vehicles, which most companies had expressed in prewar times for fear of uncontrolled expansion and narrowing margins. Thus the largest French car makers and Simca (which took over the ailing Ford-France in 1954) gradually reshuffled their manufacturing organization. This included the introduction of transfer machines at Renault in 1947, designed by a pioneering engineer of the company and then sold to the other French makers. In this field, Renault could compete with the Americans, and it took Toyota ten more years to develop and build the same type of machines, which became prominent symbols of automation.[11] It included also the introduction of job evaluation methods (of American origin) that standardized the wages and the careers of blue-collar workers. It was enhanced by the spread of decentralized factories all over France which specialized in a model or component and could get by with only a very small proportion of managers, foremen and skilled workers in their personnel.[12] Conversely, the range of models was considerably reduced, each constructor specializing in a certain segment of the market (most on low-priced, popular vehicles; Peugeot, however, on the medium segment).[13] To be fair, there were contradictory developments simultaneously, although they also came from the United States. The suggestion system for workers, which car makers had introduced in 1927, became greatly encouraged and enjoyed strong support from most workers from the mid-1940s.[14] A little later, Training Within Industry (TWI), the American programme for the further training of supervisors and of managers just above them, was introduced in French industry, notably in French car companies from 1949 onward.[15] But in contrast to Japan, the suggestion system and the TWI program did not coalesce and did not give birth to a Kaizen system. The impetus of the original Fordist production methods and product strategy, both of which enjoyed marked support from the American government in the name of productivity and from the French government in the name of growth and of democratization of consumption, was probably too strong. Also, automation did not provoke any panic, in contrast to the USA, and enjoyed all the attractiveness of modernity among Frenchmen.[16]

Only in the 1960s and early 1970s did French car makers and Simca (renamed Chrysler-France) start to build a full range of models. Nevertheless, this was not a full conversion to the strategy pioneered by Alfred Sloan at General Motors some forty years ago. Model changes were not frequent and annual modifications remained essentially cosmetic. So, against this background of softened Sloanism, Fordist production methods could only develop further. A significant proof of their full-scale, systematic implementation was the achievement in 1970 of the one-minute cycle time on the final assembly line.[17] In 1973, a former leading Renault production engineer, Pierre Debos, contemplated robotization as the only sensible solution to the repeated strikes led by semi-skilled workers. Production engineers at large had identified with Fordism.[18]

7.2. Quality, or the Contradiction between French Intellectual Advance and Limited Industrial Progress in French car Companies

The origins of the modern quality movement have long been ascribed to Americans, notably W. A. Shewhart, a researcher at Bell Telephone Company and at Western Electric, who pioneered the methods of statistical control of quality and their instrument, the quality control card, in the late 1920s. Some ten years later, Edwards Deming became one of his assistants. After World War II, he was to inspire the formidable quality movement in Japan, in which he took part till his death on 20 December 1993.[19] But only very recently has it been found out that two innovative Frenchmen had conceived similar theories at the same time.[20] So we have to ask ourselves why their methods were not grafted into industry by French firms and what substitutes were developed by French car makers when they tried to meet the quality problems engendered by the adoption of Fordism in the interwar years.

While in the USA quality control originated in industry, at the Bell Labs in 1924 with Shewhart, in France it emerged in 1925 among army officers. This was obviously a consequence of the seriousness of defective products in armament production during World War I. The military engineer Maurice Dumas, preoccupied with the quality of ammunition, discovered the theory of reception control by samples. The same year a military pharmacist, Vallery, found another application of statistical methods to the detection of frauds. Why were these discoveries not influential in French industry? Several reasons may be adduced. In the USA, it was statistical process control that was invented first, which is easier for producers to accept than reception control, which brings in the viewpoint of the customer. The relations between the army and its suppliers during World War I had not been very happy, and a new idea originating among French officers could not be very popular among industrialists after the war. At the same time, French academic mathematicians in the 1920s, although very brilliant, disdained statistical work, and even more the application of statistics to industrial matters; therefore they did not support the discoveries of the two officers. So their attempts were devoid of any practical influence for lack of powerful allies both in academia and in industry (which was also the case for a contemporary German pioneer).[21] Statistical quality control had to wait for the aftermath of World War II to penetrate French industry, and this time it was of course the American method devised by Shewhart, his colleagues, and his followers, but also, more surprisingly perhaps, its British variants—which British war production engineers devised during World War II and which were soon publicized in the *First Guide to Quality Control for Engineers*, stressing the ensuing reduction in production cost and in inspection personnel.[22]

Very early after World War I, French car makers had clearly identified the double challenge mass production posed for the quality of their manufactured products: reaching a quality at least as high as craft production but at a lower cost; ensuring the constancy of quality.[23] But there was no R & D tie-in: the design department remained almighty, and production engineers knew of its new products only when they were ready, and generally were not allowed to modify them.[24] Under these conditions and in the absence of the application of statistics as an aid in maintaining quality, car makers were left with only two options: organizing men (Taylor's idea) and organizing the supply of materials and components (Ford's idea). Yet the French companies gradually learned that these two options were not sufficient, and started to move in new directions.

Taylor and the main pioneer of his works in France, the professor of chemistry Henry Le Chatelier, believed in physical determinism: production had simply to apply the best knowledge of the laws of nature, and this was bound to increase the precision of machines. Consequently, the management of quality mainly consisted in the organization of men, i.e. constraining or motivating workers to follow a strict discipline and to be both exact and honest, accustoming them to sift out the inevitable defective components or products from the good ones.[25] This vision included a strong growth in verification and control services. It met initially with doubts and criticism from production engineers and workers. But the trend was irresistible. At Renault in 1935 there were no less than 20 structures in charge of control at central or local levels. In 1939 there were some 1,500 controllers for a total labour force of 32,000 workers. On average, the controllers were young, with a high proportion of women.[26] Team leaders at Renault (i.e. skilled workers immediately below the foremen) were instructed by articles in the in-house journal, then by a specific booklet, to exert constant and rigorous control over the output of the workers, their attention to quality, and their punctuality.[27] There was constant fear of using defective components and products, and this fear bore some fruit. In 1939 the general production manager of Peugeot, Mattern, admiringly observed that there were no stations for finishing cars at the end of Renault's assembly lines. He deplored the fact that at Peugeot they still needed areas where the defects that production controllers had not noticed could be reworked and areas where they could complete cars with missing components.[28]

Ford's policy of vertical integration and tight control over suppliers was the other model that impressed the French car makers. When Louis Renault and André Citroën came back to the United States in order to visit factories both of suppliers and of car makers (the former in 1928, the latter in 1923 and 1931), they ostensibly toured the River Rouge plant, met Henry Ford again, and praised him for having increased the integration of production. Peugeot managers also went to the US; the first trip had been in 1916 (the R & D manager); the second one was in 1926, by top managers Eugène and Rodolphe Peugeot and Lucien Rosengart, accompanied by several engineers.[29] Renault was clearly the manufacturer who carried this strategy in steel production the farthest. Management regularly invoked the bad quality, unacceptable production cost, and low output of the suppliers.[30] But its main competitors, Citroën and Peugeot, went part of the way in the same direction.[31] It is striking to see that in 1939 the general production manager of Peugeot toyed with the idea of experimenting with a higher degree of vertical integration, and admired Renault's ability to produce a significant proportion of its machine-tools. He also praised Renault for having built a 'diplomatic workshop' for roller bearers, which produced only part of Renault's needs but was used to exert pressure on specialized suppliers. His general conclusion was strongly worded: 'We can no longer be content to have with our suppliers simply the ordinary relationship of customers'.[32] In sum, in the face of the dilemmas 'make or buy', 'markets or hierarchies', the constancy of quality demanded by Fordism had been weighing intensely against market forces.

Yet these two complementary strategies derived from American theories and practices had their limitations. The French automobile makers found some of them. For instance, we have mentioned above the observation by Ernest Mattern of Peugeot that in his company production inspection was unable to detect all defects, a theme that was recognized and expanded by the American quality expert, Joseph M. Juran.[33] Management began to think of other ways to maintain quality. But here we have to treat Citroën, Peugeot, and Renault separately.

At Peugeot, manager Ernest Mattern made three main observations in 1938. To achieve quality, it was not enough to disassemble a few finished cars every day, it was becoming necessary to replace the hostility of workers to managers by cooperation, without which quality would be insufficient. Similarly, technical foremen had to be trained in such cooperation. Finally, team leaders should write down their observations about work and workers, and these should be taken into account, on the one hand by the workers of their team in order to improve quality, and on the other hand by management in order to know what was in the minds of the workers.[34] But he did not limit his remarks to the labour force. He criticized the lack of cooperation between the design department and the production department, which resulted in 'prototypes the industrial production of which had not been thoroughly studied', and called for a longer time to prepare a new model.[35] One month later, in 1939, after having toured the Renault plants, he also departed from his usual Fordist view about integration in an important respect: 'We might have closer relations with some of our suppliers. They would have the status of exclusive suppliers for a number of years according to a set formula'. He also recommended repeated inspections of all the suppliers' factories by the men of the Purchasing Department.[36] The latter suggestion had two meanings. It entailed a general application of inspections that already existed, but only for some types of suppliers: the inspection of sheet and metals in the steel firms had been in existence at the Peugeot company (and maybe in the other two larger French companies) well before 1939. It revealed a preoccupation that Mattern first expressed in 1913–14 when he imposed metallic moulds upon the smelting companies that were supplying Peugeot: the suppliers should take customer wishes into greater consideration.[37] Had the war not broken out, would he have pioneered methods parallel to Toyota's?

At Renault, management also moved towards some innovations. They instructed the team leaders, as early as 1923, to behave as models for the workers, to let the concern for quality prevail over the preoccupation with output when necessary, to prefer the prevention of defects to curative action.[38] Consequently, and in contrast to Peugeot, they became able to secure cooperation between production control and foremen. They also organized a rigorous preparation of work (by foremen acting under the supervision of the production methods department). Finally, they were very keen on the smoothness of production flows, and achieved very impressive results in that area.[39] Once again, we may hypothesize that a broader conception and practice of quality was beginning to emerge, which World War II was to delay.

At the same time in the immediate prewar years, Citroën made significant progress in quality control. This happened because André Citroën's company had become bankrupt and had been taken over in 1935 by the Michelin tyre company. The new management sent a team to review the defects of the new main model of the company, the *traction avant*. It identified no less than 600 defects. It took almost two years to eliminate them.[40] Meanwhile, the new management became sensitive to the lack of interest in the products that the survey had revealed among most workers. The decisions taken were in response to these findings. Regarding the production process, a number of operations were taken out of the main assembly line, to be either prepared or even made in neighbouring areas, whereas workers were given financial incentives when they acted to eliminate defects and took on more responsibility on machine set-up. Regarding the design process, Citroën introduced market surveys in the French car industry that enabled the company to develop new product features; it created a unit that acted as a bridge between the design department and the production process department.[41] These various departures from either the original methods of Ford or from the structural characteristics of French car manufacturing increased the problem-solving capabilities of the organization and obviously might have paved the way for larger modifications.

Thus, the overall balance of automotive quality in the first two stages of the adaptation of Fordism to France is not negative. Quality of the process and quality of the product are already a growing priority and, despite the primacy of inspection, significant results are progressively scored. The model that Peugeot introduced from 1931 onward in the lower medium segment, the 201, competed with other cars for quality and resistance. It earned the Peugeot cars their reputation of solidity, a quality that the next models confirmed.[42] Renault, which had earlier used the theme of quality against Citroën, counter-attacked. I have examined a number of Renault advertisements of the period. They all insist on quality and connect it to the long life of the cars sold.[43] It would probably have been hard to maintain this type of advertisement had they been totally false. Peugeot's Ernest Mattern became anxious in 1939 that Peugeot might lose its competitive advantage for quality to Renault, and assessed Renault's production process as often more efficient in this respect and less costly.[44] In 1930 Renault's marketing bulletin even used a phrase that the Japanese later made famous: 'customer satisfaction'. But this first encounter between Fordism and quality—typical of the usual first phase of quality management, where the focus is on quality through inspection,[45] but showing, at least in the late 1930s, signs of a new approach—was soon threatened by the expansion of the automobile market between the end of World War II and the second oil shock.

7.3. Are Quality and Quantity Incompatible?

Whereas in Japan the growth of automobile production after World War II soon went apace with the development of quality control, later total quality, I am tempted to argue that in France quality methods and institutions also progressed, but that within the car companies the logic of quality was nevertheless often sacrificed to a dominant logic of quantity and volume. At least in companies like Renault and Simca (but perhaps not so much at Peugeot and at Citroën), it was a conflict between quality and productivity, and as the latter was promoted by the top managers of the powerful operational departments, it won.[46] This was only possible because French car makers were still (until 1968) protected by high tariff barriers.

Even so, one should not forget in retrospect the advancement of quality methods and institutions, and therefore one should not fail to assess their significance. They were a response to growing quality problems, arising from hasty design, approximative tuning of the tooling of the new models, and the spread of transfer machines that relegated control to the end of the line. They were also prompted by the equally growing costs of the guarantee offered by the main car makers to their customers. So, Renault's CEO sent a young executive, an engineer from the important École Centrale, Paul Durlach, to the USA in 1954, in order to survey ways to improve product quality. He suggested the creation of a Department of Quality, directly attached to the CEO. This was done in 1956. He also suggested the generalization of American statistical process control, which—forgetting Bell and Shewhart—was then called in France 'the Ford system', and had been introduced at Renault on a small scale only in 1953, despite repeated pleas made by consultants, engineers, and the French Society of Automotive Engineers since 1949. These experts had invoked the diffusion of statistical process control in other branches and the training of specialized engineers by a French engineering school. This generalization was also accepted by top management. Durlach was bold enough to suggest the intervention of the Quality Department in the work of the Design Department when prototypes become available. His suggestion was followed. In 1956 Renault was the first automobile manufacturer in France to introduce an evaluation of the defects of each vehicle that had been produced.[47] Simultaneously, top management decided in the early 1950s to abandon the production of certain materials or components that had become uneconomical or uncompetitive in quality, a decline of vertical integration that General Motors under Alfred Sloan had pioneered in the USA during the 1930s. This major shift to contractual links with external suppliers was soon to be imitated by the other car makers in France.[48] These various initiatives modified the management of quality. It was still based on control at each stage of the product cycle, the same as in prewar times, and such an approach remained popular even among trade union leaders.[49] It was still subordinated to the logic of quantity, as was evidenced by the failure of Renault's (and Peugeot's) penetration of the US market in 1960.[50] But it was becoming more ambitious in scope, as the themes promoted by the French Association for Industrial Quality Control (AFCIQ, founded in 1957) suggest: pilot actions on productivity quality, research on reliability, relations between producers and suppliers, quality costs. Moreover, the forceful pressures that Renault exerted on the domestic producers of sheet iron in order to have them build two modern strip mills finally succeeded in the early 1950s. Thus the voice of the customer prevailed over the interests of the supplier, a change that benefited the four major French car makers and made better products possible.[51]

However, even this subordinate, though ambitious, position of the quality personnel was too much for the existing design and industrial departments. At each car maker they considered that their productivist logic and their autonomy were threatened by the burgeoning of qualiticians. At Citroën the gap became so wide that, beginning with the otherwise brilliant DS (introduced in 1955), new models presented numerous defects that the manufacturer would take one or two years to eliminate.[52] At Renault the development of Sloanism, i.e. of a full range, offered opponents to quality management an unexpected opportunity to restore the full dimensions of their powers. The first year of production of the top-of-the-range model, the R16, in 1965, was marred by customers' complaints about numerous defects. In 1966 the top managers reacted by calling for full solidarity 'between all those who take part in the design of a vehicle and those who produce it', and consequently numerous articles in the in-house journal were devoted to quality assurance.[53] Industrial departments, however, argued that a quality department was inefficient and that each operational department could thus legitimately desire full responsibility over its tasks. Their rebellion succeeded. In 1967 the department was downgraded to a secretariat, and an interdepartmental group was created to insure the maintenance and improvement of quality. This was a total victory for the Fordist barons and for Taylorist division of labour. Production policy became entirely based on quantity. This had tremendous consequences. The production dates for the introduction of new models were decided irrespective of the observations of qualiticians. The marketing department vowed a cult to new registrations, independently of customers' problems. The directors of the decentralized factories were assessed by top managers only on their ability to maintain output targets. The cost and usefulness of quality control became openly questioned inside various departments. The growing needs of customers, which Renault had taken into account with the adoption of the front-wheel drive from 1961, and which now extended to rust, noise, and braking, were neglected by the operational departments. Even value analysis, born in the USA at General Electric in 1947 in order to improve production costs, was initially interpreted by French automobile companies, first Peugeot (1965), then Renault (1967), as a weapon in the struggle against waste which was given full priority, with quality only a secondary concern.[54] Yet in the long run the loss of power of Renault's quality assurance staff may have resulted in a much stronger decentralization of quality tasks and in a diffusion of quality-oriented analyses and practices.[55]

But this was precisely the time when the voice of customers became perceptible inside the organizations thanks to new tools that managers could accept. To be fair, the existence for the first time since 1950 of an independent automobile press had given customers some voice, but it left manufacturers the option of ignoring its criticisms or of taming the journalists.[56] Now the voice of customers could be heard directly. Indeed it had to be listened to, as replacement demand took the lead, as multimotorization developed, and as trade liberalization as well as the fulfilment of the European Economic Community dramatically increased competition by foreign makes on the domestic market. The Renault Company reacted by adding to its mainly production-oriented quality tools other tools that embodied its emerging customer orientation. From 1971 the after-sale department launched a quality survey about the behaviour and performance of Renault cars after six months of use. From 1972 the product department launched an image survey pertaining to both the Renault cars and the company: it measured Renault's image on the French market and compared it to its competitors, French and foreign. From 1975, the French car makers agreed to ask a leading poll institute to survey the reliability of cars on the French market (measured, *inter alia*, by the number of breakdowns per 1,000 cars). Thereafter, quality was no more an individual perception or a rumour. It was now a statistical result, and thus an instrument of decision. Excerpts of that survey for each segment were made available at the meetings of the various committees in charge of the development of new models. By 1980 Renault decided to join a similar survey at the European level, one that was highly sophisticated and recognized by the entire automobile industry. It naturally included a customer satisfaction index. In addition, consumers' associations, both French and foreign, tested the new models more and more seriously. Quality, generally associated with safety, became one of their main criteria. However, the impact of these surveys on management in the French car companies was quite slow, partly because their message was complex to analyse, partly because the European market was still growing.[57]

Yet it was already enough to prove the inefficiency of the dual structure instituted in 1967 at Renault. In 1977 a quality department directly attached to the deputy CEO was recreated and its objectives were considerably increased. The results were mixed. On one side, it made visible a global management of quality, from design to after-sale and maintenance. New managerial tools, focusing on prevention, were conceived: training seminars, a quality handbook, and especially a quality index at Kilometre zero (called AQR, Action Qualité Renault). New procedures made the agreement of quality managers necessary to the beginning of production and then of commercialization. But the real place of quality in the company remained ambiguous. The focus of the organization on volume was still very strong. The successive directors of the quality department alternated between centralization and decentralization. Tensions grew between the quality department and the operational departments, whose relative goodwill was tempered by the real difficulties of coordinating quality and of tuning the, by now, varied quality procedures. The spectre of Japanese car companies' competitiveness and their management of quality were not enough to provide an incentive to overcome the difficulties in the transition from the quality assurance phase to a prevention phase. It would take no less than a total crisis, which successively hit Peugeot-Citroën in 1983 and Renault in 1984.

7.4. The Belated Integration of Quality into the General Policy of the French Car Companies

Both French companies thus barely escaped bankruptcy in the early 1980s. But they recovered, and a quality approach was at the heart of this recovery. Clearly it could not be reduced to a fashionable graft of the Japanese Kaizen movement. It actually involved a transformation of the two enterprises that, as of the end of 1993, was not yet finished.[58] Two questions have to be asked: was the use of quality circles *à la Japonaise* frequent at workshop level and what were the real implications of the conversion to total quality for the two companies?

Quality circles had a very uneven penetration in the French car companies and their life was precarious. The main reason was the symbolic challenge they posed to the foundations of the industrial organization that had prevailed in the French car industry since the end of World War II. As their underlying logic was heterogeneous to the classical industrial logic, either that logic could be integrated, but as part and parcel of a thorough modification of the general management system, or it could be rejected.[59] The spread of quality circles was, recently, differentiated from one car maker to the other.

The smaller maker, Citroën, which had been taken over by Peugeot in 1974, was probably the earliest to introduce quality circles in its plants. The Citroën decentralized factories in Rennes (Britanny) were thus able to modify considerably their working methods. In 1981, two years after the first missions sent by French industry to Japan to investigate the quality circles established there in the early 1960s, two self-taught cadres of the production methods control service at Rennes La Janais read an account of the missions' assessment in the bulletin of the AFCIQ (the quality association mentioned earlier). Particularly impressed by Ishikawa's diagram, they asked a student to translate the Japanese documents relating to their operation. After six months, forty-two QC circles had been formed. This local initiative received much publicity within and outside the Citroën division of the Peugeot-Citroën company.[60] By 1991 the Citroën division had 1104 QC circles. The initial target was the reduction of scrap and reworking time for production lines. After three years, in 1984, it triggered a broader perspective than worker's involvement and self-inspection. The entire Citroën division strove for the principle of 'zero breakdown, zero defect, zero stock'. It now followed a rigorous prevention strategy, including *poka yoke* (foolproof) process devices. That quality drive paid off. In 1985, the average cost of the Citroën car guarantee during one year had declined, from 2.5 per cent of the selling price to 1.9 per cent.[61] In 1993, the suggestion system (revitalized by higher bonuses, by an annual inter-factory challenge, and by a more direct involvement of foremen) saved Citroën 183 million francs.[62] The 'voluntary' participation rate in QC circle activities reached 24 per cent that same year, about the same level as the average of all the overseas Japanese subsidiaries.[63]

Peugeot also welcomed a number of quality circles. As at Citroën, the earliest ones started in 1981, on a local basis, in Dijon. But soon their diffusion was managed from the top down. Management saw them as a necessary component in the increase in automation and robotization. It also called in outside consultants, either French or American, to assist in the spread of the new institution. Such was the case at the assembly plant of Mulhouse (Alsace) between 1982 and 1985: a reduction of one level of hierarchy was an outcome of the quality strategy carried out there. Yet by 1991 the 'Voluntary' participation rate in QC circle activites was lower than at Citroën: 10 per cent in Mulhouse, and probably the same at other sites, despite higher goals. This result can be ascribed partly to internal resistance among some blue- and white-collar workers and partly to the top-down approach.[64]

But at Renault they were relatively few, essentially in a few decentra-lized factories like Flins or Sandouville that had suffered from deep qual-ity problems in the late 1970s or from their future being in question. This is partly because the battle for quality was fought at the time by people who did not occupy the highest positions in Renault's hierarchy, or who had not been educated in the highest *grandes écoles*, or who were technical qualiticians lacking more general ambitions. But this was also the result of the relative inconsistency of top management in such matters during the early 1980s. There was a proliferation of quality newspapers and quality networks, plus a two-year seminar on technological, economic, and social change in the company. In such a context quality circles were not a central priority. In June 1982, there were some fifty quality circles in the Renault company.[65]

The solution to their respective crises by the two automobile French groups opened a space for more far-searching innovations, with quality at the forefront. In the 1980s the Citroën division of the PSA (Peugeot-Citroën) group went beyond its earlier initiatives. Its managers carefully designed and enforced a global reshuffling of the methods, practices, values, hierarchy, and organization of the company, soon called Mercury Project. However, success did not come until recently, because of troubles in the selection of new models and because a quality strategy does not always yield quality.[66]

Renault's change happened in two stages. In 1985 a major breakthrough occurred. The new CEO abandoned the volume policy. He made profitability the number one target of the company, and the improvement of its products the second priority. Therefore it became possible for the company's top and middle managers to reconcile profitability and customer satisfaction. Two institutional initiatives were decided upon. At the company's headquarters, the Quality Department took charge of a permanent campaign for reliability in use, supported by computerized data. In the network, autonomous units were created to welcome the customers. The Quality Department focused on decentralized actions. It attempted experiments in total quality in various factories with the help of famous American consulting experts: Juran, Feigenbaum, Crosby. Moreover, Renault's management initiated talks with the Peugeot-Citroën group in order to jointly abandon reception control and delegate control to the supplier. The goal was twofold: to make suppliers responsible for the quality of their products and to reduce the cost of reception controls. In 1987, the two companies achieved a joint procedure for assessment of the quality potential of their suppliers. It made the actual development of partnerships between the two companies and their suppliers possible. A special procedure was developed to ensure the quality of specific components. The CEO asked leading executives to elaborate a total quality programme.[67]

After the assassination of Renault's CEO by terrorists at the end of 1986, his successor made total quality the basic tenet of his policy. The Quality Department was reorganized. A new director, Pierre Jocou, was appointed; unlike his predecessors, he came from the after-sale department and was thus considered the representative of customers within the firm. A training and research institute was established in 1988. In 1989 total quality committees and coordinators were appointed in each department, while project leaders of cross-functional teams were chosen by the CEO to pilot the preparation of new major vehicles or to improve the company's response to formidable challenges such as the management of the product life cycle, the compromises between diversity in products and standardization of components, and the selection of useful investments. Thus, the core of this approach was the improvement of human processes, in three domains that were now closely associated after having been opposed: quality, costs, and time lags. Ambitious qualitative and quantitative targets in terms of product quality and of human progress were chosen by top management. Two symbols easily materialized this new policy deployment. The Director for Quality became a member of the top executive group of the company. He twice used his newly given right to delay the production of new models as long as their quality was not fully satisfactory.[68] Total quality was one of three reasons for the conclusion in 1990 of the alliance between Renault and the Swedish company Volvo, which had a strong image of quality. It was also one of the motivations for the attempted merger, which fell through on 2 December 1993. The joint venture for quality between Renault and Volvo did not survive the failure of the merger.

How do we assess the result of this third phase, oriented both toward the prevention of problems and the integration of quality into top management strategy?

Clearly, as in other European countries and following the Japanese lead, the relationships of the two French car makers with their suppliers have undergone major changes, as we have just suggested. Under the banner of 'lasting relationships of mutual trust', car makers and suppliers now share the functions of R & D, design, production methods, training, and quality control in varying degrees. The consequences of this have been a better selection process for suppliers, a stricter hierarchy among them, and a dual policy of localization for suppliers: either the setting up of plants in the vicinity of the car makers' assembly plants (much in the Japanese way) or, in contrast, far away, in keeping with the policy of global sourcing, pioneered by Ford since 1988 and recently adopted by GM.[69] To characterize these new relationships as partnerships calls for some qualifications. In contrast, to other industries, the customers, i.e. the two car makers, are clearly the pilots. The partnership is real and efficient only if the supplier is of sufficient size and if its advantages are not counter-balanced by some drawbacks (fragility in the face of sudden changes in the environment, irreversibility of some choices, lower pressure of competition on the supplier). But under such conditions, some French suppliers, like Valeo and Bertrand Faure, have been able to internationalize.[70]

In this common trend we may distinguish striking differences between Renault and Peugeot. Partly because of its commercial and financial crisis of the mid-1980s, Renault sold some of its components subsidiaries and went further in outsourcing than Peugeot-Citroën. Peugeot-Citroën, on the contrary, first modernized and reorganized their components sector. Only recently did they sell some subsidiaries and reorganize the others into a coherent, international group.[71] Peugeot-Citroën are keener on screening the best suppliers on the basis of certification and effective productivity gains. Renault on the other hand invites its suppliers to create networks solidified by quality assurance.[72]

If we finally come to more recent quality management at the two car makers themselves, common trends are obvious, such as the centralization of buying within a strong purchasing department, closer relationships between the various departments of each firm and within each function, and more intense customer awareness.[73] But some differences in quality management remain between the two competing French makers.

In a recent interview, Renault's Quality Director gave a balanced assessment of the progress achieved. Product quality has reached the best European level and is never too far from the Japanese. But process quality has still not climbed to the level of the best competitors. The costs of quality requirements are now under control, but some managers remain sceptical regarding total quality management. Customer satisfaction has increased among both 'internal clients' and market customers, but further improvement in the company's performance is still needed.[74]

At Peugeot-Citroën, a vigorous quality management probably obtains less systematic results: it has to face two different ranges of models despite the unification of the design departments and the commonalization of parts; the position of the two central quality managers is lower in the hierarchy of top management; and the top of the range cars 605 (Peugeot) and XM (Citroën), both launched in 1989, were plagued by painful quality problems due to weaknesses in the design department, and insufficient road testing and co-ordination with production engineers.[75] These drawbacks are currently being dealt with at the local level and in general policy making.

7.5. Conclusions

The history of the French car makers' approach to quality in the course of implementing and adapting Ford's production methods leads to several general observations.

First, the size of production matters. The relatively moderate level of quantities produced during the interwar years made possible acceptable levels of quality control, largely based on inspection and vertical integration, and inspired imaginative managers to think in new directions, only to be thwarted by the war. Inversely, the period of high production growth from 1945 to 1981 brought about a divergence between the needs of quality and the demands of quantity, at the expense of the former. Conversely, the stabilization of French automotive production in the 1980s and the early 1990s may have been a favourable circumstance leading manufacturers towards total quality and greater consumer satisfaction.

Secondly, there is certainly a distinctiveness about the history of the French car makers' attitudes to quality when a comparison is drawn with the British and Japanese developments as outlined by Fujimoto and Tidd. This distinctiveness may be ascribed to the receptivity of twentieth-century French industry and society to management fads, as well as to their ability to digest them, thus making a real integration to policy deployment something that took much time and energy.[76] It also depends on the well-known positions occupied by French social actors: production engineers, alumni of the *grandes écoles,* consulting engineers, and trade unionists who gave in easily to Taylorism and Fordism for the sake of consumption. Their common interest in quantity could at certain periods counter-balance the imperative or the fashion of quality. Finally, it cannot be explained without referring to the initial primacy of the French car industrialists in the world car industry, as it both reinforced the position of production experts and in the late 1930s made the needs of the clients a familiar message.

Thirdly, can one speak today of a European way to quality that would not be a mere replica of Japanese and American models? This would be a nice extension of the Fujimoto-Tidd argument regarding both the necessary adjustment of production methods to demand and the variability of product life cycles in the car industry. It is also the current contention of Peugeot's, Citroën's and Renault's quality directors. But for the moment, as they readily acknowledge, European car makers have shown their ability at 'cherry picking' among other nations' solutions.[77] They still have to prove fully that they accept their weaknesses and apparent disadvantages and are able to overcome them by solutions adapted to European practices and volumes.

Fourthly, in the history of company structures the existence of a quality department is clearly not sufficient in itself. For the development of a quality policy real authority must be conferred on the department by the CEO and the quality targets must be adopted by the other departments. The French story might show that the strength of a quality department does not come from the number of its bureaucrats, but rather from its power of influence and its ability to let operational personnel manage quality in their own field of competence. A quality department may

TABLE 7.1. *French automobile production for selected years*

Year	No. of cars
1921	55,000
1929	254,000
1938	190,000
1945	35,000
1951	447,000
1973	2,630,000
1987	3,325,000
1992	3,768,000

Source: French Automobile Manufacturers' Association.

be a transitory organizational form, but the decentralization of quality responsibilities to operational personnel may be a permanent necessity if the recurrent problems of quality management at plant level—frontiers between functions and trades; liaison between marketing, production, and design; position and number of quality assurance personnel; flexibility of quality criteria; capitalization of learning and knowledge; different national cultures in the various industrial sites of a multinational corporation—are to be tackled adequately in the French automobile companies.[78]

NOTES

1. T. Fujimoto and J. Tidd, 'The UK and Japanese Auto Industry: Adoption and Adaptation of Fordism', paper presented at Gotenba City Conference, 1993.
2. In the absence of a general history of the French car industry, see J.-P. Bardou, J.-J. Chanaron, P. Fridenson, and J. M. Laux, *The Automobile Revolution* (Chapel Hill, 1982) and J. M. Laux, *The European Automobile Industry* (New York, 1992).
3. P. Fridenson, 'Les Premiers Contacts entre Louis Renault et Henry Ford', *De Renault Frères constructeurs d'automobiles à Renault Régie Nationale*, 4 (Dec. 1973), 247–50. J.-P. Poitou, 'Le Voyage de Louis Renault aux États-Unis d'avril 1911', ibid. 15 (Dec. 1984), 226–31. S. Schweitzer, *André Citroën* (Paris, 1992).
4. M. Wilkins and F. E. Hill, *American Business Abroad: Ford on Six Continents* (Detroit, 1964), 97, 140. A. P. Sloan, *My Years with General Motors* (Garden City, N. Y., 1964).
5. G. Declas, 'Recherches sur les usines Berliet (1912–49)', MA thesis (University of Paris I, 1977). J.-F. Grevet, 'Stratégies commerciales et développement d'une firme automobile: Berliet des origines à 1939', MA thesis (University of Lille III, 1994).
6. P. Fridenson, 'The Coming of the Assembly Line to Europe', in W. Krohn *et al.* (eds.), *The Dynamics of Science & Technology* (Dordrecht, 1978), 159–75. Y. Cohen, 'The Modernization of Production in the French Automobile Industry between the Wars: A Photographic Essay', *Business History Review* 65 (winter 1991), 758–68.
7. Y. Cohen, 'Inventivité organisationnelle et compétitivité: L'Interchangeabilité des pièces face à la crise de la machine-outil en France autour de 1990', *Entreprises et Histoire*, 3 (June 1994).
8. P. Fridenson, *Histoire des usines Renault*, vol. I (Paris, 1972). S. Schweitzer, *Des engrenages à la chaîne: Les Usines Citroën, 1915–35* (Lyons, 1982). D. Henri, 'Comptes, mécomptes et redressement d'une gestion industrielle: Les Automobiles Peugeot de 1918 à 1930', *Revue d'histoire moderne et contemporaine, 39* (Jan.–Mar. 1985), 30–74.
9. O. Cinqualbre and Y. Cohen, 'L'Usine de la grande série: André Citroën, quai de Javel', *Monuments Historiques*, 26 (Sept. 1984), 15–22. P. Fridenson, 'La

Traction avant', in P. Ory and O. Barrot, *Entre deux guerres* (Paris, 1991). Cohen, 'The Modernization of Production', 770-5.

10. P. Saint-Marc, 'Recherches sur les usines Simca de 1934 à 1958', MA thesis (University of Paris X-Nanterre, 1989). M. Wilkins and F. E. Hill, *American Business Abroad*.

11. J.-P. Poitou, *Le Cerveau de l'usine* (Aix en Provence, 1988).

12. See, inter alia, P. Caro, 'Les Usines Citroën de Rennes (1954–74)', D.E.A. thesis (École des Hautes Études en Sciences Sociales, 1993).

13. J.-L. Loubet, *Automobiles Peugeot. Une réussite industrielle 1945–74* (Paris, 1990).

14. A. Moutet, 'Introduction de la production à la chaîne en France du début de XXe siècle à la grande crise en 1930', *Histoire, Économie et Société*, 5 (Jan.–Mar. 1983), 63–82. C. Malaval, *Renault à la Une. La Presse d'entreprise Renault depuis 1945* (Paris, 1992), 43 and 58–9.

15. F. Jacquin, *Les Cadres de l'industrie et du commerce en France* (Paris, 1955). A. Kopff, 'Histoire de la Maîtrise au Centre de production Peugeot-Sochaux (1945–75)', DEA thesis (École des Hautes Études en Sciences Sociales, 1993), 52–7.

16. P. Fridenson, 'L'Industrie automobile française et le Plan Marshall', in M. Lévy-Leboyer (ed.), Le *Plan Marshall et le relèvement èconomique de l'Europe* (Paris, 1993).

17. For the Renault Company: interview by myself of a trade-unionist, Daniel Labbé, 1 Mar. 1994; C. Midler, 'L'Organisation du travail et ses déterminants. Enjeux économiques et organisationnels des réformes de restructuration des tâches dans le montage automobile', thèse de troisième cycle en gestion (University of Paris I, 1980), 40, 77, 78, 83–4.

18. P. Debos, 'L'Automation et le problème des O.S.', *Arts et Métiers*, 28 (May–June 1973). For his own overview of his action as Renault's Production Methods Manager, P. Debos, 'L'Évolution des méthodes de fabrication à la Régie de 1946 à 1967', *De Renault Frères constructeurs d'automobiles à Renault Régie nationale*, 5 (June 1974). On his Department's contract emphasis on Fordist methods, see Midler, 'L'Organisation du travail', 178–90.

19. A. Blanton Godfrey, 'History of QC at AT&T', *AT&T Technical Journal*, 1988. My thanks to Louis Galambos for this reference. ' "God of Quality Control" W. Edwards Deming Dies', *The Japan Times*, 97 (22 Dec. 1993), 12. However, in the early 1930s, Yasushi Ishida, working for the Toshiba Corporation, developed control charts for increasing the life span of light bulbs and reducing errors in their production: I. Nonaka, 'The History of the Quality Circle', *Quality Progress* (Sept. 1993), 81.

20. D. Bayart, 'Des objets qui solidifient une théorie: Le Cas du contrôle statistique de fabrication', *Sociologie du Travail*, forthcoming.

21. Bayart, 'Des objets', plus my own remarks. In the late 1930s there was a quality pioneer in the glass and chemical corporation Saint-Gobain.

22. A. Jouve, 'Considérations sur la productivité et les différents facteurs qui la conditionnent', *Journal de la Société des Ingénieurs de l'Automobile*, 24 (June 1950), 180–1.

23. F. Guilain, 'Histoire de la qualité chez Renault', MA thesis (École Supérieure de Commerce de Paris, 1991), 16. I have used this excellent thesis extensively

in the following pages, though I have not always followed its author's interpretations.

24. J. Boulogne, *Louis Renault* (Paris, 1931). R. Brioult, *Le Bureau d'études Citroën* (Paris, 1987). C. Dollet and A. Dusart, *Les Sorciers du lion. Un siècle dans le secret du bureau d'études Peugeot* (Paris, 1990).

25. Bayart, 'Des objets'.

26. Archives of the Mattern family, report by Ernest Mattern on his tour of the Renault plants, 1 Feb. 1939. Reprinted in Y. Cohen, 'Quand un homme de Peugeot visite Renault (janvier 1939)', *Renault Histoire*, 1 (Nov. 1989), 25.

27. Renault Archives, booklet of the team leader, 1923. Reprinted by Renault's current management in *Les Cahiers de la Qualité*, 1 (1990), 50–6.

28. Cohen, 'Quand un homme de Peugeot', 26–7.

29. Archives of the Renault Historical Society, account by Louis Renault of his trip to the US, June 1928. S. Schweitzer, *André Citroën*. L. Rosengart, 'Ce que nous devons apprendre de l'Amérique', *Peugeot-Revue*, 3 (June 1926), 6–9. This article was kindly supplied by Yves Cohen.

30. Fridenson, *Histoire des usines Renault*.

31. Cohen, 'The Modernization of Production', 758 and 770.

32. Cohen, 'Quand un homme de Peugeot', 19, 21, 30.

33. R. C. Wood, 'The Prophets of Quality', *The Quality Review* (autumn 1988).

34. Archives of the Mattern family, five-year plan for the improvement of the Peugeot works by Ernest Mattern, 31 Dec. 1938. I thank Yves Cohen for sharing with me this unpublished document. Peugeot enforced Mattern's ideas on the relationship between team leaders and workers: A. Kopff, 'La Maîtrise', 23.

35. Cohen, 'Quand un homme de Peugeot', 12.

36. Ibid. 30.

37. Y. Cohen, 'La Pratique des machines et des hommes: Une pensée technique en formation (1900–14)', in A. Thépot (ed.), *L'Ingénieur dans la société française* (Paris, 1985), 61–70.

38. Renault Archives, booklet of the team leader, 1923.

39. Cohen, 'Quand un homme de Peugeot', 25, 27.

40. P. Fridenson, 'La traction avant'.

41. J.-L. Loubet, 'Une sortie de crise: Citroën (1935–9)', working paper (Évry, 1994).

42. Cohen, 'The Modernization of Production', 778.

43. 'Les Principes Directeurs de Renault', *L'Illustration*, 4389, 1927. *Bulletin commercial des usines Renault*, Mar. and Dec. 1930. 'Résistance, qualité primordiale de toutes les Renault', advertisement, 1931. Renault commercial document, 1932. 'La qualité survit! achetez une Renault', *L'Illustration*, 24 Apr. 1937.

44. Cohen, 'Quand un homme de Peugeot'.

45. G. Rommel, R.-D. Kempis, and H.-W. Kaas, 'Does Quality Pay? An Empirical Study of the Automotive Supplier Industry in Europe and Japan', *The McKinsey Quarterly* (Apr.–June 1994).

46. Midler, 'L'Organisation du travail', 185 ff. Guilain, 'Histoire de la qualité'. Loubet, *Automobiles Peugeot*.

47. Jouve, 'Considérations sur la productivité', 180–1. G. Hatry (ed.), *Notices biographiques Renault*, vol. I (Paris, 1990), 44–5. M. Meuleau, 'Les H.E.C. et

l'évolution du management en France (1881–années 1970)', thèse de doctorat d'État (University of Paris X—Nanterre, 1992). Guilain, 'Histoire de la qualité'.

48. F. Picard, *L'Épopée de Renault* (Paris, 1976). J.-L. Loubet, 'La Société anonyme des automobiles Citroën (1924–68)', thèse de troisième cycle (University of Paris X—Nanterre, 1979). Loubet, *Automobiles Peugeot.*

49. *Renault Magazine* 2 (Apr. 1958).

50. Picard, *L'Épopée.* P. Dreyfus, *La Liberté de réussir* (Paris, 1977).

51. M. Kipping, 'Les Tôles avant les casseroles', *Entreprises et Histoire*, 3 (June 1994). M. Kipping, 'Competing for dollars', *Business and Economic History* (autumn 1994).

52. J. Broustail, 'Rétrospective technologique et management de l'innovation', Ph.D. thesis (École des Hautes Études Commerciales, 1991).

53. Malaval, *Renault à la Une*, 114–5.

54. Guilain, 'Histoire de la qualité'.

55. A hypothesis suggested to me by my colleague Christophe Midler in Mar. 1994.

56. D. Gaucher, 'La Presse automobile en France depuis 1945', MA thesis (University of Paris X—Nanterre, 1977).

57. Guilain, 'Histoire de la qualité', 38 and 77, plus Christophe Midler's thoughtful comments to me in Dec. 1993.

58. C. Midler, *L'Auto qui n'existait pas. Management des projets et transformation de l'entreprise* (Paris, 1993) J.-L. Loubet, 'PSA Peugeot-Citroën, 1973–92. Histoire d'un groupe automobile dans les années de crise', *Actes du GERPISA* 10 (Apr. 1994), 126–47.

59. F. Chevalier, *Cercles de qualité et changement organisationnel* (Paris, 1991).

60. C. Moire, 'Du sur mesure au prêt à porter. La Diffusion des cercles de qualité dans les entreprises françaises entre 1979 et 1984', Ph.D. thesis (École Polytechnique, 1985), 60, 62, 68–70, 73, 157–8, 167.

61. Loubet, 'PSA Peugeot-Citroën', 141.

62. C. Lévi, 'Citroën offre une prime à la suggestion retenue', *Le Monde*, 49 (13 Apr. 1994), IV.

63. I thank my colleague Tetsuo Abo for this comparison.

64. P. Jansen and L. Kissler, 'Direkte Arbeitnehmerbeteiligung und Wandel der betrieblichen Arbeitsbeziehungen als Managementaufgabe. Erste Untersuchungsergebnisse aus der deutschen und französischen Automobilindustrie. Das Beispiel Peugeot-Mulhouse', in L. Kissler (ed.), *Management und Partizipation in der Automobilindustrie. Zum Wandel der Arbeitsbeziehungen in Deutschland und Frankreich* (Frankfurt am Main, 1992), 155–91. A. Coffineau and J.-P. Sarraz, 'Partizipatives Management und Unternehmensberatung. Das Projekt ISOAR bei Peugeot-Mulhouse', ibid. 145–53.

65. Archives of the Centre de Gestion Scientifique de l'École des Mines de Paris, accounts of interviews with Mr Belard and Mr Trembelland, 3 June 1982.

66. X. Mercure, *Citroën, une nouvelle culture d'enterprise* (Paris, 1989). A. Routier, 'Citroën: La Résurrection', *Le Nouvel Observateur* 20 (31 Mar. 1994), 72–3.

67. Guilain, 'Histoire de la qualité'.

68. See the first three annual issues of Renault's *Les Cahiers de la Qualité*, 1991–3; F. Lucas and P. Jocou, *Au coeur du changement. Une autre démarche de management:*

La Qualité totale (Paris, 1992); Midler, *L'Auto qui n'existait pas*. Interview of Antoine de Vaugelas, head of the organization department, by Gilles Garel and myself, 26 Apr. 1994.

69. V. Devillechabrolle, 'Automobile: L'Excellence à marche forcée', *Le Monde*, 49 (16 June 1993), 31. A. Gorgeu and R. Mathieu, 'Les Fournisseurs de l'industrie: Nouveaux Impératifs, nouvelles localisations', *La lettre d'information du C.E.E.*, 4 (Sept. 1993), 1–8. G. Rommel, et al., 'Does Quality Pay?'.

70. X. Eloy, 'Le Dialogue constructeur-fournisseur: L'Exemple des pneumatiques chez Renault', *Annales des Mines*, 197 (Oct. 1991), 99–103. M. Barbier de la Serre, 'Bertrand Faure: La Stratégie d'un équipementier', ibid. 104–6. Y. Blanc, 'La Stratégie industrielle et financière d'un équipementier indépendant: Valeo', ibid. 107–9. Midler, *L'Auto qui n'existait pas*. F., Pallez, 'Le Partenariat dans l'industrie', *Revue Française de Gestion Industrielle*, 12 (Jan. 1993), 5–15.

71. Loubet, 'PSA Peugeot-Citroën', 121–3 and 131–3. Renault's annual reports, 1981–93.

72. Devillechabrolle, 'Automobile: L'Excellence', 31.

73. Gorgeu and Matheu, 'Les fournisseurs', 3. C. Guéry and L. Corbi, 'Acheteur: L'Emergence d'un poste-clé pour l'entreprise', *Le Figaro*, 163 (25 Apr. 1994), 51. J.-C. Moisdon and B. Weil, 'Groupes transversaux et coordination technique dans la conception d'un nouveau véhicule', *Revue Française de Gestion Industrielle*, 11 (Apr. 1992).

74. P. Jocou, 'Renault est-il une entreprise de qualité?', *Avec*, 12 (10 Dec. 1993), 10–12.

75. *PSA Peugeot Citroën. Le groupe en bref*, Paris, 1993 I. C. Bonnaud, 'Projet Z-8: La Peugeot du siècle', *L'Auto-Journal*, 45 (15 Mar. 1994), 36–40. Routier, 'Citroën: La Résurrection', 72.

76. C. Midler, 'Logique de la mode managériale', *Gérer et Comprendre*, 2 (June 1986), 74–85.

77. P. Jocou, 'Postface', *Les Cahiers de la Qualité*, 3 (1993), 121–6. A further proof is in the recent emphasis on Renault's unit that, as in the good old days, is in charge of dismantling competitors' models: 'Le Centre d'Analyse de la Concurrence', *Avec*, 13 (10 Mar. 1994).

78. See two fine monographs on the Renault plant of Sandouville (Normandy) in 1988: J.-C. Croiger, 'Le Traitement de la non-qualité dans une usine de montage automobile: Analyse du rôle du service qualité en bout d'usine', DEA thesis (Institut d'Études Politiques de Paris, 1988). H. Barry, 'Étude des deux groupes de montage d'une usine d'automobiles dans un contexte de forte contrainte qualitative', DEA thesis (Institut d'Études Politiques de Paris, 1988). Also Guilain, 'Histoire de la qualité', 122. On the impact of a different national culture on product quality, see T. Globokar, 'Gérer en Slovénie ou les difficultés de la communication interculturelle', working paper, Paris School of Management, 1994 (about the Renault plant at Ljubljana, 1991–3).

8

A Note on the Origin of the 'Black Box Parts' Practice in the Japanese Motor Vehicle Industry

TAKAHIRO FUJIMOTO

8.1. Introduction

The purpose of this paper is to explore the origin and historical development of the 'black box parts' practice in the Japanese automobile and parts industry. The black box parts system refers to a certain pattern of transactions in which a parts supplier conducts detailed engineering of a component that it makes for an automobile maker on the basis of the latter's specifications and basic designs. In a sense, this is a kind of joint product development between a system maker and a component supplier, in that the latter is involved in the former's new-product development process.

Although the system is referred to variously as 'black box parts', 'grey box parts', 'design-in', and 'approved drawings', it has been known to be a prevalent practice in the Japanese automobile industry and one of the sources of its competitive advantages in recent years. It was also one element of the Japanese supplier system that many of the US motor vehicle suppliers tried to adopt in order to narrow the competitive gaps between them and the leading Japanese motor vehicle makers.

Although the black box parts practice has attracted the attention of both researchers and practitioners internationally, its historical origin had not been studied much by business historians. The published company histories of the major Japanese motor vehicle makers do not seem to have touched upon this aspect of supplier management either.[1] Thus, research on the historical evolution of the black box parts system seems to be

This chapter is an abridged version of a paper presented at the 21st Fuji Conference in Jan. 1994. For the full text and reference, see T. Fujimoto (1994). 'The Origin and Evolution of the "Black Box Parts" Practice in the Japanese Auto Industry.' Tokyo University Faculty of Economics Discussion Paper 94-F-1.

important for those who wish to understand, transfer, or adopt the practice.

In the broader context of the Japanese-style production system in the postwar automobile industry, a historical study of this practice would seem to provide an important insight into how the Japanese car makers built core capabilities within their highly competitive system. A study of black box parts might enable us to better understand how historical imperatives, entrepreneurial visions, and learning from other firms have impacts on the evolutionary patterns of production systems and capabilities.

On the basis of the above reasoning, this paper will first describe the current practice of supplier involvement in product development in the major Japanese motor vehicle makers, such as Toyota and Honda, as well as giving a brief literature survey. It will then examine the origin and historical evolution of the black box parts practice at Toyota and Nissan. The last section will focus on the supplier side by analysing some preliminary results of a questionnaire and field surveys.

8.2. Current Practice of Supplier Involvement in Product Development

8.2.1. Basic Concepts

Let us first define some key concepts. Clark and Fujimoto (1991, ch. 6) classified transactions between automobile makers and their suppliers into three broad categories, depending on suppliers' levels of involvement and capability in product development. Figure 8.1 illustrates typical examples of supplier involvement in product development by using a simplified information asset map. Three basic categories were identified here: supplier proprietary parts, black box parts, and detail-controlled parts. This classification is basically the same as that by Asanuma (1984, 1989): marketed goods, drawings approved (*shoninzu*), and drawings supplied (*taiyozu*). The magnitude of supplier involvement in engineering is higher in the former classification and lower in the latter.[2]

(1) *Supplier Proprietary Parts*: The supplier develops a component entirely from concept to manufacturing as its standard product (*shihanhin*). Some highly standardized components, such as batteries, may belong to this category.

(2) *Black Box Parts*: Developmental work for the component is split between the assembler and the supplier. In a typical case, the former creates basic design information such as cost/performance requirements,

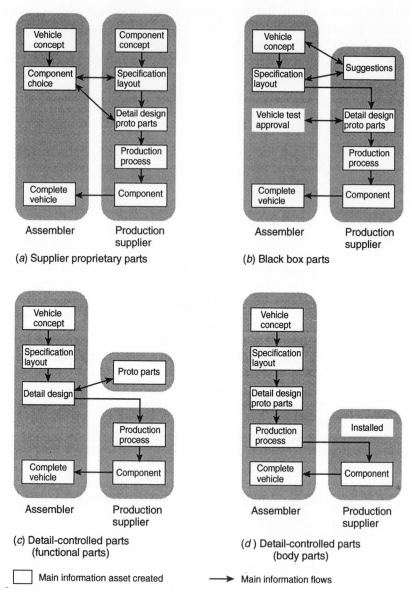

Fig. 8.1. *Typical information flows with parts suppliers*
Source: Clark and Fujimoto (1991: 141).

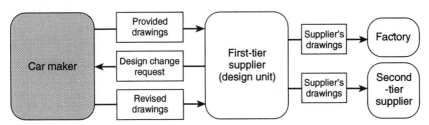

(a) Detail-controlled parts (provided drawings)

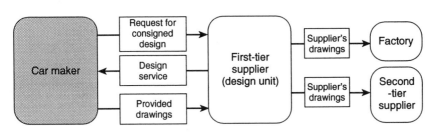

(b) Black box parts (consigned drawings)

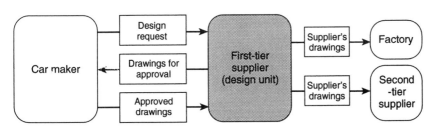

(c) Black box parts (approved drawings)

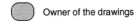   Owner of the drawings

Fig. 8.2. *Basic paper flows for each mode of transaction*
Source: Company A (revised and translated by Fujimoto).

exterior shapes, and interface details based on the total vehicle planning and layout, while parts suppliers do detailed engineering. As shown in Fig. 8.2, there are two sub-categories of black box parts: approved drawings (*shoninzu*) and consigned drawings (*itakuzu*).

(2a) Approved drawings: in this case, the drawings are eventually owned by the supplier, which assures design quality and patent rights over the parts in question. That is, the supplier has to make engineering actions in response to field claims related to the parts. In exchange for this responsibility for quality, the suppliers enjoy a greater degree of design discretion for better manufacturability and cost reduction.

(2b) *Consigned drawings*: unlike approved drawings, final drawings are owned by the car maker, but detailed engineering work is subcontracted out to the suppliers. The former pays the design fee to the latter as a separate contract, and is free to switch suppliers at the manufacturing stage. It is the car makers that take responsibility for quality assurance, though.

(3) *Detail-Controlled Parts*: The third category is the case in which most of the component engineering work, including parts drawing, is done in-house. In this way not only basic engineering but also detailed engineering is concentrated in the hands of the car maker, although the suppliers can make requests for design changes for better manufacturability and cost reduction. It is called the 'provided drawings (*taiyozu* or *shikyuzu*) system' in Japan.

There are some other ways for suppliers to get involved in new car development projects, including an arrangement called 'guest engineer', or 'resident engineer', in which component engineers are dispatched from the supplier to the car maker to work jointly with the latter's engineers.

8.2.2. Literature Survey

Although there have been many studies on manufacturing practices and subcontracting systems in the Japanese motor vehicle industry, studies of the product development side of the supplier system have been relatively scarce. One of the important exceptions has been a series of systematic studies by Asanuma (1984, 1989), which described and analysed the inter-firm flow of design information between assemblers and suppliers. Asanuma defined various categories of suppliers in terms of a combination of suppliers' relation specific skills, and made some predictions on how the level of relational quasi-rent would differ among the categories (Asanuma, 1989). In this way, Asanuma gava a clear economic interpretation of the effectiveness and stability of the black box parts system. Asanuma also pointed out that the approved parts system may have originated either from marketed goods or from provided drawing parts in different ways, and that black box parts (approved drawings) are found more often in the Japanese automobile industry than in the electric machinery industry.[3]

Nishiguchi (1989, 1993) thoroughly studied the practice of the Japanese-style subcontracting system and described it as 'strategic dualism', a new form of contractual relations that was based on 'problem-solving-oriented collaborative manufacturing'. He also argued that the system was the evolutionary product of a complex historical interaction. Nishiguchi suggested that the black box design concept evolved from what he calls 'bilateral design' in the late 1950s and early 1960s, in which suppliers made VA (value analysis) and VE (value engineering) proposals, as well as supplier-driven innovations, for the motor vehicle makers.

Clark and Fujimoto (1991), by collecting data from 29 product development projects from Europe, the USA, and Japan in the late 1980s, found that the average Japanese project in our sample relies much more on black box parts than the average American project, which relies heavily upon detail-controlled parts. Europeans are positioned in the middle. Together with supplier proprietary parts, the average Japanese project of the 1980s relies on supplier's engineering in roughly 70 per cent of purchased parts, as compared with 20 per cent in the USA and 50 per cent in Europe. Note that the average fraction of procurement cost in total production cost is about 70 per cent in Japan, 70 per cent in the USA (this includes parts from component divisions of each company), and 60 per cent in Europe in the same sample.[4] Clark and Fujimoto also estimated that the lower in-house development ratio of the Japanese makers contributed to their lower engineering hours per project (i.e., higher development productivity).

In the second-round survey conducted in 1993, the same authors found that the US car makers increased the ratio of black box parts significantly. This is considered to be part of their catch-up efforts to narrow the competitive gaps with better Japanese makers.

Cusumano and Takeishi (1991) studied the parts suppliers of four different component categories, and found that the Japanese car makers relied more on the black box parts system in the parts studied.

Takeishi, Sei, and Fujimoto (1993), by collecting data from about 120 parts suppliers in Kanagawa Prefecture, which included not only first-tier but also second-, third-, and fourth-tier parts suppliers, indicated that a majority of the first-tier suppliers in the sample (59%) made either black box or supplier proprietary parts, while the fraction was much smaller (23%) in the second-tier suppliers and zero in the third- and fourth-tier suppliers. This implies that the black box parts practice is concentrated heavily among the first-tier suppliers.

Thus, the current practice of black box parts has attracted the attention of both practitioners and researchers in recent years, and there have already been several important works on this subject. However, there has been hardly any literature exploring historical aspects of the system. As well, to the author's knowledge there has been virtually no description of this topic in the official company histories of major Japanese auto companies and parts suppliers. Against this background, the current paper will examine how the system evolved over time by relying mostly on first-hand data obtained from the companies in question.

8.3. Hypotheses on the Emergence of the System

8.3.1. The Logic of System Emergence versus System Stability

When we observe a certain stable pattern of organizational activities and capabilities in a group of firms, such as the black box parts practice among the Japanese motor vehicle makers, there are at least two ways of explaining the phenomenon: explaining how the pattern emerged (the logic of system emergence), or explaining how it was sustained (the logic of system stability). Although they are intertwined in the real world and what we observe now is a system that emerged *and* has been sustained up till now, the two kinds of logic can, in theory, be discussed separately.

8.3.2. Generic Hypotheses of System Emergence

A study of the emergence of a system is essentially a historical analysis. It tries to explain why a certain economic organization chooses a particular set of trials for system changes. Generally speaking, there are several alternative logics explaining the emergence of a new pattern (Fig. 8.3).

Rational calculation. An organization deliberately chooses a new course of action that satisfies or maximizes its objective function by examining a feasible set of alternatives based on its understanding of environmental constraints and limits of capabilities. The neoclassical decisions further assume that the economic actors are equally capable and face an identical environment.

Random trials. This logic assumes that it is a matter of pure chance for an organization to choose a particular trial. A lucky one gets a better system, an unlucky one gets a poor one.

Environmental constraints. An organization detects certain constraints imposed by *objective* or *perceived* environments, and voluntarily prohibits a certain set of actions. The constraints may be objective (e.g., laws and regulations), or they may be self-restraints based on its perception of the environments. Obedience to other organizations is also included in this category.

Entrepreneurial vision. A desirable set of activities is directly chosen by entrepreneurs of the organizations on the basis of their visions, philosophies, or intuitions, without much analysis of the capabilities and constraints of those activities.

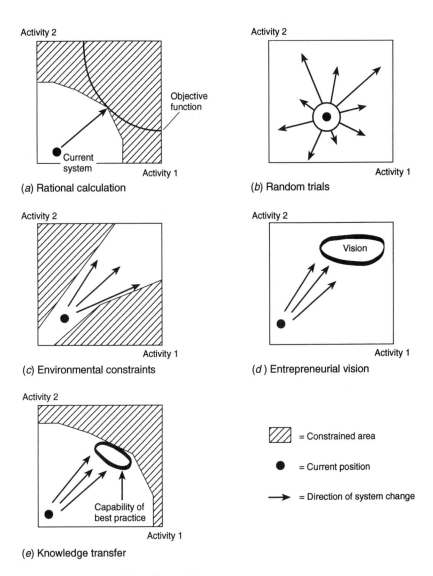

Fig. 8.3. *Some generic hypotheses of system emergence*

Knowledge transfer. A certain pattern is transferred from another organization to the one in question. The transfer may happen within the industry (competitor, supplier, customer, etc.) or across industries. Also, the transfer may be a *pull* type, where the adopter-imitator of the system takes the initiative, or it may be a *push* type, where the driving force behind the transfer comes from the source organizations.

This may not be an exhaustive list of possible logics of system emergence. Nor is it likely that we can explain a certain system evolution by a single logic. A combination of different logics would normally be needed.[5]

8.4. Origin of the Black Box Parts System

8.4.1. The Change in Procurement Policy at Toyota after the War

Soon after World War II ended, Toyota changed its procurement policy again,[6] this time emphasizing outsourcing. Kiichiro described it thus:

I want to change the parts manufacturing policy drastically. In the past, for various reasons, Toyota made many parts in-house and thus could not concentrate on parts procurement. From now on, we should encourage specialization among our suppliers, have them do research in their specialty, and nurture their capability as specialist factories. We will ask such specialist manufacturers to make our specialist parts.[7]

Although there are some signs that Kiichiro already had this vision of creating a group of specialist vendors in the prewar era, it was only after the war that Kiichiro declared it an official policy.

8.4.2. The Birth of Nippondenso (1949)

In 1949 Nippondenso separated from Toyota and became an independent company, in accordance with a restoration plan (*saiken seibi keikaku*) that Toyota submitted to the government. The electric parts factory of Toyota was suffering a deficit.[8]

The new company, Nippondenso, had design and engineering capability from the beginning; virtually all of Toyota's electric component engineers, who were making parts drawings based on sketches of parts in the US cars (e.g. Delco and Lucas), were transferred to the new company to form an engineering department there.[9] It naturally followed that the transactions between Toyota and Nippondenso were based on a black box parts system from the beginning. Engineers of Nippondenso made their parts drawings on the basis of specification drawings (rough assembly drawings) provided by Toyota, which in turn gave approval of the former's drawings. In other words, an approved parts (*shoninzu*) system has existed at Nippondenso since 1949.[10] In 1951 a post was created to handle approved drawings when the number of such drawings increased. In 1952 Nippondenso expanded transactions with non-Toyota makers such as Mazda (Toyo Kogyo), Mitsubishi, and Honda, which virtually meant diffusion of the black box parts practice outside the Toyota group. In 1956, formal rules for handling approved drawings were established. Drawings with direct Bosch influence continued until the late 1960s.

8.4.3. Origin of Toyota's Procedures on Approved Drawings

Assuming that Toyota is one of the first mass-producing motor vehicle makers that systematically used the black box parts system, it is necessary to identify when Toyota formally started this practice. According to a survey of Toyota's purchasing and engineering procedures, the oldest documents available that use the word 'approved drawings' at Toyota are Appproved Drawings Rules (*shoninzu kitei*) and Approved Drawings Handling Rules (*shoninzu shori kitei*) from 1953.[11] These internal rules of the design office, in turn, mention that they replaced similar rules from 1949. The 1949 documents themselves no longer exist at Toyota, according to our survey. It is not clear whether any internal rules existed prior to 1949, but given the fact that systematic formats for Toyota's documents were established in 1948, it is likely that the 1949 rule was the first formal rule to define the approved drawings.

The coincidence in timing makes us suspect that the separation of Nippondenso from Toyota (also in 1949) might have been closely related to the origin of the black box parts (approved drawings) system. It is not clear if this separation itself made Toyota establish the black box parts rule, though.[12]

In any case, Toyota's approved parts rules back in 1953 were rather simple: only one page a few sentences long. The rules specified that the approved drawings be submitted to Toyota's engineering department via the purchasing department, that the engineering department was to process the drawings according to the Approved Drawings Handling Rules, and that approved drawings be sent back to the supplier via the purchasing department. The rules have been essentially unchanged since then.

8.4.4. Possibility of Knowledge Transfer in the Prewar Era

Although it is likely that the transaction between Toyota and Nippondenso was the origin of the Toyota-style approved drawing system, it does not necessarily mean that it was the origin for the Japanese motor vehicle industry in general. In fact, according to the survey that the author conducted in 1993 (see Fujimoto, 1994a), six of the sample firms (3% of the 177 respondents) answered that they had established the approved drawing system in the prewar era (before 1945). The six companies were three Nissan group suppliers and three independent (supplying parts to both Toyota and Nissan) parts makers.

Looking at these early adoptions of the approved drawing system more closely, one may hypothesize that the approved drawing system was in fact imported from other industries, such as the prewar locomotive and/ or aircraft industries. In fact, three of the above six companies had close connections with one or other of those two industries. Many of the other companies supplied relatively generic parts (i.e., merchandise goods), such as tyres, paint, oil filters, cables, and gaskets.

Thus, it is possible that the black box practice was transferred from the locomotive and aircraft industries, which were more advanced than the car industry.[13] For example, Sawai (1985) argues that some locomotive and component suppliers accumulated technological capabilities and started joint engineering with the customer, the Ministry of Railroad, during the 1920s. The aircraft industry might have started the black box parts practice later (1930s), but its influence on the postwar motor vehicle industry might have been more direct. As some major aircraft parts suppliers were located in the eastern part of Japan, Nissan, with most of its plants located in the same area, might have learned this practice more directly than Toyota at this initial stage.

The above survey results also indicate the possibility that there is another path to the black box parts system: suppliers of relatively generic parts (e.g., tyres, paint, etc.), or supplier proprietary parts, converted themselves to black box parts suppliers.[14]

In summary, it would be safe to say that Nippondenso was one of the very early adopters of the approved drawing system, but it was not 'the earliest' one. It is nevertheless important to note that Toyota's approved parts system turned out to be significantly more systematic and effective than that of the other motor vehicle makers, including Nissan (see Section 8.5). Thus, the Toyota-Nippondenso transactions might not have been the earliest of the black box parts practice in general, but it is likely to be one of the origins of the relatively effective version of the system (i.e., Toyota-style black box parts).

To sum up, the historical imperative (i.e., environmental constraints) in terms of technological capabilities, caused by the hiving off of Nippondenso, seems to be a major reason explaining the Toyota-style black box parts system. As for the approved drawings in general, transfer of knowledge from other prewar industries such as locomotives and aircraft may be important. In any case, it should be noted that, unlike many other elements in automobile production systems, this was not a practice that was adopted directly or indirectly from the Ford system.[15]

8.5. Diffusion of the Black Box Parts System

8.5.1. Toyota and Nissan in the 1960s: Model Proliferation

Triggered by the 'motorization' in the domestic market, both production volume and the number of models increased rapidly in the 1960s.[16] If the Japanese motor vehicle companies could not recruit sufficient engineers and workers to deal with the rapid expansion of production, it would be reasonable to infer that pressures to subcontract out a larger fraction of their operational tasks to the parts suppliers increased during the 1960s.

A similar logic may hold in the case of product development and black box parts, although the pattern of model proliferation may have been somewhat different from that of production volume expansion. The number of basic models in 1960 (excluding foreign models produced by licence agreements) was 8 in 1960 (2 from Toyota), 24 in 1965 (4 from Toyota), 37 in 1970 (8 from Toyota), and 46 in 1980 (10 from Toyota).[17] Thus, proliferation of basic models occurred mostly in the 1960s (the latter half in particular), but it slowed down in the 1970s.

Figure 8.4 shows a rough estimate of the product development work-load for Toyota and its suppliers since the 1960s. It shows that there were

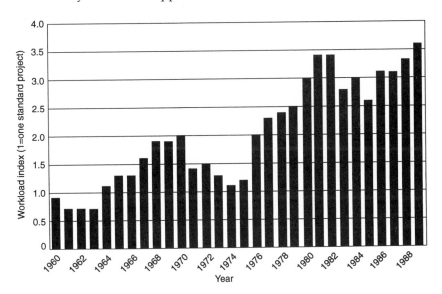

Fig. 8.4. *Estimated workload of Toyota's car development*

a few waves of development workloads: the late 1960s (Japan's motorization period), around 1980 (between the two oil crises), and the late 1980s (the 'bubble' era).[18] Although the result is very preliminary, the figure seems to indicate that there may have been significant pressures for Toyota and other major Japanese motor vehicle makers to alleviate the problem of growing development workload by asking the parts suppliers to do a part of the product development tasks.

In the area of product development, few researchers have pointed out that the black box parts system (i.e., subcontracting of engineering tasks) was diffused rapidly during the 1960s. However, the author's 1993 survey clearly shows that the late 1960s was, in fact, the peak period for adoption of the approved drawing system among the Japanese first-tier suppliers.

An interview conducted by the author at Nissan also indicates that the black box parts system prevailed during the late 1960s at that company.[19] The diffusion process is said to have been quite informal, though. When an informal process of spreading the black box parts system was beginning at Nissan, Toyota was already refining its formal process for the system. In 1961, the internal rules on approved parts became company-wide rules (*sekkei kenkyu kitei*), seven pages in length. By the late 1970s, the rules had evolved into a very detailed procedure of nearly fifty pages. Nissan lagged behind Toyota in formalizing and systematizing the black box parts practice.

8.5.2. Toyota in the 1980s: Continued Diffusion of the Black Box Parts Practice

Diffusion of 'approved drawings' continued at Toyota during the 1980s. According to company data, the fraction of approved drawings in the total number of engineering drawings increased from 30 per cent in 1980 to 37 per cent in 1992 (in the case of the Mark II model).[20] Proliferation of product variations after the mid-1980s is likely to have created pressures to rely more on the engineering resources of the suppliers (see Fig. 8.4).

8.5.3. Nissan in 1980s: Benchmarking and Institutionalization

Although Nissan's black box parts system may have already originated before the war and was diffused in the 1960s, as was the case at Toyota, the system is said to have lacked a coherent and effective system of formal procedures. It was in 1986–87 that Nissan finally established a system of formal procedures for black box parts, which Nissan called the 'New Approved Drawing System'.[21] This was apparently a part of Nissan's efforts to renovate its organization in the late 1980s.

Preparation for this shift started in the early 1980s, when the Japanese motor vehicle industry was recovering from the second oil crisis. This was a period when Nissan switched from restrictive to expansive strategies in production and product mix. Thus, Nissan was predicting a rapid increase in product development workload. At the same time, Nissan's managers were concerned about the fact that its approved drawing system was not as developed as that of Toyota. After a benchmarking study of Toyota's practice, for example, it was revealed that Toyota's inputs to black box parts suppliers (a few pages of specification documents with rough sketches) were much simpler than those of Nissan (a roll of fairly detailed specification drawings). This implied that Nissan's black box parts system consumed more in-house engineering resources, and that it put more restrictions on suppliers' efforts to make their parts design easier. It was also pointed out that the timing of the specification freeze at Nissan was late compared with Toyota.

After a series of analyses, Nissan's engineering administration department proposed the following revision of the black box parts system in the mid-1980s (see Fig. 8.5).

- The old approved drawing system used 'specification drawings' or *shiyozu* (incomplete drawings that the suppliers were expected to complete) as inputs to the suppliers.[22] The new system abolishes specification drawings and introduces a 'specification instruction form' (*shiyo teiansho*), which was closer to Toyota's 'Design Request Form'. In this way, suppliers could enjoy more design discretion to apply their technological expertise to the designs.
- Division of responsibility for development tasks between Nissan and the suppliers was not clear in the old system. In the new system, it is proposed that a 'Design Task Assignment Table' and a 'Check Sheet' clarify the responsibility of each party. This measure was expected to eliminate duplication of development work between Nissan and the suppliers.
- Responsibility for quality assurance was not clear in the old system, either. In the new system, it is proposed that the approved drawing suppliers take full responsibility for the functionality of the component as a unit.
- In the old system, timing of supplier selection was relatively late. In the new system, supplier selection is made much earlier (around the end of the product planning phase); this gives the supplier a longer lead time for component design.
- Specification drawings were unilaterally submitted to the suppliers in the old system. The new system adds a period during which Nissan and the supplier jointly develop specifications.

Overall, the new approved drawing system adopted some of the prac-

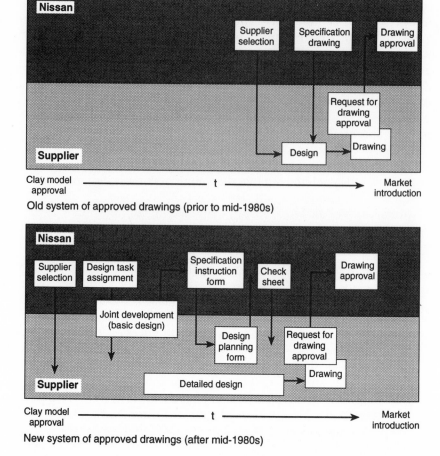

Old system of approved drawings (prior to mid-1980s)

New system of approved drawings (after mid-1980s)

Fig. 8.5. *Nissan's adoption of new approved drawings (mid-1980s)*
Source: Nissan.

tices that Toyota had already had: a relatively simple specification form, clear division of responsibility for quality assurance, early supplier selection, and joint development of specifications. At the same time, Nissan managers expected that the new system would save engineering person-hours for its in-house engineers.

There was some controversy inside Nissan over the shift to the new system. Some argued that technological hollows might be created in the new system, in that Nissan's engineers would cease to be real engineers if they stopped making engineering drawings (Japanese design engineers tended to believe that drafting capability was a prerequisite to being a good engineer). However, the argument that Nissan would have needed thousands of additional engineers to maintain the old system when it had to expand its product line overshadowed such concerns, according to a Nissan manager.[23]

In this way, Nissan introduced in the 1980s a new black box parts system, in many ways similar to that of Toyota. As discussed later, Nissan group suppliers adopted the approved drawings system almost as early as their Toyota counterparts, but their system significantly differed in content from Toyota's. In fact, it might be that Nissan was among the last of the Japanese car makers to adopt the Toyota-style approved drawing system because there had been a tacit agreement that Toyota group suppliers (e.g., Nippondenso) could supply parts to any Japanese car maker except Nissan group assemblers, while Nissan group suppliers took the opposite step of excluding the Toyota suppliers (see 'type 1' and 'type 2' in Fig. 8.6). Although independent and neutral suppliers ('type 3' in Fig. 8.6) might have acted as intermediaries that transferred the essence of the Toyota-style system to Nissan, it is likely that the separation of Toyota and Nissan supplier groups was at least partially responsible for Nissan's lateness in adopting a more effective black box parts system. In other words, the partial separation of supplier networks between Toyota and Nissan may have hampered the transfer of a Toyota-style black box parts system to Nissan.

According to the study by Clark and Fujimoto (1991), the fraction of black box parts in total procurement cost was consistently high among the Japanese motor vehicle makers.[24] It is thus likely that the diffusion of the approved drawing system among the Japanese motor vehicle makers came to an end by the late 1980s.

8.5.4. *Diffusion of the Black Box Parts System: Survey Results*

Let us now turn to the result of the author's survey mentioned earlier. The survey asked approximately when the companies started various activities related to black box parts practices. The results are generally consistent with the above description of the diffusion process (Fig. 8.7):

- The peak period for the institutionalization of the approved drawing system, as well as actual diffusion of the system within each supplier was the late 1960s. The next wave came in the 1980s (Figs. 8.7 (*b*) and 8.7 (*c*)).
- The peak time for the start of informal requests from the motor vehicle makers was the early 1960s. The pattern of informal requests tended to precede that of formal institutionalization (Fig. 8.7(*a*)).

Starter

Assembler	Supplier	A	B	C	D	E	F
Toyota group	Toyota	▨					
	Daihatsu	▨					
	Hino	▨	▨				
others	Mitsubishi		▨			▨	
	Suzuki		▨	▨		▨	
	Honda				▨		
	Mazda	▨					
	Isuzu	▨			▨		
Nissan group	Fuji (Subaru)						▨
	Nissan Diesel		▨				▨
	Nissan						▨

Radiator

Assembler	Supplier	A	B	C	D	E	F	G
Toyota group	Toyota	▨						
	Daihatsu		▨					
	Hino			▨				
others	Mitsubishi			▨				
	Suzuki				▨			
	Honda			▨	▨			
	Mazda							▨
	Isuzu				▨			
Nissan group	Fuji (Subaru)							▨
	Nissan Diesel					▨		
	Nissan						▨	

Type 1: Toyota's and Nissan's suppliers are separated

Seat

Assembler	Supplier	A	B	C	D	E	F	G	H	I	J	K	L	M	N	O
Toyota group	Toyota	▨														
	Daihatsu		▨													
	Hino			▨											▨	
others	Mitsubishi					▨										
	Suzuki						▨	▨								
	Honda								▨							
	Mazda									▨						
	Isuzu				▨				▨			▨				
Nissan group	Fuji (Subaru)												▨			
	Nissan Diesel													▨		
	Nissan															▨

Type 2: many dedicated suppliers

Head Lamp

Assembler	Supplier	A	B	C
Toyota group	Toyota	▨		
	Daihatsu	▨		
	Hino	▨		
others	Mitsubishi	▨	▨	
	Suzuki	▨		
	Honda	▨		▨
	Mazda	▨		▨
	Isuzu	▨		
Nissan group	Fuji (Subaru)	▨		
	Nissan Diesel	▨	▨	
	Nissan	▨		

Shock Absorber

Assembler	Supplier	A	B	C	D
Toyota group	Toyota	▨			
	Daihatsu	▨			
	Hino	▨			
others	Mitsubishi		▨		
	Suzuki	▨			
	Honda		▨		
	Mazda	▨			
	Isuzu	▨			
Nissan group	Fuji (Subaru)	▨			
	Nissan Diesel	▨			
	Nissan	▨			

Type 3: oligopoly by neutral suppliers

▨ Transactions exist as of 1990.   In-house production is omitted for simplicity.

Fig. 8.6. *Patterns of assembler and supplier relationship (1990)*
Source: Industry experts.

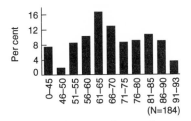

(a) Informal requests for supplier participation in product engineering started to come from makers

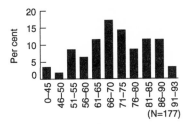

(b) Approved drawing system was formally institutionalized as company procedure

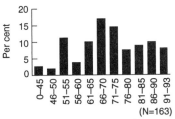

(c) A majority of the transactions with the main customer became approved drawing parts

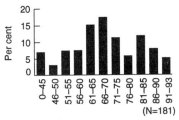

(d) Started to hire product engineers regularly from colleges

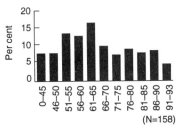

(e) Engineering design section was established in the factory

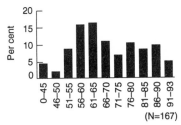

(f) Product engineering department was established separately from the factory

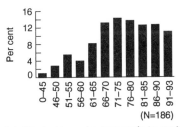

(g) Started to make proposals to customers on product concept and technology

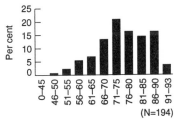

(h) Started full-scale value analysis and value engineering activities

Fig. 8.7. *The first timing of supplier's engineering activities*
Source: A questionnaire survey by the author in 1993.

- Many of the first-tier suppliers started to regularly hire college graduates for engineering jobs during the 1960s. Establishment of formal engineering sections or divisions in and outside the factories tended to precede full-scale activities in product engineering (Figs. 8.7(*d*), (*e*), and (*f*)).
- The suppliers started to make engineering proposals and conduct VA-VE activities mostly after the 1970s (Figs. 8.7 (*g*) and (*h*)).

These results seem to be consistent with the foregoing argument that the adoption of the approved parts system on the supplier side was influenced by the rapid increase in engineering workload on the side of the motor vehicle makers (Fig. 8.4), and that the formal establishment of the system may have been preceded by informal participation of suppliers in automobile product development.

What about the difference between Toyota and Nissan? Figure 8.8 compares the timing of institutionalization of the approved parts system between Toyota-related suppliers (including the independent type supplying both to Toyota and Nissan) and Nissan group suppliers. Apart from the difference in the content of the black box parts system between the two groups, discussed earlier, the pattern of timing in introducing some kind of approved drawing system is fairly similar, although the Toyota-related suppliers tended to be slightly earlier.[25] Also, there is a small peak in the early 1950s in the case of Toyota-related suppliers. This, however, does not seem to be the diffusion effect from Nippondenso. Again, the origin of the black box parts system (Toyota version) and its diffusion seem to be generally separate both in logic and timing.

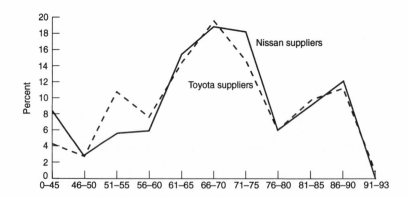

Fig. 8.8. *Comparison of Toyota-related suppliers and Nissan suppliers in the timing of institutionalization of approved drawing system*

Note: Toyota-related suppliers include those that supply both Toyota and Nissan.
Source: A questionnaire survey by the author in 1993.

8.6. A Case of Evolution of Black Box Parts: Company B

8.6.1. Company B (Interior Parts): From the 1960s to the 1990s

In order to examine more closely how each supplier's design capability evolved over time, company B, a supplier of interior parts, was chosen as a case.[26] This company maintains a close tie with Toyota in terms of capital participation and transaction.[27] Established in the early 1960s, the company itself is fairly new, and it started with virtually no design capability. Besides, interior parts are relatively difficult to fit into the black box parts system because of their close interdependence with total vehicle design. Thus, the case seems to provide an informative example of how a 'latecomer' to the black box parts system accumulated design capability step by step.

8.6.2. Width of Design Capability

Let us now examine the case more closely. Figure 8.9 shows the product line of Toyota for which company B adopted the black box parts system in parts X and/or Y.

- Starting from parts X for the first-generation Publica, Toyota's entry class model, company B expanded its width of business as Toyota expanded its product line.
- Company B's business with Toyota continued, but there was a switching of suppliers at the level of individual transactions for various reasons.[28]
- Company B started its black box parts arrangement with a single product, the Crown, and gradually expanded the range of Toyota models that adopted this arrangement.
- The company also expanded those of its products that adopted the black box parts system form parts X to parts Y.

8.6.3 Depth of Design Capability

The depth of company B's design capability also increased in thirty years. Figures 8.10 and 8.11 explain how the company acquired skills necessary for black box parts, such as detailed parts design, prototyping, detailed assembly design, testing, industrial design, basic design, and product planning. A brief history of how it built up capability is presented below.

1. In the early 1960s, company B started its business with a small engineering department of three or four people. The unit had the capabil-

Fig. 8.9. *Step-by-step acquisition of engineering capability at Company B*

Note: The fork lift business is omitted for simplicity.

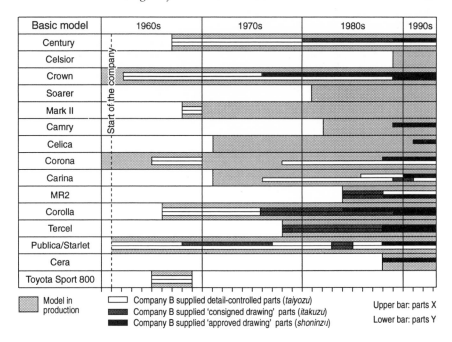

Fig. 8.10. *Company B's transactions with Toyota (passenger car)*

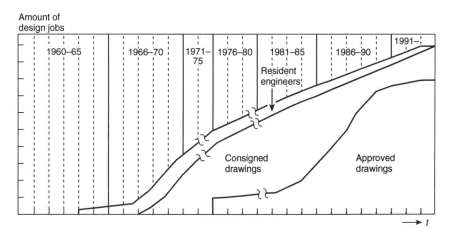

Fig. 8.11. *Workload of product design at Company B by types of drawing*

Note: This diagram represents the impressions and recollections of company staff rather than actual data.

Source: Company B.

ity for engineering administration (i.e., handling given drawings), basic production engineering (i.e., making bills of materials based on Toyota's parts drawings), and prototyping, but it had virtually no capability for designing and drafting its products. The company was fully dependent on detailed drawings provided by Toyota.

2. In the mid-1960s, three years after the establishment of the company, company B started to send its young engineers to Toyota's interior engineering department for one or two years each. They worked on Toyota premises, learning not only product technologies but also Toyota's engineering process itself. The young engineers from company B at first did miscellaneous jobs assisting Toyota's interior engineers, but Toyota soon started to give them a set of jobs related to company B's target product. They acquired the capability to draw assembly drawings of parts X around that time.

3. In the mid-1960s, as mentioned earlier, company B started a job for Toyoda Automatic Loom Works on an approved drawing basis.

4. In the late 1960s, company B started to get some jobs from Toyota on a consigned drawing basis. Toyota still took final responsibility for quality assurance, but the arrangement was more like a joint development, in that the engineers from both companies often worked on Toyota premises. The amount of work under this arrangement grew rapidly in the early 1970s.

5. In the 1970s, company B established a system to test and evaluate prototypes, following Toyota's testing standards. As Toyota still took quality assurance responsibility, though, the testing conducted by company B was 'consigned testing'.

6. In the same period, company B's engineers started to participate in the process of setting detailed specifications, basic designs, and test standards.[29] Thus, company B gradually acquired the capability not only of detailed design but also basic design (i.e., specification setting).

7. Also, in the early 1970s, company B's engineers became capable of making assembly drawings of parts Y, ten years later than was the case with parts X.

8. In the mid-1970s, fifteen years after the establishment of the company and eight years after it started consigned drawings, company B finally started to do business with Toyota on an approved drawings basis. It started this system in one model. The switch from consigned to approved drawings meant that the company had to take a risk as to quality assurance, but it acquired more discretion to design manufacturable parts. For Toyota, this also meant further reduction of its own engineering workload.

9. During the same period, company B's engineers working on Toyota premises started to be treated as full-fledged engineers, which seems to reflect improvements of the former's engineering capabilities.

10. In the late 1970s, company B won the Toyota Quality Control Prize (Toyota's version of the Deming Prize), which implied that company B

was now ready to move to the approved drawings system and become a specialist parts supplier. The prize thus triggered a rapid growth in company B's design jobs. Also, the company started up its research activities around this time. Its initial research efforts were concentrated on claims regarding the functionality of parts X.

11. Also, in the late 1970s, the company created a quality assurance department, which included test engineers, separate from its engineering department. It also started to dispatch test engineers to Toyota.

12. In the early 1980s, company B extended its engineering capability further upstream, participation in some of Toyota's product planning. It also started market studies on the durability and functionality of parts X independently or jointly with Toyota.

13. In the early 1980s, company B started to hire industrial designers. A few years later, it started to dispatch designers to Toyota, as it had design engineers and test engineers.

14. In the mid-1980s, jobs based on approved drawings started to grow rapidly. This was partly because the approved drawing system was adopted for the company's two major products: parts X for Toyota Corolla and Crown. The shift was completed in the early 1990s.

15. In the late 1980s, company B started to do structural planning for Toyota on a consigned drawing basis. This virtually meant working jointly with Toyota's engineers to make basic designs, from which company B made detailed drawings on an approved drawing basis. The number of resident engineers was reduced during the same period.

In this way, it took company B over twenty years to acquire full capability for black box parts with dozens of engineers and designers. Interactions between Toyota and company B were carefully managed with long-term perspectives. For example, resident engineers meant that the engineers of company B had access to the engineering rooms of Toyota (i.e., beyond the reception area). Of about two hundred suppliers that Toyota dealt with, less than one hundred had the privilege of dispatching resident engineers.

It should also be noted that consigned parts (*itakuzu*) functioned as a kind of 'rehearsal' that paved the way to company B's transformation to an approved drawing (*shoninzu*) parts supplier.

8.7. Summary and Implications

8.7.1. Summary: Logic for the Emergence of Black Box Parts

The case of the black box parts practice among Japanese motor vehicle makers provides a good example of how a group of manufacturing firms build new capabilities over time. The foregoing discussion seems to indi-

Logic	Empirical Evidence
Rational calculation	● Before the trial: benefits of black box parts are not well predicted. ○ After the fact: Toyota was quick in institutionalizing what it inadvertently tried and found to be effective.
Environmental constraints (technological dependence)	● Few suppliers had engineering capability in the 1930s and 40s. ○ Toyota - Nippondenso case: Toyota lacked electric parts engineering capability as it separated from Nippondenso in 1949.
Environmental constraints (insufficient number of engineers)	● Black box parts practice originated before the model proliferation period. ○ Survey result: the peak time of diffusion of the black box parts system coincided with model proliferation period in the late 1960s. ○ Respondents of the survey tended to agree that high engineering workload of the car makers triggered the shift to black box parts. Interviewees at Nissan, Company A, and Company B agreed with this hypothesis, at least partially.
Knowledge transfer	● Ford and other US mass producers did not have black box parts, and thus they were not the source of the practice. ◉ Prewar aircraft industry in Japan may have been a source of the black box parts practice. ○ Indirect knowledge transfer (benchmarking) from Toyota to Nissan in the 1980s. ○ Rapid diffusion of black box parts practice among the Japanese first-tier suppliers (survey results).
Entrepreneurial vision	○ No clearly stated comments on the black box parts system by Toyota's executives were found in the formal company history. ● Kiichiro Toyoda's vision of growing specialist parts vendors after the war might have facilitated the development of black box parts practice.

Note: ○ = Evidence consistent with the hypothesis   ● = Evidence contradicting the hypothesis

◉ = Inconclusive   'Random trials' omitted from the table.

(a) Black box parts system is the joint result of various historical factors

Fig. 8.12. *Summary of the emergence logic for black box parts system*

cate that the system emerged and evolved over time by a complicated combination of driving forces surrounding the industry. As the case of company B suggests, the process of developing a black box design system requires step-by-step collaboration and tenacious capability building by both the assembler and the supplier. Also, as shown in Fig. 8.12, the present case seems to indicate that a combination of aspects of system

Logic		Empirical Evidence

Logic of system emergence	Logic of system origins	*Knowledge transfer (inter-industrial):*
		   From the prewar aircraft and locomotive industries
		
		Environmental constraints (historical imperatives):
		   forced by Toyota's separation from Nippondenso

	Logic of system diffusion	*Environmental constraints (forced growth)*
		   Limited engineering resources of the car makers during
		   the product proliferation period
		
		Knowledge transfer (inter-industrial):
		   Accelerated by competition by design capability builiding
		   among the suppliers

Logic of system stability		*Competitive rationality* Saving in product development cost Saving in production cost through manufacturing Suppliers' monopoly rent seeking is limited

(b) Logic of the systems origin, diffusion, and stability varies

Fig. 8.12. *Continued*

emergence, such as environmental constraints (historical imperatives), entrepreneurial visions, and knowledge transfer, partly explains how the practice originated and was developed, and that the logic of system origin, diffusion, and stability may be altogether different. In any case, it should be noted that the black box parts system was not a practice that the Japanese motor vehicle makers adopted from the Ford system in America, unlike many other elements of the Toyota-style system (Fujimoto and Tidd, 1993).

8.7.2. Origins

As for the origin of the black box parts system, historical evidence makes us infer that the transactions between Toyota and Nippondenso in 1949 were probably one of the origins (if not the only origin) of the black box parts practice, particularly its Toyota version.

Historical imperatives, or technological constraints, seem to have played an important role here: first, before the war, Toyota could not find decent electric parts suppliers in Japan, so it was almost forced to design and make such parts in-house; secondly, after the war, Toyota had to hive off the electric parts factory for its own survival; thirdly, when Nippondenso was created in 1949 as a result of the separation, Toyota found that it had to rely on the engineering capability of Nippondenso, as virtually all the electric engineers had moved to the separated company. In this way, the historical imperative that Toyota lacked the technological capability for electric parts appears to have forced Toyota to apply the approved drawing (i.e., black box parts) system to its transactions with Nippondenso from the beginning.

This story implies that the black box parts practice emerged not because of rational calculation by the automobile companies (at least before the fact), but because of a set of constraints or historical imperatives imposed on the motor vehicle makers. The motor vehicle companies apparently realized the benefit of this practice much later, when their engineering workloads started to soar as a result of product proliferation in the 1960s.

Inter-industrial transfer of the practices in the prewar locomotive and/or aircraft industries may also partly explain the origin of the black box parts system. Transformation of supplier proprietary parts (e.g., tyres and paints) to black box parts may be another passage to the new system. Further investigation is needed to examine these hypotheses, though.

Entrepreneurial visions (e.g., conceptualization and visions by Kiichiro Toyoda, Taiichi Ohno, etc.) apparently did not play a decisive role in the case of black box parts, unlike some other cases (see Fujimoto and Tidd, 1993). As indicated in the foregoing discussion, there is some evidence that the founder of Toyota Motor Manufacturing had a vision of nurturing parts specialist suppliers even prior to the war, when Toyota had to make many parts in-house. Although Kiichiro's remarks did not refer to a black box parts system, it is possible that what he meant by 'specialist' was a parts maker who also had research and development capability for designing certain types of parts. To the extent that Kiichiro's 'specialist' concept implied capabilities for parts design, it is possible that Kiichiro's vision of the supplier system served as a catalyst, if not driving force, for subsequent development of the black box parts practice. However, in the case of black box parts, entrepreneurial visions were not unequivocally and explicitly put into words by the executives and managers, unlike the case of Just-in-Time and Total Quality management.

8.7.3. Diffusions

The foregoing case also indicates that the *historical imperative* of high growth with limited resource inputs in the product engineering area of the motor vehicle companies (i.e., shortage of in-house engineers at model proliferation time) created constant pressures to subcontract detailed component engineering wherever possible.

From the suppliers' point of view, the black box parts arrangement meant a great opportunity to develop design capability, build up a technological entry barrier against the motor vehicle makers' efforts to make the parts in-house, gain some quasi-rent, and survive as a first-tier parts supplier, although it was accompanied by the risk of taking full responsibility for quality assurance. Competitive pressures from rival suppliers also accelerated their efforts to build up design and engineering capability in order to match their competitors' efforts. Thus, once the motor vehicle makers started to offer the opportunities of shifting to the black box parts arrangement, the suppliers tended to have rational reasons to accept the offers, as long as they had enough managerial capability to do so.

Transfer of knowledge and managerial resources between suppliers (e.g., from Nippondenso to other suppliers), between assemblers (e.g., from Toyota to Nissan), as well as between an assembler and a supplier (e.g., from Toyota to company B), was a key engine for diffusion of the black box parts practice. Thus, originating presumably from the transactions between Toyota and Nippondenso in the late 1940s, the black box parts practice grew rapidly during the high growth era of the 1960s, and the diffusion process came to an end within Japan in the 1980s. The diffusion is now in progress across national borders: from Japan to the USA.

The diffusion of the black box parts system in general, in terms of timing, did not differ much between Toyota and Nissan. However, it was in the content and effectiveness of the system that the two companies differed for a long time. Thus, it was only in the 1980s that Nissan modified its black box parts system to make it more like the Toyota version.

To sum up, the origin and evolution of the black box parts practice, which is one of the sources of the competitive advantages of the Japanese motor vehicle industry of the 1980s, may be explained by a combination of the system emergence logic discussed in Section 8.3, including historical imperatives and inter-firm and inter-industrial knowledge transfers, rather than rational calculation or deliberate strategic choices prior to the event. While the system, after it was established, might have turned out to contribute to competitive advantage, the benefits were not clearly recognized by the firms when it was emerging. After the companies learned from their unplanned trials, however, some of the firms started to adopt the new system more intentionally, deliberately, and institutionally.

Thus, to the extent that an ex-post rational system emerged out of unintended imperatives or constraints, it may be meaningful to explore and analyse the logic of system emergence. More importantly, it was this capability of systematizing the new elements after the trials, rather than that of rational calculation, that seems to explain the difference in the content and effectiveness of the black box parts systems of Toyota and

Nissan up to the mid-1980s. This leads us to raise the last question: what is the source of the inter-firm differences in performance and capabilities?

8.7.4. Implication: Patterns of Capability Building

The foregoing case of the evolution of the black box parts system may provide some insights from the veiwpoint of dynamic aspects of manufacturing firms' core capabilities (Teece, Pisano, and Shuen, 1992).

Long-term evolution of capabilities. The suppliers accumulate capability of component engineering through long-term collaboration between assemblers and suppliers. The process needs step-by-step enhancement of design and engineering capabilities both in width and depth.

Competition based on capability building. As the case of company B indicates, suppliers making similar components competed with each other in terms of gaining component design capability. Thus, short-term price competition (i.e., bidding) is not the only mode of competition. In the long run, competition in building engineering capabilities becomes crucial. In other words, companies compete to build inter-firm differences in capability and performance in their favour.

Limit of capability in rational calculation. The foregoing cases indicated that Toyota was more effective than some other makers in designing and implementing the black box parts system. Is this because Toyota was more capable of predicting the effectiveness of the system than others? It seems unlikely. The case studies in this paper seem to indicate that even Toyota could not predict the effectiveness of the black box parts system prior to its trials, and that Toyota, forced by historical imperatives and environmental constraints in many cases, tended to make initial trials in an unplanned manner. Thus, the inter-firm difference in the results does not seem to be caused principally by inter-firm differences in rational calculation.

Inter-firm gaps in ex-post capability of systematizing the successful trials. What Toyota could do better than its rivals seems to be not so much rational calculation before the trials as systematization and institutionalization *after* the trials.[30] Opportunities for making certain trials for new systems came to other companies as well. For example, Toyota and Nissan both faced the same kind of historical imperative by which they were virtually forced to rely partially on suppliers' engineering resources. They both made trials in this direction. However, as Nissan itself found in the mid-1980s, the effectiveness of the resulting systems was significantly different. This seems to be because Toyota had the capability to learn quickly from what it inadvertently tried, understanding the core benefit of the trials, institutionalizing them as a set of formal procedures, and diffusing them within the company and throughout the supplier network better than its rivals.

To sum up, inter-firm differences in certain *dynamic capabilities* seem to have created competitive differences among the motor vehicle makers in terms of utilizing suppliers' potential in component engineering. However, it was not so much the ability to make better choices before the trials, as the ability to learn from the trials already made, for whatever reason. That is, *post-trial capability*, rather than pre-trial capability (e.g., rational calculation or visions), seems to have been crucial in the foregoing case of the emergence of the effective system. Although the current paper focused only on the emergence of the black box parts practice, there seem to be other elements of the Toyota-style production-development system as well, in which post-trial capability was the key for their evolution.

There are various ways in which a company differentiates its capability for competitive performance in a dynamic way. In some cases, differences in entrepreneurial visions may create a competitive advantage for a particular firm (Okouchi, 1979). Capability of rational calculation may be critical in other cases. In still other cases, pure luck may give some firms certain advantages. In the current case of the black box parts practice, though, it seems to be the capability of systematizing unplanned trials that created Toyota's advantages over its rivals (see Fig. 8.13). Dynamic capa-

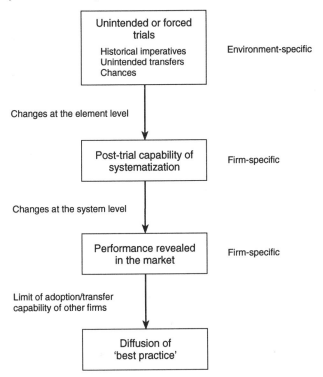

Fig. 8.13. *A typical pattern observed in the case of black box parts evolution*

bilities of coming up with visions, calculating rationally, or imitating from others tended to be onlookers in this particular case.

The concept of 'capability' in the dynamic aspect of manufacturing firms has attracted much attention among both academic researchers and practitioners in recent years, but so far the linkage between the concept and empirical findings does not seem to be that clear. In this regard, historical studies on detailed aspects of manufacturing capabilities, such as the present paper, may be able to contribute something to further development of the evolutionary theories of capability building.

NOTES

1. An exceptional case is the company history of Kojima Press (Kojima Press Kogyo Co. Ltd. [1988], 248–50), which briefly describes the origin of the black box parts practice at this Company. What is illustrated here is somewhat similiar to the case of company B described in this chapter.
2. On the strength and weakness of each category, see Clark and Fujimoto (1991), ch. 6. For details of current practices at Toyota, see Fujimoto (1994a).
3. See, also, Helper (1991) for the history of the US supplier system.
4. Regarding the component divisions as inside the companies, Mitsubishi Research Institute (1987) estimates average US outside parts ratio to be from 52 to 55%.
5. For specific hypotheses, see Fujimoto (1994a).
6. For prewar history, see Fujimoto (1994a) and Fujimoto and Tidd (1993).
7. *Toyota Jidosha 30 nenshi*, 253–4.
8. *Toyota Jidosha 30 nenshi*, 277. There is a hint that Kiichiro Toyoda had planned to encourage separate electric parts specialists, while *Toyota Jidosha 20 nenshi* (300) indicates that the electric parts unit of Toyota was losing money and was separated for better management. Which is the stronger motive for the separation is not clear.
9. This paragraph is based on interviews with Yoshihiko Furuya, Director, and Michihiro Ohashi, General Manager, of Nippondenso on 16 Sept. 1993.
10. This fact was confirmed by Kazuyoshi Yamada, former engineer of Nippondenso.
11. The contents of this paragraph are based on an interview with Toshihito Kondo, Masami Komatsu, and Shoji Kasama, Purchasing Planning Department of Toyota, 4 Aug. 1993.
12. Another possibility is that Toyota Auto Body, separated from Toyota in 1945, built cabs for Toyota on the basis of approved drawings. The appendix of the 1953 Approved Drawings Rules specifies how to handle the cabs by Toyota Auto Body as an exceptional case, though. This makes us suspect that approved drawings from Toyota Auto Body did not become an issue prior to 1953.
13. The author is grateful for some valuable insights from Haruhito Takeda and Kazuo Wada of Tokyo University on this issue.

14. This path was already predicted by Asanuma (1989).
15. For the elements of the Toyota-style production systems adopted originally from the Ford system, see, for example, Fujimoto and Tidd (1993).
16. For the evolution of supplier systems in postwar Japan, see Fujimoto (1994a).
17. Clark and Fujimoto (1992).
18. For the definition of the workload index, see Fujimoto (1994a).
19. An interview with Ryo Hatano, General Manager of Material Purchasing Department, Nissan (8 May 1993).
20. Note that these data are based on the number of engineering drawings, as opposed to purchasing cost.
21. This section is based mainly on the interview with Ryo Hatano, General Manager of Material Purchasing Department, Nissan (8 May 1993).
22. Note that the black box parts transactions at Nippondenso at the initial stage were also based on such drawings, or *'shiyozu'*, supplied by the can makers. The inputs from the automobile makers were subsequently simplified to specification documents, though. See S. 4 of this paper.
23. Nissan's purchasing staff were not concerned about the possibility that early selection of suppliers would deprive them of their price negotiation power, because they were confident of the reliability of their target cost system.
24. The average fraction of black box parts in procurement cost among the Japanese samples ($N = 12$) was 62%, with a standard deviation of 17%. Toyota was not among the highest in this figure; Nissan was above the Japanese average.
25. Note that some Nissan suppliers started the approved drawings system before the war. Considering that many of the Nissan suppliers are located in the eastern part of Japan, where there some major aircraft makers (e.g., Nakajima), they may have been aircraft parts suppliers before and during the war. Further investigation is needed on this point. See also S. 4 of this paper.
26. The author greatly appreciates the cooperation of anonymous managers from company B. The author also conducted a detailed survey at company A. See Fujimoto (1994a) for details.
27. Toyota holds a minority share of company B's stock. Virtually all of company B's business is related to the Toyota group.
28. As for parts X, company B was one of several suppliers (and one of two major ones) for Toyota.
29. Such a specification is called structural planning; it consists of a number of documents and rough drawings and is made by translation of product planning, an upstream document.
30. For the concept of ex-post capabilities or post-trial capabilites, see Fujimoto (1994b).

REFERENCES

Asanuma, B. (1984), 'Jidosha sangyo ni okeru buhin torihiki no kozo' [Structures of parts transactions in the automobile industry], *Kikan gendai keizai*, summer, 38–48.

——(1989), 'Manufacturer-Supplier Relationships in Japan and the Concept of Relation-Specific Skill', *Journal of Japanese and International Economies*, 3, 1–30.

Clark, K. B. and Fujimoto, T. (1991), *Product Development Performance* Boston.

——(1992), 'Product Development and Competitiveness', *Journal of Japanese and International Economies*, 6, 101–43.

Cusumano, M. A. and Takeishi, A. (1991), 'Supplier Relations and Management: A Survey of Japanese-Transplant, and U.S. Auto Plants', *Strategic Management Journal*, 12, 563–88.

Fujimoto, T. (1994a), 'The Origin and Evolution of the "Black Box Parts" Practice in the Japanese Auto Industry', Tokyo University Faculty of Economics Discussion Paper 94-F-1.

Fujimoto, T. (1994b), 'Reinterpreting the Resource-Capability View of the Firm: A Case of the Development-Production Systems of the Japanese Auto Makers'. Paper presented to Prince Bertil Symposium, Stockholm, June.

Fujimoto, T., Sei, S., and Takeishi, A. (1994), 'Nihon jidosha sangyo no supplier system no zentaizo to sono tamensei' [The total perspective and multifaceted nature of the supplier system in the Japanese auto industry], *Kikai Keizai Kenkyu* [Studies of Machinery Economics], 24, May, 11–36.

Fujimoto, T. and Tidd, J. (1993), 'The U.K. and Japanese Auto Industry: Adoption and Adaptation of Fordism', Imperial College working paper. The Japanese version: 'Ford system no donyu to genchi tekio: Nichi-ei jidosha sangyo no hikaku kenkyu', *Kikan keizaigaku ronshu* [The Journal of Economics, Tokyo University], 59/2, 36–50, and 59/3, 34–56.

Helper, S. (1991), 'Comparative Supplier Relations in the U.S. and Japanese Auto Industries', *Business and Economic History*, second series, 19, 153–62.

Kojima Press Kogyo Co. Ltd. (1988), *Okagesama de 50 Nen Minna Genki de* [Thanks to you we have all been well for 50 years].

Mitsubishi Research Institute (1987), *The Relationship between Japanese Auto and Auto Parts Makers* Tokyo.

Nishiguchi, T. (1989), 'Strategic Dualism', Ph.D. thesis (Oxford). (Published by OUP in 1993, *Strategic Industrial Sourcing*, Oxford University Press, New York.)

Okouchi, A. (1979), *Keiei kosoryoku* [Entrepreneurial imagination], Tokyo.

Penrose, E. T. (1968), *The Theory of the Growth of the Firm*, Oxford.

Sawai, Minoru (1985), 'Senzenki Nihon Tetsudo Sharyo Kogyo no Tenkai Katei- 1890 Nendai–1920 Nendai,' [The development of the prewar Japanese locomotive industry—from the 1890s to the 1920s], *Shakai Kagaku Kenkyu* [Social Science Studies], 37/3, Oct.

Teece, J. T., Pisano, G., and Shuen, A. (1992), 'Dynamic Capabilities and Strategic Management', University of California Berkeley working paper.

Toyota Motor Corporation (1957), *Toyota Jidosha 20 nenshi* [20 years of Toyota Motors].

—— (1967), *Toyota Jidosha 30 nenshi* [30 years of Toyota Motors].

—— (1978),*Toyota no ayumi* [History of Toyota].

PART III

THE FORD SYSTEMS AND SOCIO–ECONOMIC SYSTEMS

9

Men and Mass Production: *The Role of Gender in Managerial Strategies in the British and American Automobile Industries*

WAYNE A. LEWCHUK

Mr Ford's business is the making of men, and he manufactures automobiles on the side to defray the expenses of his main business.

Revd S. S. Marquis, Director, Ford Sociology
Department, 1915–21.[1]

9.1. Introduction

Until very recently, the shop floors of the leading automobile manufacturers, and the unions that bargained for these workers, were the domain
of men. At the Ford Motor Company's plants in Detroit, women represented less than 2 per cent of the total workforce during the 1920s and
1930s.[2] At its massive River Rouge complex outside Detroit, Ford employed over 70,000 men but not one woman in a production department,
in the early 1940s.[3] In Britain, at Austin's Longbridge shops men also
dominated the workforce, but women were found in larger numbers,
representing between 7 and 10 per cent of the workforce between the
wars.[4] This paper examines why men dominated both workforces and
offers a preliminary hypothesis to explain why British employers might
have been more willing to employ women than American employers.

I would like to thank Haruhito Shiomi, Kazuo Wada, and other participants at the
21st Fuji Conference, Ava Baron, Les Robb, Robert Storey, and Pam Sugiman for
their comments and encouragement. Dale Brown was instrumental in ensuring
clarity and focus. The research was partially funded by the Social Sciences and
Humanities Research Council. This paper is a continuation of work originally
published in the *Journal of Economic History*, 53/4, 1993.

Research during the 1970s and 1980s directed at understanding the experiences of women at work is now having a profound impact on our understanding of men at work.[5] Two themes in this literature are relevant to the arguments to be made below. The first is the argument that gender norms are socially constructed and change over time as men and women interact in their daily lives. The second is the evidence that a deeper understanding of the role of women within the family, kin networks, and the 'informal economy' can illuminate the relationship of women to work and work-based organizations. This literature encourages us to think of men and women bringing economic and gender interests to the workplace. For example, in the case of men, their role within the production process plays an important role in how they define their masculinity and has implications for their status and authority in society and within the family. While male managers and male workers may have opposing economic interests in some contexts and complementary ones in other cases, they are all men. The possibility of male workers participating in the restructuring of work contrary to their economic interests in order to protect their gender interests must be considered.

From this perspective, managerial strategies are gendered strategies, in the sense that they reflect and shape gender norms in the pursuit of economic objectives. The equilibrium between gender norms, managerial strategies, and the technical organization of work is likely to be particularly fragile during periods of significant technical change. Ava Baron has argued that technical change can create a crisis in masculinity, 'a crisis for men workers both as men and as workers'.[6] Building on these insights, this paper will suggest that a similar crisis occurred during the transition to mass production in the automobile industry as employers struggled with the conversion of labour time into effort. How the crisis was resolved may shed light on why women represented such a small component of the new labour force and why some automobile producers, especially those in Europe, had difficulty matching Ford's success.

9.2. The Automobile Industry as a Male Domain

One of the critical points made in the literature on women at work is the social construction of gender norms that give substance to concepts of masculinity and femininity and how these gender norms influence the allocation of work between men and women. Ava Baron, Sonya Rose, and Elizabeth Faue have shown how lived experiences, interpreted through the lenses of language and discourse, shape and give meaning to gender identities and how men and women come to understand their roles in society.[7] Recent research on nineteenth-century Britain has shed new light on how gender norms gave men an advantage in securing skilled work outside the home.[8] Notwithstanding the bias in labelling men's work as skilled, regardless of the tasks actually performed, there is evidence that many nineteenth-century occupations required workers to make decisions based on specialized knowledge and that these occupations were dominated by men.[9] Men monopolized such work in part thanks to social norms that identified control, independence, and the ability to make decisions as inherent masculine traits.

Men performing such tasks were fulfilling the preconceptions of existing norms of masculinity. David Montgomery argued that, for nineteenth-century skilled men, the concept of masculinity was 'embedded in a mutualistic ethical code'. In both Britain and North America, the 'manly' worker 'refused to cower before the foreman's glare—in fact, often would not work at all when a boss was watching'. He accepted group-imposed quotas or stints, even when management offered attractive incentives to speed up production. The skilled male worker was expected to behave in a 'manly' fashion towards his fellow male workers, which implied not accepting management deals that might undermine a fellow brother's job.[10]

Keith McClelland's study of the mid-nineteenth-century British 'representative artisan', suggested that such a worker was a man in control of his life and worthy of respect.[11] In the words of a British miner in 1873, a man wanted, '... the independence of the workshop, and he wanted to be able to pursue his work in such a manner and under such a condition that it should not be a degradation to him in his eyes. He wished to be independent in following his ordinary daily occupation.'[12] Many of these values were taught during an apprenticeship when boys learned the mysteries of a trade, appropriate attitudes towards employers, male codes of sexual conduct, and male social responsibility. In the process, a boy went through a period of 'unfreedom' from which he emerged a free and independent man, that is, 'one of the lads'.[13]

The gender division of labour was never fully determined by gender norms. Employers were under pressure to minimize costs, making the employment of low-wage women and children tempting, despite tensions created by gender norms. The supply of labour, although arguably itself a product of gender norms, was also important. British men were more likely to pursue the training needed to master the mysteries of nineteenth-century trades. This imbalance reflected differences in the expected return to training for men and women given the gender bias in allocating work. It also reflected the norms of male and female behaviour that ensured that only men pursued training.[14] During the first stages of the Industrial Revolution in both Britain and the United States, women readily found employment in industries such as textiles.[15] The late-nineteenth-century shift to the production of standardized goods with less skilled labour opened new opportunities for women in many industries.

Between 1903 and 1930, the automobile industry in Britain and the United States underwent major changes in technology as mass production became the dominant mode of production. At first, a number of Detroit firms did turn to low-wage female workers during this shift. Hudson began employing women in 1910 when they built special facilities for them in their new plant and may have employed as many as 5,000 women by the mid-1920s. In 1927, women trimmers were employed at Murray, which supplied bodies to Ford. At Studebaker, women were employed in large numbers on drill presses, lathes, and internal grinders, and they could be found on the assembly line at the Piquette plant in the 1920s.[16]

However, by the late 1920s, before the onset of the depression, many American automobile employers lost their early enthusiasm for employing women. A study by the Women's Bureau in 1928 charged that American employers showed little interest in exploring how the labour of women could be more effectively used in automobile plants.[17] By 1941, only 2,700 of the approximately 200,000 workers employed by the six leading Detroit vehicle producers were women.[18] In 1942, when women were brought into the plants in large numbers on war work, they tended to work in isolation from men in predominantly female departments. Once the war ended, women were quickly 'invited' to leave the automobile plants and return to the home, thereby maintaining the masculinity of the industry.[19] In Britain, the initial shift to female employment in the automobile industry was not reversed. Data from firms belonging to the Engineering Employers Federation indicate that by 1923 female employment in the automobile and parts industry had reached over ten per cent and by 1934 was over 17 per cent.[20] By contrast, in the American automobile and parts industry women represented 6.2 per cent of the workforce in 1923, but only 4.7 in 1934.[21]

The failure of women to make greater inroads into British and especially American plants is even more puzzling if one accepts contemporary opinion that trends in technology after 1910 reduced the strength and knowledge requirements for working in automobile plants, opening possibilities for employing more women.[22] A mid-1920s review in the United States claimed:

Quickly—over night as it were—the machine, gigantic, complex and intricate, has removed the need of muscle and brawn. . . . Instead we have a greater demand for nervous and mental activities such as watchfulness, quick judgments, dexterity, guidance, ability and lastly a nervous endurance to carry through dull, monotonous, fatiguing rhythmic operations.[23]

The author went on to argue that such work was perfectly suited for women. An internal survey of Ford jobs in 1913 or 1914 concluded:

At the time of the inquiry . . . there were then 7,882 different jobs in the factory. Of these, 949 were classified as heavy work requiring strong able-bodied, and practically physically perfect men; 3,338 required men of ordinary physical development and strength. The remaining 3,595 jobs were disclosed as requiring no physical exertion and could be performed by the slightest, weakest sort of men. *In fact, most of them could be satisfactorily filled by women or older children.*[24]

A number of recent studies have pointed to the role of labour laws and factory legislation in restricting industrial wage work for women. However, a study of the automobile industry in the United States in 1928 dismissed this explanation. Conditions in the industry already exceeded most of the standards imposed by law, and there were no restrictions in Michigan on night work by women. Instead, it was argued, 'In no other industry studied was there found such violent prejudice [by employers] against women's employment because of the mere fact that they were women. It is this prejudice that is a stronger force than any legislative requirements in closing opportunity to women in such occupations.'[25]

Ruth Milkman has argued that Ford's decision to follow a high-wage and high-effort strategy, and to abandon at least temporarily a more traditional strategy of employing the least costly labour time available, precluded the employment of women. She suggests that even during periods such as the 1930s, when firms were more concerned about minimizing costs, the dominant ideology that in times of high unemployment men should be offered jobs first shaped hiring policies.[26] There is little doubt that the general acceptance of male priority in employment had negative consequences for women seeking work, especially in times of high unemployment. It is less obvious why the decision to pursue a high-wage policy in 1914 should close the door to women seeking jobs. Ford could have increased the wages of women relative to their alternative opportunities and extracted more effort from them as well. They were almost certainly available, even outside war years, given their employment in the thousands by other Detroit assemblers mentioned above. Had Ford employed women at high wages, there is no reason to believe that the effort-to-earnings ratio of women would have been less attractive than that of men on many jobs.[27]

In the next section I show that at Ford women were excluded from production, not because men were being paid a high wage, but rather because it was unclear if time could be converted into effort as efficiently in a mixed-gender workforce. The exclusion of women was part of a broader strategy by Ford to reshape masculinity along lines more consistent with conditions in a mass production factory. Ford consciously excluded women from the workplace and created a fraternal system, a men's club, to help male workers adjust to a world of monotonous repetitive work. In the process, Ford shifted gender norms at work and standards of labour productivity, and also remodelled the family and the role of working-class men in society. The Ford strategy gave real meaning to Marquis's claim quoted above that 'Mr Ford's business is the making of men.'

9.3. Masculinity and Mass Production at the Ford Motor Company

In the pre-mass-production period, Ford relied heavily on skilled male workers and had little choice but to leave them a degree of control over how vehicles were assembled and the rate at which time was converted into effort.[28] The spread of mass production and the transition to an unskilled workforce created both new problems and new possibilities for converting time into effort. In 1910, about 60 per cent of Ford's workforce was classified as skilled mechanics. By 1913, skilled workers comprised 28 per cent of the workforce and by 1917 they represented only 8.6 per cent of all employees.[29] The skilled workers had been replaced by unskilled assemblers and machine tenders, many of whom were machine paced after the introduction of the moving assembly line in 1913.[30]

For men, the new conditions of work undermined any sense of control or independence at work, the key characteristics of manly work amongst nineteenth-century skilled workers. For male managers, the rising capital investment and the organization of production along flow lines increased the importance of extracting as much effort as possible from labour and, as argued by Daniel Raff, at as regular a pace as possible.[31] Ford and his managers pointed to the removal from workers of virtually all decision making power as a great social advance. Cruden claimed that, in the 1920s, Ford tour guides boasted about the simple and repetitive nature of Ford work. They claimed that Ford workers were well suited to such an arrangement, and that 'most of the workers have had little or no schooling. . . . They have never been taught to think; and they do not care to think. . . . All of which means that they get to like their monotonous jobs.'[32] Ford argued that 'the vast majority of men want to stay put. They want to be led. They want to have everything done for them and to have no responsibility.'[33]

While Ford may have believed that his male workers were indifferent to the loss of skill and autonomy, one of the few accounts by a contemporary car worker paints a different picture. In 1914, Frank Marquart, then a young factory hand, but eventually a key player in the rise of the UAW, recounted how the boys and men employed in the industry would pass their time in the local saloon talking shop, each trying to impress on the others how important his job was and how much skill it required.[34] Other evidence of male disenchantment with the new conditions of work was the growing labour relations crisis at Ford associated with the shift to mass production. As early as 1908, as employment levels grew, Ford began experiencing problems converting labour time into effort and experimented with time studies and profit sharing plans.[35] By 1913, daily absenteeism reached 10 per cent of the workforce and the annual turnover rate was 370 per cent.[36] Arthur Renner, who began working at Highland Park in 1911, recalled that, during this period, relations between male workers were fractious. It was not unusual to see five or six fist fights in progress while walking from one department to another.[37] Detroit workers were also looking to unions to defend their interests. Raff argues that at least part of Ford's motivation in doubling wages in 1914 was to ensure that the Wobblies did not gain a foothold in his shops.[38] The high rates of turnover, the turmoil on the shop floor, and the growing interest in unions suggest that existing male workers were less than enthusiastic about the new conditions of work under mass production.

In 1914, Ford moved to stabilize labour relations by effectively doubling real wages to five dollars a day. Turnover rates fell to a fraction of their pre-1914 levels. Daniel Raff has argued that the payment of five dollars a day was more than simple compensation for accepting boring and monotonous work.[39] Instead it is seen in part as an efficiency wage given in anticipation of higher effort norms. However, perhaps of even more importance for explaining the timing and the magnitude of the pay increase, according to Raff, was the desire to keep newly active unions such as the Wobblies out of the Ford plants.[40]

There remain, however, troubling questions regarding forces behind the five-dollar day, the high-effort norms associated with this period, and the absence of women at Ford. One of the problems is that many of the factors that Raff suggests motivated the payment of a higher wage in 1914 remained in place throughout the 1930s, by which time the premium offered by Ford had eroded in real and relative terms. In 1914, the many Ford workers earning the daily minimum plus bonus made 92 per cent more than the average worker in the industry. By 1925 the advantage was less than 3 per cent, and, by 1934, the Ford worker was actually earning less than the industry average.[41] It is also puzzling why Ford would raise wages 100 per cent to combat unions that rarely succeed in raising wages more than 20 or 30 per cent. In addition, the payment of a premium to men does not help us explain the absence of women. All of the reasons given by Raff for paying men more money apply equally to women. If keeping the union out was the main reason for paying the higher wage, one might have expected Ford to employ women, which would have compounded the task of organizing a union. Any thesis explaining the reasons behind the five-dollar day needs to explain why only men could profitably be employed at a wage above their opportunity cost and why a male-only workforce was the most effective way of preventing unionization.

A reinterpretation of changes taking place at Ford during this period suggests that the five-dollar day was one component of a broader strategy to revise norms of masculinity in keeping with the new conditions of monotonous and repetitive work. The campaign to promote such work as masculine began in earnest with the introduction of the moving assembly line. The new norms of masculinity at work would depend less on the intrinsic qualities of work and the skill needed to execute it, and more on the ability to work hard and the manufacture of useful products. An appropriately titled but short-lived company publication, *The Ford Man*, suggested in 1917 that

the nobility of labor rests in the practical merit of the product it makes. . . . A common complaint against modern labor is that it breeds dissatisfaction. This complaint cannot be lodged against Ford workers. . . . Even when the extreme specialization of the work obliges the man to make the same motion over and over again, the monotony of the action has not smothered the consciousness of the importance of his task. . . . He is building something that is a benefit to humanity and why should not his work be a pleasure.[42]

In 1919, the paper again stressed the importance of making useful articles, claiming, 'Anyone working for an organization that has good working conditions, is paying good wages, and is making a thing that people need can be sure that he is in a good calling.'[43] This was a theme Ford would stress in his autobiographies, where he wrote, 'There is one thing that can be said about menial jobs that cannot be said about a great many so-called more responsible jobs, and that is, they are useful and they are respectable and they are honest.'[44] In focusing on the usefulness of the Model T, Ford was tapping an undercurrent of American populist culture that stressed the usefulness of work and was in part a reflection of pressing material demands in a developing economy.[45]

The strategy calling for more work and pride in producing a utilitarian product could have been applied equally to men or women. However, other evidence suggests that these characteristics were being promoted as components of a new masculine respectability. Effort and speed came to be identified as laudable features of Ford work, contributing to the positive self-image of men in these positions. Those who could not work at the expected pace were ridiculed and characterized as suitable only for women's work, such as cutting ribbons. An early Ford manager wrote, '. . . many of them—often who have been measuring ribbon, perhaps and doing such things in life—find they can not, in the language of the shop, "stand the gaff" and they immediately resign. . . . They would rather go back to the ribbon counter for $12 or $15 a week than to stand what is necessary to become efficient in the Ford Motor Co.'[46]

The campaign to glorify and masculinize effort needs to be read in conjunction with the implementation of the eight-hour five-dollar day. A careful reading of this strategy reveals that more was at work than simply motivation through the cash nexus. Not only was Ford raising wages, he was also promoting a particular vision of men and women at work, in the home and in society, a vision that he hoped would enable and motivate his male workers to work harder.[47] When the new wage standard was introduced in early 1914, it applied only to men.[48] Its objective was, in the words of a contemporary Ford manager, 'to help men to be better men and to make good American citizens and to bring about a larger degree of comfort, habits and a higher plane of living among our employees by sane, sound and wholesome means'.[49] Marquis, who had taken over the Sociology Department in 1915, argued that the intention was, '. . . to uplift the community; make for better manhood and character of his employees . . . and fix in their minds such ideas of right living as go to make better American citizens'.[50] He suggested:

We have in mind a man who is right in his relations toward his employer, his family and towards the community in which he lives. This is the kind of man we have in view. This is the human product we seek to turn out, and as we adapt the machinery in the shop to turning out the kind of automobile we have in mind, so we have constructed our educational system with a view to producing the human product we have in mind.[51]

To J. R. Lee, Ford's first personnel manager, the five-dollar day was a comprehensive strategy that linked together monetary inducements to effort, the home as a nurturing centre enabling high effort, and the redefinition of masculine work as work requiring abnormal effort. In 1914 Lee wrote, 'Mr Ford believes, and so do I, that if we keep pounding away at the root and the heart of the family in the home, that we are going to make better men for future generations, than if we simply pounded away at the fellows at their work here in the factory.'[52] Charles Reitell, a contemporary journalist, argued, 'To attain a normal day's production the worker is timed so as to keep up an energetic gait for eight hours a day—this can only be done when a well-regulated living is carried on by the worker in his home life.'[53] This is a point recently made by May in her analysis of the five-dollar day. She wrote:

Ford's family wage implicitly recognized the contribution of women's domestic labor to a stable and secure family life. In all likelihood, Ford believed that women's contribution was greatest in their emotional, nurturing, and motherly roles. This emphasis on psychological rather than material comfort parallels the arguments of many Progressive reformers, who saw the female emotional, affective role as a necessary aspect of family life which should be support by adequate wages.[54]

Some Ford managers made a direct link between stability at home and effort at work. Marquis argued, 'Almost immediately after the profit share plan was inaugurated, there was a voluntary increase in efficiency of about 20 per cent. . . . It stands to reason that a man who comes out of a home well fed, whose children are properly clothed, whose wife is happy, and where there is a freedom from debt and all its worries, cannot help but be a great deal more efficient.[55] In another paper Marquis argued, 'We have made a discovery. You have all heard that the family is the foundation of church and state. We have found that the family is the foundation of right industrial conditions as well. Nothing seems to lower a man's efficiency more than wrong family relations.'[56]

With the higher wages and shorter hours, men were expected to provide for their families and participate more fully in family life. In return their status within the household would be enhanced. Ford argued, 'If a man feels that his day's work is not only supplying his basic need, but is also giving him a margin of comfort and enabling him to give his boys and girls their opportunity and his wife some pleasure in life, then his job looks good to him and he is free to give it his best.'[57] Men were encouraged to buy homes and life insurance, and were discouraged from drinking outside the home or taking in boarders. Ford's promotion of men as responsible heads of households is evident in their policy of allowing a divorced wife to request the Sociology Department to deduct a set amount from the husband's pay. If he objected he was fired.[58] This new focus on home life, made possible by both the five-dollar day and the shorter hours, represented a distinct departure from the culture of nineteenth-century skilled men.[59] As the focus of male activity outside the workplace, the family was to replace the working men's societies, the Mechanics Institutes, and perhaps even trade unions, which all played an import role in the lives of respectable nineteenth-century skilled working men and which gave meaning to their identities as men.

Accompanying the increase in wages and reduction in hours was a campaign to promote the Ford Highland Park plant as a fraternal community, a male club. *The Ford Man* was a key contributor to this new campaign; its stated objective was 'to cultivate and establish the broadest fellowship among Ford workers through understanding each other'.[60] It stressed that 'the Ford shop is a community in which all kinds of men' could be found.[61] Marquis, in a piece on the 'Eight-Hour Day' wrote, 'The attitude of the Ford Motor Company toward its employees is not paternal, but fraternal. The "Help the Other Fellow" spirit runs through all we do. The men catch this spirit and are practicing it in their relation toward one another.'[62] A pamphlet distributed to Ford workers in 1915 claimed that the objective of the five-dollar day was 'to better the financial and moral standing of each employee and those of his household. . . . To implant in the heart of every individual the wholesome desire to Help the Other Fellow, whenever he comes across your path.'[63]

The 'Help the Other Fellow' philosophy was a regular component of the preachings of *The Ford Man*. In 1918, when the newly introduced Ford suggestion plan was not living up to expectations, the editors admonished the men and foremen to pull together and to 'speedily learn the solid truth that "helping the other fellow" we can't miss helping ourselves'.[64] It was the central idea of what was described as the spirit of Ford. 'The Ford spirit is toward Achievement! . . . "Help the Other Fellow!" That's the Ford spirit, a splendid spirit of co-operation. . . . It is a spirit that brings out the best there is in every man. . . . Heed then, the spirit of Ford!—be dependable, be thrifty, be efficient, be all good things to all men.'[65]

Employment policies also reflect the climate of community and fraternalism being developed at Ford. The changes implemented went much further than the rationalizing of hiring in 1913 and the removal of the foreman's autocratic power. In 1914, when demand for vehicles fell, married men were given preference over single men for employment.[66] Ford's stated policy was 'to give employment to those who are in the greatest need. As a result, a slight preference is given to married men, or single men who have others totally dependent upon them for support, although single men with no dependents are not barred.'[67] According to William Baxter, who worked in the Sociology Department, renamed the Employment Department in the 1920s, if a worker selected for lay-off by a foreman was 'in dire financial straits which could be substantiated, we would make arrangements to return the man to his department and ask the foreman to replace him with some other employee in more fortunate circumstances'.[68] Over time this practice created ill feeling, and opportunities for abuse, and was one of the reasons the unionization campaign was eventually successful.[69]

Ford also had a policy of hiring men who traditionally found it difficult to find employment in manufacturing shops, including convicts, disabled men, and petty thieves. It was suggested that Ford was 'the first corporation to undertake on a large scale the work of securing efficiency from defectives by the proper adjustment of the man to his work. In this instance the inception of the idea came through the motto, "Help the Other Fellow" originated by Ford.'[70] By November of 1917, Ford employed 6,095 crippled or physically substandard men, representing 18 per cent of the workforce. By 1928 there were 2,600 male convicts working in the plants.[71] In 1926, Ford declared that he would help solve Detroit's crime problem by hiring 5,000 boys to keep them out of mischief.[72] Even the limited employment of women after 1919 contributed to the restructuring of Ford as a fraternal community, where men supported each other for the common good. 'Like the substandard men, the women are employed not because they are women. Most of the Ford women are wives or daughters of Food men, who have been in some way temporarily or permanently disabled. A woman whose husband is an active worker in the factory cannot obtain employment in the Ford plants.'[73]

Collectively, the campaign to glorify hard work and generate pride in making useful articles, the promotion of a fraternal community through selective hiring policies, the reduction in hours worked and the granting of a family wage that raised the status of men within their households and the community, created a package whose intention was to make monotonous and repetitive work manly and hence respectable, if not enjoyable, for Detroit men. Together, these changes replaced the paternalist strategy employed by Ford in the pre-mass-production era with a fraternalist strategy, one more in tune with the new conditions of work. Such a fraternal community might also have been resistant to organizing by outsiders and hence help secure the labour peace essential to an integrated production process. It was a strategy that left little room on the shop floor for women.

Was Ford successful in creating a fraternal community and making monotonous repetitive work desirable to men? The telling of this story has depended heavily on statements by Henry Ford and Ford managers. It is difficult to get evidence regarding the actual impact on workers. It is likely that Ford workers were neither as joyful about monotonous work, nor as enamoured of the Ford community spirit—which rarely extended beyond the age of 40—nor as thankful for a wage packet that was eroded by inflation and employment instability as Ford, Lee, and Marquis would have us believe. Reports suggest that in 1918 absenteeism and lateness were again problems.[74] The pages of the union papers such as *Auto Workers News* and *The Ford Worker* are filled with complaints regarding labour speed up and intolerable working conditions in the Ford plant, especially during the 1930s.[75] For instance, one contributor to *The Ford Worker* complained, 'Such is the wonderful thing called the "Ford Spirit". It makes sure that we workers feel like slaves, we are kicked from pillar to post. The star men, the foremen, the straw bosses, all of them have but one reason to keep their jobs and that is more slave driving, more speed-up. We can only stop this when we put up a united resistance. I for one say let's organize.'[76] When given a chance, Detroit workers, some of them from Ford, did organize and joined fledgling unions in 1919 and again in the mid-1920s.[77] Racial and ethnic tensions within the male workforce also speaks to the limits of the fraternal model as proposed by Ford. For some Ford workers, the Ford system was simply a massive intrusion into one's life. In the words of a young married polisher in the 1920s, 'You've got to stand for a lot of things. . . . You just have to see that the boss gets nothing on you, and if he does just let him bawl you out.'[78] Fraternalism itself was an evolving strategy. The shift to the River Rouge Plant in the late 1920s, the decline of the Model T, and the ascendancy of Charles Sorensen mark the rise of a much harsher regime, one more dependent upon fear rather than pride, to get the work out.[79] Nonetheless, the gender die had been cast and women remained outside the Ford shops.

There is evidence, however, that Ford was at least somewhat successful in redefining masculinity and, through this, labour productivity. Ford attracted male workers easily to his shops, and many stayed for a long time. In an industry with a history of high turnover rates due to its seasonal nature, Ford had 25,000 workers on the payroll in 1945 with 20 or more years of seniority.[80] There is also ample evidence that from the point of view of supervisors, a community spirit existed amongst Ford workers, at least at Highland Park, until the late 1920s. P. E. Haglund, who began working for Ford in 1915 and became a supervisor in the open hearth in the 1920s, recalled, 'The employees worked willingly. They would tackle jobs and they tried to get results without any particularly excessive pressure. There was an elective drive in the men to do a good job at Highland Park.'[81] K. C. Klann, another early Ford worker who became the supervisor of motor assembly in 1912, referred to the 'help the other fellow spirit' at Highland Park.[82] In a similar vein, Alex Lumsden who began working in the tool room in 1913, eventually becoming head of the department in 1923, recalled, 'Our infatuation of the Ford Motor Company in those days was something that I could write a story about. . . . It created a tremendous loyalty on the part of a large, large group of men in the shop. . . . One of the common remarks was, Now listen, that's not the Ford spirit.'[83]

Absolute measures of effort norms are difficult to find, but by most anecdotal accounts, Ford was more successful than most employers in enforcing high-effort norms prior to the 1930s. The wife of one Ford worker stated, 'The chain system you have is a slave driver! . . . That $5.00 day is a blessing—but oh they earn it.'[84] Another worker returned to a position at Dodge after a period at Ford, claiming he was 'too fatigued after leaving the Ford factory to do any serious reading or attend a play or concert'.[85] A Yale student on a work assignment at Ford in the 1920s reported, 'You've got to work like hell in Ford's. From the time you become a number in the morning until the bell rings for quitting time you have to keep at it. You can't let up. You've got to get out the production . . . and if you can't get it out, you get out.'[86] Stories are also told of Ford River Rouge workers riding the Baker streetcar so exhausted after their shifts that they fell asleep, whether sitting or standing.[87]

During the 1920s, Ford was viewed by some of his sternest critics with begrudging admiration as somehow different from other employers. One critic wrote:

Really, up until that summer (1927), the workers of Detroit had set up Henry as a little tin god, to whom the poor forlorn workers could always run in times of stress and unemployment, and always get a job at five dollars a day. Dodge might close; Hudson might shutdown; even General Motors might lay off—but Henry Ford, never! . . . Had not Ford himself said that he was interested in the lives of his workers![88]

Another wrote of Ford, 'The conviction that character and manhood are requisites for employment has spread throughout the entire shop; in fact throughout the whole city of Detroit.[89] To some at least, Ford was successful in 'making men'.

9.4. Masculinity and Mass Production in the British Automobile Industry

Our hypothesis that gender played a central role in Ford's strategy for converting time into effort under mass production provides an opportunity to speculate on the role of gender in the British automobile industry. While much more work needs to be done to confirm our initial hypothesis, we will suggest that in Britain the nineteenth-century norms of masculinity were more deeply entrenched and that British employers opted to adapt working conditions under mass production to these norms, in contrast to Ford's strategy of adapting the gender norms of his workers to the new technology. Specifically, male British employers were unwilling to undermine the participation of their male workers in decision making on the shop floor.

British managers were willing and able to implement many of the new technical innovations associated with mass production, such as grinding and milling machines.[90] They were less willing and less able to transfer authority for decision making from male shop-floor workers to a new class of managers. Many factors help explain the persistence of nineteenth-century authority patterns within British establishments. The collective resistance of workers and the slower pace of technical change due to smaller markets were clearly important. We may also want to consider the role of gender and the reluctance of male British managers to strip British male workers of decision making power and affront their norms of masculinity.

The Fordist production system came to Britain before World War I.[91] During 1912 and 1913, Ford had introduced American methods to his Manchester assembly plant. A prolonged strike in 1912 ended in defeat for labour. Ford raised wages dramatically in the aftermath of the strike. For most workers wages doubled to 10*d.* an hour between early 1912 and September of 1913. In April of 1914, three months after its introduction in Detroit, profit sharing came to Manchester in the form of a 5*d.* bonus for eligible workers. Unlike Detroit, where a large percentage of workers were denied their profit sharing bonus initially, in Manchester it was virtually universal. Of the 1,400 workers employed at the time only 50 or 60 were denied their bonus and by October of 1914 they were no longer employed by Ford.[92] British investigation of workers' homes was less judgemental. Only 49 were identified as living in bad homes. Of these most could be explained by periods of illness or unemployment before working for Ford. Again unlike the Americans, who saw the home as the centre of a patriarchial society, the British had a less overtly gendered vision. To British Ford management Ford workers need '. . . a wider outlook, a keener sense of what a home and its comforts should be, and a stronger desire for those refinements for themselves and their families, which constitutes the mainspring of most human endeavour'.[93]

Most British managers held serious reservations about American methods of production and the Fordist system. Their criticisms were often couched in gendered terms and the need to respect male desires to participate in decision making at work. Bayley, before the British Institute of Automobile Engineers, argued:

In America, I understand, the labour available is much more amenable to systematised working. In England there is difficulty in getting a man to do exactly what he is told, because he is apt to think a great deal more for himself than do his fellows in America. Therefore, a system in this country has to be more elastic and less precise than many American systems are said to be.[94]

Perry Keene, from Austin, had the following observation on American labour management methods:

In America you have to employ methods which a crowd can carry out, but the British individual will not have that . . . the Britisher will not have 'herd' methods. He has the individualistic tendency, and it is a British tendency that you have to allow for.[95]

To convert time into effort, many British employers, especially those in the motor vehicle industry, turned to a variety of incentive payment systems with large bonus rates, a system of indirect control of effort norms. This left sellers of labour time greater control over the level of effort but insulated profits from variation in effort by automatically adjusting wage payments.[96] These techniques had been pioneered in British steel plants and in the coal mining sector.[97] Similar strategies were used in many American firms, but the extent to which British management relied on British labour to coordinate shop floor activity and self-enforce effort norms under incentive payment systems seems extraordinary.[98]

Austin managers saw the use of payment by results as an alternative to the Ford system for controlling labour.[99] They argued:

There are still a few employers who object to piecework on principle. Their standpoint is that an efficient management ought to be able to get the same results at an agreed rate of wage without having to pay more money to encourage the men to work harder. . . . The daily task system at fixed wages may perhaps be workable in American, or even Continental factories, but the necessary . . . driving works policy would not be acceptable either to English Labour or Management.[100]

The thesis that the right to participate in decision making remained an important attribute of British notions of masculinity at work warrants further investigation.[101] This might help explain why women were accepted in British automobile plants in larger numbers and why overall female participation rates were higher in British manufacturing at the turn of the century. The strategy of creating a fraternalist 'male environment' by excluding women in order to encourage men to work hard does not appear to have been a central part of the British system of mass production. As long as women were allocated jobs that British male managers and British male workers agreed did not require decision making authority, the conversion of male labour time into effort would not be undermined.

This went hand in hand with the persistence in Britain of traditional strategies that held that paying the lowest wages was the least expensive way of producing goods, a strategy that made women workers and boys attractive employees. In an attempt to regulate the conditions under which such workers entered the plants, British unions and the Engineering Employers Federation signed collective agreements in the early 1920s that allowed for special grades and pay rates for boys and women. When Rover challenged the differentials and tried to raise the wages of women in the early 1930s, they were forced to live by the agreement and lower the wages of women to the old standard.[102] In the mid-1920s, one quarter of the Austin workforce was classified under the agreement as either female or boys under the age of 21 and hence payable at the lower rate.[103]

For the men who did make up the majority of the British workforce, the inability of British management to completely dislodge them from their nineteenth-century craft roots allowed them to retain an important aspect of how they defined their masculinity. This stabilized both labour relations in the industry and the conversion of time into effort. Even though the British male worker retained the appearance of control of shop-floor decision making, the structure of the payment system made it financially unattractive to employ this control to reduce effort norms. This was a strategy that British male managers, who prided themselves on their nearness to the shop floor, may have felt comfortable with as men.

It is important not to misinterpret the British system of production in the interwar years as being based on nineteenth-century craft technology. As early as the 1920s many elements of mass production were entrenched in British shops and the control male British workers had over their jobs was more myth than fact. However, this myth survived well into the post World War II period. Thompson has documented the existence of this mythical craft world in the modern assembly line era, a myth that served the short-run interest of labour and management, but that paved the way for long-run decline.[104] In the United States, the increased authority of management over effort levels, and through these the productivity of capital equipment, paved the way for future capital investment in productivity-enhancing innovations. In Britain, the decision to allow male workers some say over effort levels made capital intensive investments in productivity-enhancing innovations less attractive. Instead, British managers turned to changes such as the gang system perfected at Standard in the 1950s, which tried to discipline workers to work harder without undermining their nominal control of shop-floor decision making.[105] But in the end, time caught up with the British industry. New capital equipment was installed in the late 1950s and early 1960s, but lack of managerial control over effort left British firms vulnerable to workers trying to even the score after decades of instability and low wages. In the words of Thompson 'Neither they [labour] nor their masters had any answer when the international capitalist market blew the final whistle.'[106]

9.5. Conclusion

The implementation of mass production technology brought with it a heightened managerial focus on effort levels in both Britain and the United States. A capital-intensive integrated system could be profitably employed only by sustaining high and stable levels of effort. However, enforcing such an effort norm was problematic. In its American form, the new technology generated tensions for some men as it removed many of the attributes of work prized by skilled men in the nineteenth century. The results presented in this paper indicate that Ford went beyond raising wages and tightening up supervision to ensure an adequate supply of effort. There was a conscious attempt to revise notions of masculinity. Through company papers, public speeches, and employment policies, a new vision of man at work, within the family and in society, was promoted. Working hard in the company of other men on a useful product and being paid well was supposed to make Ford work manly, even though it was repetitive, boring, and devoid of many of the control elements characteristic of nineteenth-century skilled work. The Ford strategy balanced economic and gender interests as it reached beyond the workplace to shape relations of men in the household and in society at large. The loss of control, independence, and status at work was compensated by gains in the household and in the rest of society. The exclusion of women from the Ford shops was a component of this strategy.

In Britain, male workers retained at least the appearance of greater control over decision making for most of the interwar period and continued to define their masculinity using norms reflecting the experiences of their nineteenth-century craft brothers. Even after World War II, when most researchers accept that work in British automobile plants left male workers little real say over decisions, the myth of craft work survived. As long as the real or imaginary exercise of skill was the basis of masculinity and the conversion of time into effort in British shops, the constraints on employing women would be looser compared with the Ford shops, where the exclusion of women was a central component of a high-effort fraternalist strategy.

While further research on the changing notions of masculinity within early mass production shops is needed to see the extent to which men accepted this new vision of masculine work, and how widespread it was in the industry as a whole, such an approach opens up new possibilities for understanding the success of Ford relative to his competitors and the different experiences of American and British firms in the early period of mass production.

NOTES

1. Revd S. S. Marquis, Director, Ford Sociology Department, 'The Ford Idea in Education', (1916), 12, Ford Archives, Acc. 293.
2. On the gender composition of the Ford workforce, see 1912: Edward Levinson Papers, Reuther Archive, Acc. 85, Box 1, File Automobile History 1910–15, *Free Press* clipping, Dec. 1912; 1914: Arnold and Faurote, *Ford Methods and the Ford Shops*, 58; 1917 & 1921: Ford Archive, Acc. 940, Box 16, File Labor Employment Totals; 1925–9: Ford Archive, Acc. 732, Factory Counts; 1941: UAW Research Department, Employment of Women and Negroes; Apr. 1943, Reuther Library, Acc. 350, Box 11, File 11.
3. UAW Research Department, Employment of Women and Negroes, Apr. 1943, Reuther Library, Acc. 350, Box 11, File 11. On the role of women in early automobile unions, see Gabin, *Feminism in the Labor Movement*, ch. 1.
4. Basement Engineering Employers Federation, File Number Employed 1910.
5. See the collection by A. Baron (ed.), *Work Engendered*, esp. chs. by Baron, 'An "Other" Side', and Blewett, 'Manhood and the Market'. See also Parr, *Gender of Bread Winners*; Blewett, *Men, Women and Work*; Valverde, 'Giving the Female'; Cockburn, *Brothers*; Pollert, *Girls, Wives, Factory Lives*, Introduction; Milkman, *Gender at Work*, 1–26.
6. See Baron, 'Contested Terrain Revisited', 61.
7. Baron, 'An "Other" Side'; Faue, *Community of Suffering and Struggle*, 69–99; Rose, *Limited Livelihoods*, 1–21.
8. See Rose, *Limited Livelihoods*, 22–31; Valverde, 'Giving the Female', 621–5.
9. See Phillips & Taylor, 'Sex and Skill'.
10. Montgomery, *Workers' Control in America*, 13–14. See also Kimmel, 'The Contemporary Crisis of Masculinity'; Clawson, *Constructing Brotherhood*, 145–9.
11. McClelland, 'Some Thoughts on Masculinity'.
12. Cited in McClelland, 'Some Thoughts on Masculinity', 172.
13. Ibid. 170.
14. Ibid. 169.
15. See Kessler-Harris, *Out to Work*, 20–22; Dublin, 'Women, Work, and the

Family'; and Rose, 'Gender Segregation'; Rose, 'From behind the Women's Petticoats'.

16. *News* (27 Oct. 1910) in the Edward Levinson papers, Reuther Library, Acc. 85; Misc. clippings, R. W. Dunn Collection, Reuther Library, Acc. 96; *Auto Workers News*, June 1927, in R. W. Dunn Collection, Reuther Library, Acc. 96.

17. US Department of Labor, *The Effects of Labor Legislation*, 233.

18. Data is for Ford, Chrysler, General Motors, Hudson, Briggs, and Continental. See UAW Research Department, Employment of Women and Negroes, Apr. 1943, Reuther Library, Acc. 350, Box 11, File 11.

19. Milkman, *Gender at Work*, 99–127; Kossoudji & Dresser, 'The End of a Riveting Experience', 519–25.

20. The British data is for member firms of the Engineering Employers' Federation and includes most of the major British producers. Ford and Vauxhall (GM) were not members, and only the Morris engine works in Coventry is included. See EEF Basement, Number Employed 1910.

21. See US Dept. of Labor, *Variations in Employment*; 'Wages, Hours, Employment, and Annual Earnings in the Motor-Vehicle Industry, 1934', *Monthly Labor Review* 42 (Mar. 1936), 523.

22. On the success of women in metal shops during World War I see US Department of Labor, *The New Position of Women*, 93–112.

23. Reitell, 'Machinery and its Effects', 43.

24. Ford, *My Life and Work*, 108.

25. US Dept of Labor, *Effects of Labor Legislation*, 220.

26. Milkman, *Gender at Work*, 12–26.

27. There is no evidence anywhere in the Ford archives of an attempt to hire women in this period, or their unsuccessful use in production departments. It was an option that was simply not considered.

28. Arthur Renner, Reminiscences, Ford Archives, 9; Ford Old Timers, Notes on George A. Brown, Ford Archives, Acc. 616.

29. Meyer, *Five Dollar Day*, 48–51.

30. For details on these changes see Hounshell, *From the American System*; Lewchuk, *American Technology*.

31. See Raff, 'The Puzzling Profusion', 1–28.

32. R. L. Cruden, 'Ford's Flimflammery', *Labor Age* (June 1928), in R. W. Dunn Collection, Reuther Library, Acc. 96, Box 2, File 2–21, 15.

33. Ford, *My Life and Work*, 99.

34. Marquart, *Auto Worker's Journal*, 12.

35. Letter from Couzen to employees announcing profit sharing, 29 Dec. 1909, Ford Archives, Acc. 683, Box 1.

36. Meyer, *Five Dollar Day*, 83.

37. Renner, Reminiscences' 14.

38. Raff, 'Wage Determination Theory', 387–99.

39. See Raff & Summers, 'Did Henry Ford Pay Efficiency Wages?'.

40. Raff, 'Wage Determination Theory'.

41. For industry average, see 1914: Census, US Motor Industry, Ford Archives, Acc. 96, Box 10; 1925: handbook of Labor Statistics 1924–6 (US Dept. of Labor, Bureau of Labor Statistics, 1927) No 439; 1934: 'Wages, Hours, Employment and Annual Earnings in the Motor-Vehicle Industry, 1934',

Monthly Labor Review 42 (1936), 523. For Ford minimum plus bonus see Hours Worked and Wage Rates, Ford Archives, Acc. 572, Box 32.

42. *The Ford Man* (20 Sept. 1917), 2–3.

43. *The Ford Man* (3 Sept. 1919), 2.

44. Ford, *My Life and Work*, 278.

45. See Rodgers, *The Work Ethic*, 9–13.

46. G. Bundy, *Work of the Employment Department* (Bureau of Labor Statistics, Bull. 196, May 1916), in Frank Hill Papers, Ford Archives, Acc. 940, Box 6, File Labor—Racial & Group Discrimination, 68.

47. See May, 'The Historical Problem of the Family Wage', 399–424.

48. Shortly after its introduction, public pressure forced Ford to allow women heads of households to qualify for the higher wage. It was not until late 1916 that Ford allowed all women the right to participate in profit sharing. See *New York World* (25 Oct. 1916), in Frank Hill Papers, Ford Archives, Acc. 940, Box 16.

49. Letter dated 26 Jan. 1914 to C. L. Gould, Omaha branch, Five Dollar Day, Reuther Library, Acc. 683, Box 1.

50. Untitled manuscript, Marquis Papers, Ford Archives, Acc. 293, p. 1.

51. Marquis, 'The Ford Idea In Education', 12.

52. Cited in Meyer, *Five Dollar Day*, 124.

53. Reitell, 'Machinery', 38.

54. May, 'Historical Problem', 416.

55. Untitled manuscript, Marquis Papers, Ford Archives, Acc. 293, p. 6.

56. S. S. Marquis, 'The Eight Hour Day', Marquis Papers, For Archives, Acc. 293, p. 15.

57. Ford, *My Life and Work*, 120.

58. James O'Connor, Reminiscences, Ford Archives, 31.

59. On the tension between long hours of work in the nineteenth century and the role of men within their family, see Cross, *Quest for Time*.

60. *The Ford Man* (20 Sept. 1917), 2.

61. Ibid. 3.

62. Marquis, 'The Eight Hour Day', 20.

63. *Helpful Hints and Advice to Employees to Help Them Grasp the Opportunities Which are Presented by the Ford Profit-Sharing Plan* (1915), Vertical Files, Reuther Library, Box 14, Folder titled Ford Motor Company 1920s, p. 3.

64. *The Ford Man* (17 May 1918), 1.

65. *The Ford Man* (3 May 1918), 2. See also *The Ford Man* (17 Jan. 1919), 2.

66. *Journal* (16 May 1914), Edward Levinson Papers, Reuther Liibrary, Acc. 85, Box 1, File Industry History, 1910–15.

67. Memo titled, 'Applicants', Marquis Papers, Ford Archives, Acc. 293. See also Rumely, 'Mr Ford's Plan', 669.

68. William P. Baxter, Reminiscences, Ford Archives, 29. See also Willis Ward, Reminiscences, Ford Archives, 26–7.

69. Renner, Reminiscences, 35. According to Willis Ward, who worked in Employment Department during the period when Marshall was a prominent figure, Marshall would secure employment for prospective purchasers of Ford cars referred to him by local dealers. In return, Marshall earned a handsome commission on these sales. See Ward, Reminiscences, 15–16.

70. J. E. Mead, 'Rehabilitating Cripples at Ford Plant', *Iron Age* 112 (26 Sept. 1918), 739–43.
71. *New York Times* (9 May 1928).
72. Dunn, *Labor and Automobiles*, 71. At the time, there was little growth in the Ford workforce and so the net result of this was to bring in 5,000 fresh young low-paid male workers in place of 5,000 older higher-paid men.
73. Li, 'A Summer in the Ford Motor Company', R. W. Dunn Collection, Reuther Library, Acc. 96, Box 1.
74. *The Ford Man* (18 Mar. 1918).
75. See *Auto Workers News* (5 May 1934); *The Ford Worker* (Oct. 1926, Dec. 1929, and 20 Apr. 1932).
76. *The Ford Worker* (Dec. 1929), 3.
77. Meyer, *Five Dollar Day*, 171 and 197.
78. R. L. Cruden, 'No Loitering: Get Our Production', *The Nation* (12 June 1929), 698.
79. See the account of working at the Rouge by Walter E. Ulrich, published by the League for Industrial Democracy, 1929, p. 8, Vertical Files, Reuther Library, Box 14.
80. Length of Service Report (as of May 1945), Ford Archives, Acc. 616.
81. P. E. Haglund, Reminiscences, Ford Archives, 62.
82. K. C. Klann, Reminiscences, Ford Archives, 88.
83. Alex Lumsden, Reminiscences, Ford Archives, 10–12.
84. Cited in Russel, 'The Coming of the Line', 45.
85. Cited in Raff and Summers, 'Did Henry Ford Pay Efficiency Wages?', 74.
86. Cited in Meyer, *Five Dollar Day*, 41.
87. Marquart, *An Auto Worker's Journal*, 30.
88. Cruden, 'Ford's Flimflammery', 15.
89. Rumely, 'Mr Ford's Plan to Share Profits', 668.
90. Owen Linley, 'Manufacturing on a Medium Scale', *Motor Trader* (8 July 1914); A. A. Remington, 'Some Possible Effects of the War on the Automobile Industry', Presidential Address, *Proceedings of the Institute of Automobile Engineers* (1918), 7.
91. For details of Ford's British operations see Lewchuk, *American Technology*, 152–8.
92. Report on Profit Sharing from Manchester, 12 October 1914, Papers of the Henry Ford Office, Reuther Library, Acc. 62.
93. Ibid. 3.
94. Comments on a paper titled 'Works Organisation', *Proceedings of the Institute of Automobile Engineers* 11 (1916/17) 396. Other criticisms of American practice can be found in, 'Mass Production', *Machinery*, (27 Nov. 1919); 'Robert Hadfield's Toast to the London Association of Foremen', *Managing Engineer*, (May 1916), 7; 'Applied Time Studies', *Automobile Engineer*, (Dec. 1920), 502.
95. A. Perry Keene, 'Production—A Dream Come True', *Journal of the Institute of Production Engineer*, 7 (1928), 31.
96. Lewchuk, *American Technology*.
97. Wilkinson, 'Industrial Relations'.
98. For an American example of payments by results see Schatz, *The Electrical Workers*.

99. For statements by Engelbach and Keene see Ward Papers, MRG1, Organisation Section, w/8/29-34/13/476, pp. 2–14, housed at the Business History Unit, London School of Economics.
100. EEF Archives, W(3)129, Piece Work in the Toolroom, 1 Feb. 1934, 26–8.
101. See L. Grant, 'Women in a Car Town: Coventry 1930–45' (unpublished paper).
102. See Downs, 'Industrial Decline'.
103. EEF, Number of Work People Employed, Basement EEF.
104. Thompson, 'Playing at Being Skilled Men'.
105. For details see Lewchuk, *American Technology*.
106. Thompson, 'Playing at Being Skilled Men', 69.

REFERENCES

Arnold, Horace L. and Faurote, Kay L. *Ford Methods and the Ford Shops* (New York, 1916).
Auto Worker (Chicago, various dates).
Auto Workers News (Detroit, various dates).
Bairoch, P., *La Population active et sa structure* (Belgium, 1968).
Baron, Ava, ed., *Work Engendered: Towards a New History of American Labor* (Ithaca, 1991).
—— 'An "Other" Side of Gender Antagonism at Work: Men, Boys, and the Remasculinization of Printers' Work, 1830–1920', in Baron, ed., *Work Engendered*, 47–69.
—— 'Contested Terrain Revisited: Technology and Gender Definitions of Work in the Printing Industry, 1850–1920', in B. Wright, ed., *Women, Work and Technology* (Ann Arbor, 1987), 58–83.
Blewett, Mary H., *Men, Women and Work: Class, Gender, and Protest in New England Shoe Industry, 1780–1910* (Chicago, 1988).
—— 'Manhood and the Market: The Politics of Gender and Class among the Textile Workers of Fall River, Massachusetts, 1870–80', in Baron, ed., *Work Engendered*, 92–113.
Clawson, M. A., *Constructing Brotherhood: Class, Gender and Fraternalism* (Princeton, 1989), 145–9.
Cockburn, Cynthia, *Brothers: Male Dominance and Technological Change* (London, 1983).
Cooper, Patricia, 'The Faces of Gender: Sex Segregation and Work Relation at Philco, 1928–38', in Baron (ed.), *Work Engendered*, 320–50.
—— *Once a Cigar Maker: Men, Women, and Work Culture in American Cigar Factories, 1900–19* (Urbana, 1987).
Cross, Gary, *A Quest for Time: The Reduction of Work in Britain and France, 1840–1940* (Berkeley, 1989).
L. L. Downs, 'Industrial Decline, Rationalization and Equal Pay: The Bedaux Strike at Rover Automobile Company', *Social History*, 15 (1990), 45–73.
Dublin, Thomas, 'Women, Work, and the Family: Female Operatives in the Lowell

Mills, 1830–60', *Feminist Studies*, 3 (1975) 30–9.

Dunn, Robert W., *Labor and Automobiles* (New York, 1929).

Engineering Employers Federation, Modern Record Centre, University of Warwick, Coventry.

Faue, Elizabeth, *Community of Suffering and Struggle: Women, Men and the Labor Movement in Minneapolis, 1915–45* (Chapel Hill, 1991).

Ford Archives, Greenfield Village and Henry Ford Museum, Dearborn, Michigan.

Ford, Henry, *My Life and Work* (New York, 1923).

Ford Man, The (Ford Motor Company, various dates).

Ford Times, The (Ford Motor Company, various dates).

Ford Worker, The (The Ford Shop Nuclei of the Worker [Communist] Party of America, Detroit, various dates).

Fordson Worker (Ford Motor Company, various dates).

Gabin, Nancy, *Feminism in the Labor Movement: Women and the United Auto Workers, 1935–1975* (Ithaca, 1990).

Gartman, David, *Auto Slavery: The Labor Process in the American Automobile Industry, 1897–1950* (London, 1987).

Goldin, Claudia, *Understanding the Gender Gap* (New York, 1990).

Hounshell, David, *From the American System to Mass Production* (Baltimore, 1984).

Humphries, Jane, 'Protective Legislation, the Capitalist State and Working Class Men: The Case of the 1842 Mines Regulation Act', *Feminist Review*, 7 (1981).

Iron Age (various dates).

Kessler-Harris, Alice, *Out to Work, A History of Wage-Earning Women in the United States* (New York, 1982).

Kimmel, Michael S., 'The Contemporary Crisis of Masculinity in Historical Perspective', in H. Brod, ed., *The Making of Masculinities: The New Men's Studies* (Boston, 1987), 121–53.

—— 'The Cult of Masculinity: American Social Character and the Legacy of the Cowboy', in M. Kaufman, ed., *Beyond Patriarchy: Essays by Men on Pleasure, Power and Change* (Toronto, 1987), 235–49.

Kossoudji, Sherrie A. and Dresser, Laura J. 'The End of a Riveting Experience: Occupational Shifts at Ford after World War II', *American Economic Review, Proceedings*, 82 (1992), 519–25.

Labor Age (various dates).

Land, Hilary, 'The Family Wage', *Feminist Review*, 6 (1980), 55–77.

Lewchuk, Wayne, *American Technology and the British Vehicle Industry* (Cambridge, 1987).

Marquart, Frank, *An Auto Worker's Journal: The UAW from Crusade to One-Party Union* (University Park, 1975).

Matthaei, 'History of Women', 18–28.

May, Martha, 'The Historical Problem of the Family Wage: The Ford Motor Company and the Five Dollar Day', *Feminist Studies*, 8 (1982), 399–424.

McClelland, K., 'Some Thoughts on Masculinity and the "Representative Artisan" in Britain, 1850–80', *Gender and History*, 1 (1989), 164–77.

Meyer III, Stephen, *The Five Dollar Day: Labor Management and Social Control in the Ford Motor Company, 1908–21* (Albany, 1981).

Milkman, Ruth, *Gender at Work: The Dynamics of Job Segregation by Sex during World War II* (Urbana, 1987).

Montgomery, David, *Workers' Control in America* (Cambridge, 1979).

Nation, The (various dates).

Nevins, Alan, *Ford: The Times, the Man, and the Company* (New York, 1954).

New York Times (various dates).

Parr, Joy, *Gender of Bread Winners: Women, Men, and Change in Two Industrial Towns: 1880–1950* (Toronto, 1990).

Phillips, Anne and Taylor, Barbara, 'Sex and Skill: Towards a Feminist Economics', *Feminist Review*, 6 (1980), 79–88.

Pollert, Anna, *Girls, Wives, Factory Lives* (London, 1981).

Raff, Daniel M. G., 'The Puzzling Profusion of Compensation Systems in the Interwar Motor Vehicle Industry' (unpublished paper, 1992), 1–28.

—— 'Wage Determination Theory and the Five-Dollar Day at Ford', *Journal of Economic History*, 48 (1988), 387–99.

—— and Lawrence H. Summers, 'Did Henry Ford Pay Efficiency Wages?', *Journal of Labor Economics*, 5 (1987), 57–86.

Reitell, Charles, 'Machinery and its Effects upon the Workers in the Automobile Industry', *Annals of the American Academy of Political and Social Science*, 116 (1924), 37–43.

Reuther Library, Walter P., Wayne State University, Detroit, Michigan.

Rodgers, Daniel T., *The Work Ethic in Industrial America, 1850–1920* (Chicago, 1974).

Rose, Sonya O., *Limited Livelihoods: Gender and Class in Nineteenth-Century England* (Berkeley, 1992).

—— 'Gender Segregation in the Transition to the Factory: The English Hosiery Industry, 1850–1910', *Feminist Studies*, 13 (1987), 163–84.

—— 'From behind the Women's Petticoats: The English Factory Act of 1874 as a Cultural Production', *Journal of Historical Sociology*, 4 (1991), 32–51.

E. A. Rumely, 'Mr Ford's Plan to Share Profits', *World's Work*, 27 (1914).

Russel, Jack, 'The Coming of the Line', *Radical America*, 12 (1978).

Schatz, R. W. *The Electrical Workers: A History of Labor at General Electric and Westinghouse, 1923–60* (Urbana, 1983).

Thompson, Paul, 'Playing at Being Skilled Men: Factory Culture and Pride in Work Skills among Coventry Car Workers', *Social History*, 13 (1988), 45–69.

US Department of Labor, *The New Position of Women in Industry* (Bulletin of the Women's Bureau, No. 12, Washington, 1920).

—— *The Effects of Labor Legislation on the Employment of Women* (Bulletin of the Women's Bureau No. 65, Washington, 1928).

—— *Variations in Employment Trends of Women and Men* (Bulletin of the Women's Bureau, No. 73, Washington, 1930).

Valverde, Marianna, 'Giving the Female a Domestic Turn: The Social, Legal and Moral Regulation of Women's Work in British Cotton Mills, 1820–50', *Journal of Social History*, 21 (1988), 619–34.

Wilkinson, F. 'Industrial Relations and Industrial Decline: The Case of the British Iron and Steel Industry, 1870–1930', (Department of Applied Economics, Cambridge, Working Paper No. 8915).

10

Production Methods and Industrial Relations at Fiat (1930–90)

STEFANO MUSSO

10.1. Introduction

Industrial relations were deeply connected with the implementation of the Ford system at Fiat. Fordism was seen as a response to long-standing difficulties in labour relations resulting from the strength of the labour movement in northern Italy, and particularly in Turin, where trade unions and left-wing political parties were deeply rooted in the workers' neighbourhoods.

On the eve of World War I, after Fiat managers had made their first study trips to America and before Taylor's works were translated into Italian,[1] the journal of the Turin-based Employers' Association wrote that traditional piecework systems were ineffectual in increasing the pace of work, and that the problem of low productivity because of normal working habits would only be solved when technical innovation minimized the influence of the skill of the workers on the production process.[2]

Of course, Fiat's main reason for introducing Fordist methods was not the control over work rules. the Ford system was primarily linked to market strategies and to production volumes. Nevertheless, control over workers was considered a secondary but important reason, and the new methods were thought to offer a useful opportunity in this regard.

But Italian employers were suffering from an illusion. From the very beginning, their attempts to rationalize the production process and to speed up work rates were hindered by worker resistance. The only time the new methods could be implemented, as it turned out, was when the labour movement experienced a period of defeat and weakness, because management never achieved any form of worker consent. As a result, the existing confrontational model of labour relations persisted and was even aggravated. This was due both to the centralization and politicization of the trade unions as well as to the reluctance of the employers to accept any level of worker participation beyond the merely passive cooperation permitted by the Taylorist organization.

By the late 1960s the fully implemented Fordist system, with its subdivision of tasks, direct shop floor control, and machine pacing, triggered off a rebellion against such unskilled mechanical work. The new disruptive attitude eroded the values of senior workers, values based on professional pride and work ethics, and hindered the transmission of these values to younger generations.

By focusing on labour relations, I hope to illuminate various aspects of the adoption of Fordism by the major (and at present the only) Italian car manufacturer, as well as to point out, because of the importance of shop-floor relations in the performance of any company, the most important factor of competitive disadvantage for Italy.

Since the Italian labour movement has always laid stress on the quality of working conditions, any change to the industrial relations system can only be implemented by a dramatic shift from Fordist methods to a multi-skilled and participatory production system. Such a shift is under way, as I briefly discuss in the last section of this article before my conclusions, even though numerous obstacles still have to be removed.

10.2. The Introduction of Scientific Management Techniques in the 1930s: The Bedaux System at Fiat Lingotto

The first implementation of certain Taylorist methods in the Italian engineering factories dates back to World War I, when the large-scale output of firearms and shells, along with the massive hiring of women and semi-skilled workers, made the production process specially suited to experimentation with the new American principles. The process of reorganization, however, was at that time in its earliest stages. In the old plant of Fiat Centro, a real *Ufficio analisi tempi e metodi* (time and motion study bureau) did not yet exist; control over the production process, by using a stopwatch for timing machine operations and worker motions, was experimental and occasional, and the foremen were still in charge of fixing piecework rates. The postwar industrial and social unrest hindered the introduction of time and motion study on a regular basis. Such an introduction took place only after the defeat of the workers' struggles, the seizure of power by fascism, and the creation of a corporative system of industrial relations in which strikes and free trade unions were banned.

The turning point of this period was the setting up of the new plant at Lingotto, which was one of the most advanced in Europe. After the movement of the assembly line had been mechanized in 1925,[3] a period of intense time and motion study was started in order to enhance productivity through new standard times for machine operations and new piece rates. The Fordist chain could not be balanced without the Taylorist time and motion study. The whole process of rationalization was thus carried through under the banner of the Bedaux system.[4]

From the mid-1920s to the outbreak of World War II, the main technical innovations, which had a strong influence on the qualification and classification of the workers, were the assembly line and the semi-special-purpose machine tools in the machine shops.

Subdivision and simplification of tasks both in assembling and manufacturing led to a considerable use of semi-skilled workers in direct production departments, but highly skilled workmen were still required in casting, forging, maintenance, tool-making, and the set-up of machinery. The indirect labour force (skilled workers like maintenance men, toolmakers, set-up men, plus labourers for internal transport) was 45 per cent of the total personnel at Fiat in 1921 (36 per cent at Citroen in 1927, and 39 per cent at Renault in 1936).[5] Compared with big engineering factories like Ansaldo, Fiat employed more highly skilled workers (specialists, following the classification fixed in the national agreements; see Tables 10.1 and 10.2); this comparison provides evidence that where semi-skilled labour was consistently used, a higher quota of specialist workers was necessary to change the set-up of machinery.

The rationalization process, therefore, even if it meant the loss of professional content in many workplaces, did not imply the declassifying of individual craftsmen, by compelling them to do less skilled tasks. Such a de-skilling did not take place in the Italian engineering industry. The growing weight of this branch in the national economy, with an increase in employment that was much more rapid than in the industrial sector as a whole, suggests that the demand for skilled workers was not being reduced. Rather, the labour market remained largely favourable to them, and their number was growing, although their percentage in the total

TABLE 10.1. *Contractual hourly minimum wages (in Italian lire) in the engineering industry by categories of male adult worker (province of Turin, 1929)*

Categories of male adult worker	Hourly minimum wage
Highly-skilled workers	3.50
Skilled workers	2.75
Low-skilled workers	2.45
Unskilled workers	2.25

Source: L'Informazione industriale, 29 Mar. 1929.

TABLE 10.2. *Male adult workers by categories (percentage of the total), Fiat (mass production) and Ansaldo (big engineering construction), 1938–51*

Categories of male adult worker	Fiat (1938)	Ansaldo (1938)	Fiat (1951)
Highly-skilled workers	10.8	2.4	11.3
Skilled workers	20.5	58.5	24.6
Low-skilled workers	62.3	31.2	61.0
Unskilled workers	6.4	8.0	3.1

Source: Figures for Fiat are in G. Zunino, 'Struttura industriale, sviluppo tecnologico e movimento operaio a Torino nel secondo dopoguerra', in E. Passerin *et al.*, *Movimento operaio e sviluppo economico in Piemonte negli ultimi cinquant' anni* (Turin, 1978). Figures for Ansaldo are in P. Rugafiori *Uomini, macchine, capitali. L'Ansaldo durante il fascismo 1922–45* (Milan, 1981).

number of the workforce was decreasing. In the province of Turin, for example, the metal workers numbered 34,000 in 1911, 49,000 in 1927, and 98,000 in 1939; at Lingotto the workforce grew from 6,927 in 1923 to 16,250 in 1938. Skilled workers maintained a strong position in the labour market. By 1921, the average skilled worker was receiving 40 per cent higher pay than the labourer, and this difference remained up to the end of the 1930s. Payment by results reduced the disadvantages for semi-skilled workers, since the highly skilled workers were normally excluded from piecework; yet for the latter increases outside collective bargaining on the basis of individual merit were more frequent. In conclusion, although certain trades declined, the new machinery and the new production methods also created new skilled tasks that, generally speaking, prevented the senior workers from considering the changes as threatening.

Following the example of Ford, in the late 1920s, along with the introduction of the Bedaux system, Fiat fostered company welfare programmes, in order to gain the consensus of its personnel. But the intensified labour effort resulting from the new piece rates led to ongoing claims from both skilled and semi-skilled workers in direct production. Yet the protests could not go far beyond being mere complaints, even if the claims of the workers were often taken up by the fascist union papers.

Collective bargaining under fascism gave birth to national agreements that subdivided blue-collar workers into categories (four for adult males in the metal and engineering trades) according to skill, and fixed the corresponding minimum wages (by province). As a result of a very centralized bargaining system, the minimum level of wages was tightly bound to the political and economic purposes of the fascist dictatorship.

Increases and reductions in wages were settled by Orders in Council or agreements at the national level under the supervision of the Corporations' Office.

The workers' claims about Bedaux led to a set of regulations in the national contracts concerning piecework, in order to avoid cut-off in the absence of technical improvement, and put piece rates under a certain amount of union control. These rules were quite favourable to workers, in their spirit at least, but the letter of the law could not stop the numerous devices used by the time and motions study offices.

The local unions were left without any initiative except for control over the enforcement of agreements. But this task was very difficult to fulfil, since the unions had no bargaining autonomy, nor were they supposed to initiate any form of struggle. The Corporations' Office, for its part, did not make any serious attempt to monitor the real situation and curb defaulting employers. All it did was provide conciliatory procedures that had little efficacy in the eyes of most workers.[6] Nor did rank-and-file pressure have more effect. Employers had all the advantages on their side in any negotiations at the national level, since they could take their time, without the threat of strikes, and put off questions that required complex political bargaining within the Corporations' Office.[7]

The individual strategies of reaction to the increasing pace of work taken in other countries where democracy held out could not take place to a comparable extent in Italy, because of the strict social control imposed by the fascist regime and the consequent weakness—both political and economic—of the workers. The rationalized factory did not suffer from absenteeism or a high turnover due to voluntary departures, nor was factory discipline hindered to any significant degree by worker resistance, even if work rules were somehow transmitted by senior workers. However, submission or acquiescence did not mean consensus. As the promises of the regime to improve the condition of the working class through the new corporative order failed to materialize, workers became more and more distrustful of the official propaganda and less willing to bear the sacrifices required by the fascist search for a rapid increase in the power of the nation. Within the factory, discipline could be kept only through an exaggerated control by the foremen, in order to prevent sly transgressions.

The first rationalization through the Bedaux system did not eliminate the importance in the production process of the old skilled, unionized, and militant workers, who in the postwar struggles had occupied Fiat plants and tried to manage the factories themselves. In the 1930s the intensification of work rates could be carried out because the workers were not allowed to react. But Fiat's failure to achieve worker consensus just deferred a major confrontation.

10.3. The War Years: One Step Forward and Two Steps Backward in the Rationalization Process

When World War II broke out, various rationalization proposals were put forward,[8] but they were destined to clash with the emergency situation created by the war. The main problems over efficient production stemmed from a lack of continuity in state orders, owing to military indecision over the features desired in the tanks, guns, and aeroplanes. The production of both aeroplanes and tanks at Fiat, and even more so in other automobile factories that were used on much more limited scales of production, remained organized along the lines of handicraft production.

The rationalization programmes therefore ended up being largely confined to techniques for stepping up work rates. They were in any case destined to fail, starting from late 1942, in the wake of the growing economic and manufacturing disorganization caused by the heavy Allied bombing. The list of problems was a long one: destruction and repair work, decentralization of production to other plants or to outside the urban area, inefficient transportation, absenteeism or tardiness by employees, and, last but not least, reduced physical stamina in the workers caused by food shortages and the cold in poorly heated homes.

Under these circumstances the development of time and motion study was impossible. The number of idle periods increased, as did the number of abnormal situations affecting worker performance. The old practices of adjusting the piecework at the end of the shift, through informal negotiation on the shop floor between workers and foremen, were restarted and consolidated. The foremen were also responsible for coping with a whole myriad of problems, obstacles, and unexpected circumstances; these also proved to be impediments to the proceduralization of the work carried out by the technical offices. Thus to some extent the foremen took their revenge on the offices: the technical responsibilities that Taylorist-style rationalization had tended to take away from the foremen (relegating them to a primarily disciplinary role) were returned to them. In practice what resulted was an old workers' claim, also made by the rank-and-file members of the fascist trade union in its arguments against the Bedaux system: the return of decision-making authority to the foremen as far as piecework rates were concerned.[9]

At the end of the war Vittorio Valletta, the managing director of Fiat, clearly pointed out what Fiat's strategy in the succeeding decades would be: the mass production of small cars for Italian and European markets. But the relaunching of mass production implied an enormous amount of reorganization work and technological modernization, and this had to take into account a long period, determined by the economic situation and the war, during which traditional job routines had been strengthened, output had often been interrupted or slowed down, a strong use of indirect manpower had been established, and a flexible management of work rates had been arranged. This situation continued long after the end of the war and was in some ways accentuated by renewed labour activities, at least up until 1948. The reorganization and relaunching of production had to wait till the end of the 1940s, when aid provided by the Marshall Plan would permit technological renewal to be introduced into the larger companies, and especially into Fiat, which enjoyed the largest quota of American aid.

10.4. Defeat of the Left-Wing Union in the 1950s and the Implementation of Fordist Methods at Fiat Mirafiori

In the summer of 1945 the situation at Fiat was one of great political uncertainty. Valletta had been excluded from the management of the company, which had been entrusted to a commissioner appointed by the Comitato di Liberazione Nazionale (National Liberation Committee). There was a serious market problem, and the workforce was too large, as a result of the freeze on dismissals decreed for reasons of public order. Between 1939 and 1945 the manpower of the entire Fiat group had grown from 46,000 to 57,000 workers, while over the same period output had suffered a steep decline.[10] The workers who in the Fiat factories in Turin had gone on strike in March 1943 against fascism and the war, mostly supported the socialist and communist positions. The trade union representatives in the factories (the CI or commissioni interne, 'internal commissions') also had a lot of power. On the one hand they helped maintain a certain amount of discipline because they were loyal to the line taken by the parties of the left, which upheld the need to cooperate in the reconstruction of the country; on the other hand they acted as spokesmen for the workers' claims. Their protests were not over conditions of work, since for a long time the pace of production had been slowed down and, after the Liberation, the piece rate had been abolished as a symbol of capitalist exploitation (the companies had done nothing to retain it, given the disorganization of manufacturing; at Fiat a bonus system was reintroduced in March 1946, and it was, at least initially, quite favourable to the workers). The workers' mobilization and strikes were directed against the high cost of living and the poor supply of basic necessities. The CI at Fiat took part in the running of company health and recreational programmes, even after Valletta's return to the helm of the company at the beginning of 1946.

The disorganization in production that lasted for about five years, from 1943 to 1947, led to the consolidation of workers' practices; at the same time the new unitary trade union demanded that workers should take part in the economic reconstruction and fed hopes for profound social change. In the judgement of Fiat management the situation within the plants had become one of lax discipline and little sense of duty; abuses ranged from special hours agreed to on the basis of individual needs, to workers arriving late or rushing off early to the locker rooms at the end of the shift. Most of all, though, management complained about work rates that enabled many workers to reach the standard output well before the end of their shifts, and then to hang around in the workshops. Management had little tolerance for the arrogance of the militant trade unionists who moved freely around the factories spreading propaganda and seeking political recruits. The workers, however, defended the flexibility over standard times, as it represented a more human concept of work that ought to be upheld against any attempts to take work rates back to the fascist period, when a concept of work as exploitation prevailed.[11]

More generally, the workers wanted to defend a way of life in the factory, which had become a social space open to self-organization by the workers, a space that accommodated veterans' and partisans' associations, shops, bicycle repair centres, and so on.

The growth of production was recognized as a primary target by the workers' unions even after the parties of the left had been excluded from the government (1947). The most active militants were skilled workers whose concept of work was that of the professional worker. Their belief in productivity was founded on an idea of work as a moral and social duty, on a professional pride, and on an attitude that was basically in favour of technological progress. But the industrial relations model that the mainly socialist and communist trade unions were striving to achieve could not but be in sharp contrast to the Taylorist and Fordist model. In the former the workers' organization guaranteed the self-discipline of the workers and assumed powers of control and decision making, pressing for a production management that would not jeopardize worker skills and work rules. The latter envisaged a factory run by a centralized authority acting through a hierarchical chain of command in which everyone had to stay in his own place, obey, and perform a subordinate function directed toward definite management objectives.

With the implementation of the modernization programmes financed by the Marshall Plan, Fiat intended to impose a great and decidedly rapid change in the control over workers and work rules. The backward steps taken by scientific management in Italy during World War II had helped to make the industrial conflict more acute at the very time Fiat was trying to impose, and was in fact succeeding in imposing, the only rationalization model that it knew and recognized as suitable, a Fordist model based on the complete control by management over workers and work flow.

The socialist and communist trade union (FIOM-CGIL) lost its absolute majority in representation of the workers in the 1955 CI elections. The defeat came at the end of a slow erosion of worker power. The counter-offensive by the company management had begun in 1948, after the Christian Democrat electoral victory over the Popular Front and the start of internal reconstruction with the aid of the ERP. There were two methods for imposing the rationalization: a reduction in the freedom of action of CI members and of militant politicians and trade unionists; and a return to employer management of the company's part of employees' wages (the extra-minimum contractual wages that the company used to pay under union control from 1945 to 1948). Management's victory on both fronts was facilitated by a split within the unions (1948), and also by the centralized negotiating system, inherited from the fascist period and taken over by the socialist-communist CGIL.

Just like the acquired rights put forward by the members of the CI (a certain number of hours off for union activities, the right to move around the workshops, etc.), the wage conditions and norms at Fiat (in terms of basic pay, production bonuses, and piece-work systems) were more favourable to the workers than those stipulated by inter-confederal agreements and labour contracts at a national level. The CI therefore found that there were no legal pretexts for their claims: internal negotiation was based on largesse on the part of the company rather than on contractual rights. And in actual fact it was Fiat that asked, little by little as the pendulum swung in its favour, for rules and conditions to be nearer to national labour contracts, or left to the discretion of the company, as the case might be. Wage expenditures not set down in contracts were gradually removed from the control of the CI and targeted at rewarding discipline and individual merit.

The conflict, which took on overtly political tones, saw Fiat use a range of ploys: output-linked wage incentives, company welfare programmes, and the firing of hundreds of union activists. The company's main objective was to re-establish internal discipline. In the eyes of management, restoration of the traditional hierarchical authority had to precede technological modernization. It was considered the necessary starting-point for the innovation of work cycles, without which the technology of mass production, with its machine pacing, could not have guaranteed the expected increase in output.

The defeat of the antagonistic FIOM trade unionism marked the advent of a system that was rigidly Taylorist and bureaucratic, in which there was place for only a single authority in the company and no room for mediation. Even the collaborationist company union (SIDA), which had helped in the defeat of the FIOM, was given by and large a peripheral position. At the same time, the CI were deprived of their role; they no longer had to mediate a conflict of interests, in that conflict was not supposed to exist any longer. Despite their moderate majority, the CI were considered an obstacle or potential danger.

Although order was thus re-established, once again it was done without achieving consensus. Fiat management's victory was won at the cost of an internal laceration that was not healed—to the extent that when the workers' struggle restarted in the late 1960s, even though it was in a radically changed historical situation, it was hailed as revenge for the defeat suffered in the fifties by those workers who, inside or outside the company, had remained militant.[12]

The question had often been raised as to why Fiat did not try to use the workers' ideas of productivity and professionality within a different work organization in order to encourage their active cooperation in improving output, by the creation of a participatory system of industrial relations that would temper their opposition.[13] Just as often, the centralism and politicization of the Italian trade union movement have been remembered, along with the movement's dependence on political parties, as well as the rigidness of employer associations and their reluctance to accept any kind of worker participation. Fiat was dominated by an organization that believed in the principles of Taylorism and Fordism, which held that making workers share in the responsibilities would have no significant impact on the results of a production process that hinged on bureaucratic control and engineering rationalization. In those years of great economic growth and the labour movement's defeat, management attention was focused on output and its own ability to seize the opportunities offered by the market. Worker motivation was not seen as a problem.[14]

The setting up of a strictly hierarchial and very centralized management system was accompanied by increased rationalization and a subdivision of workers' jobs that was more intense than in other European—especially British and West German—automobile companies. With the political firings of the fifties Fiat lost many skilled workers, yet the company coped with the expansion in demand and production that began by the mid-1950s by taking on massive numbers of unskilled workers from southern Italy.

Turin, along with other big industrial centres in northern Italy, received a large number of immigrants in the fifties and sixties. Resident population in the city increased by 65 per cent between 1951 and 1971, leaving many social problems unsolved, especially housing and the lack of services. By the late 1960s the situation was explosive, and it no doubt was one of the causes behind the outbreak of the workers' conflict. In 1968 a twelve-year cycle of agitation started in Italy, and Fiat was especially hard hit. The bitterness of the struggle and its duration when compared to other parts of Europe can be traced back to the fact that the workers doing the least skilled jobs and the monotonous and repetitive work on the assembly line were not *Gastarbeiter*, as in other European countries, but Italian citizens.

10.5. Implementation and Modification of the Ford System

In the interwar period, roughly 80 per cent of Italian passenger cars were produced by Fiat. In the 1920s, the majority were exported (the most important foreign market being Great Britain). Fiat cars were either medium or large in size, and Fiat's major internal competitors—Lancia and Alfa—produced even larger sized cars or sports cars. The domestic market was limited, the automobile being a luxury product for well-to-do people. Fiat introduced hire-purchase in 1926 (through Sava, an automobile sales joint-stock company), and in 1936 launched the Topolino (Mickey Mouse), the very first Italian small car (80,000 were produced in the five years up to and inclusively 1940). Given its relatively high price, however, the Topolino model was destined for the lower middle classes, not for the working classes.

Despite these attempts to broaden internal demand, in Italy there was only one automobile for every ninety-nine people, compared to one car for every four persons in the United States. In the 1930s, the fascist policy of national self-sufficiency (autarchy) hindered exports (see Table 10.3). At the same time, the slow economic recovery after the Great Depression did not permit the domestic market of consumer goods to grow; recovery was fostered only by state expenditures for armaments. Market preconditions for the mass production of cars did not exist before the late 1950s.

The implementation and selective adaptation of Fordist methods at Fiat in the interwar period can be traced back to the attempt to keep up with competition in foreign markets, which became stronger with the growing protectionist tendencies in the 1920s and 1930s. The Italian car industry was basically an export-led trade. At the same time, especially after the Great Depression, the limited domestic market was completely closed to foreign manufacturers (see Table 10.3). The Ford Motor Company tried to produce in Italy through a joint-venture with Isotta Fraschini, but every attempt was stopped by Mussolini in 1930.[15] Thus the Italian car industry did not have any subsidiary abroad. Strong protection persisted until the early 1960s, so that the growing internal demand in the 1950s was reserved for the national industry.

Quality, variety, and design were emphasized up to World War II, although some attention was given to the possible expansion of internal demand for small cars by the late 1930s. The selective adaptation of Fordist methods between the wars was primarily linked to competition in foreign markets; yet a certain role was played by technology-oriented managers who associated competitiveness with the capability to catch up with the most advanced techniques, and who wished, moreover, to enhance control over workers and work rules.

The decision to build the new Lingotto plant was taken during World War I, when expansion in production highlighted serious inefficiencies in

TABLE 10.3. *Number of cars and trucks produced by Fiat, total national production in Italy, and percentage of exported automobiles, 1920–61*

	Fiat	Total national production	Percentage of exports
1920	14,314	21,100	53.6
1921	10,326	15,200	68.5
1922	10,466	16,390	69.4
1923	13,629	22,820	56.0
1924	25,113	37,450	50.6
1925	39,720	49,400	58.8
1926	25,756	63,800	53.6
1927	47,513	54,300	61.3
1928	47,765	57,600	49.1
1929	46,187	55,100	43.0
1930	35,120	46,400	44.7
1931	18,169	28,400	42.0
1932	22,012	29,600	22.2
1933	32,406	41,700	17.9
1934	36,527	45,402	21.0
1935	37,355	50,493	29.5
1936	41,071	53,144	38.5
1937	64,157	77,708	43.3
1938	56,053	70,777	28.6
1939	55,821	68,834	37.2
1940	34,194	47,856	31.9
1941	26,347	38,798	19.3
1942	21,850	30,407	15.5
1943	15,686	21,134	11.1
1944	7,911	13,781	10.6
1945	6,956	10,290	—
1946	5,792	28,983	9.9
1947	40,144	43,736	25.0
1948	53,340	59,953	25.3
1949	79,800	86,054	20.4
1950	120,079	127,847	17.1
1951	132,153	145,553	22.2
1952	124,612	138,446	19.1
1953	161,502	174,320	18.1
1954	192,794	216,715	20.4
1955	250,072	268,766	27.8
1956	293,564	315,802	27.6
1957	319,646	351,815	33.9
1958	359,240	403,560	41.9
1959	443,968	500,784	44.2
1960	542,300	644,633	31.6
1961	623,178	759,140	32.3

Source: Fiat in cifre (Turin, 1974); ANFIA, *Automobile in cifre* (Turin, 1962).

the old Fiat Centro plant. In the latter, the old-style layout of machinery by the type of machine (functional layout) made the transporting of material and work in progress very costly. At Lingotto the layout of machinery was designed according to process in production flow (linear layout). The plant was subdivided into units (departments) corresponding to sub-assemblies of component parts (motor, rear axle, front axle, transmission, and so on). Each department was equipped with the necessary machine tools, benches, assembly line, and testing rooms, so that it could furnish its component of the car (diversified for the various models) ready for further assembly (non-specialized linear layout).

The general movement of the work in process was upward, from the ground floor to the testing track on the roof. On the top floor were two assembly lines for engines (one for four-cylinder motors, one for six-cylinder motors), as well as two corresponding final assembly lines (one for cars equipped with four-cylinder engines, and one for cars equipped with six-cylinder engines).[16]

The production system at Lingotto was studied in order to adapt Fordist methods to lot production of different component parts for a wide range of car models (in the years between 1928 and 1938, Fiat had 42 models of car—15 of which were basic models—and 22 models of truck).[17]

Internal and external demand was neither large nor standardized enough to allow economies of scale. Single-purpose machines found little utilization, because of the reduced production volumes; the prevailing kind of machine tools were semi-special, or multi-purpose machines, which were mainly made in America. These flexible machines were set up for the production of a certain amount of identical pieces, and then again adapted (with jigs, fixtures, and other attachments) to manufacture a new lot of different component parts. A limited number of skilled workers adjusted the machines, then it was the task of semi-skilled machine tenders to manufacture the lot of pieces.

The introduction of the assembly line was accompanied by a period of intense time and motion study, in order to improve the coordination of production flow and to speed up work rates. Through new standard times for machine operations and new piece rates Fiat aimed to enhance productivity. In its American counterparts productivity was four times higher, while the technological gap was limited more to quantitative than to qualitative aspects.

The Ford Plant in Highland Park, which had been the reference model for Lingotto, produced 3,000 cars a day, compared to the latter's 200 a day. Differences in size of output and in productivity were due, at least partially, to the fact that Highland Park was mainly an assembly plant, unlike Lingotto, which was a factory producing most of the component parts. This difference also explains Fiat's decision to maintain the incentive of piece-work—which was one of the most important modifications of the Ford system. Given the prevailing kind of machine tools, which required frequent set-up and adjustment, it was not easy for Fiat to impose fixed work rates.

Giovanni Agnelli himself was President of the Italian Bedaux Society, founded in 1927, which was one of the first important Italian consulting companies in production management. The Bedaux system was intended not merely as piece-work, but as a means of rationalizing and synchronizing production processes through a general analysis of machine and workers' operations. The Bedaux system was first introduced, and experimented with, at the Officine di Villar Perosa, a Fiat-owned ball-bearing factory located on the outskirts of Turin. It was a technologically advanced plant, operating on a high-production basis and utilizing special automatic machinery (about 70% of its output was exported to America). After a couple of years, starting in 1929, the Bedaux system was extended to the automobile plants in Turin.[18]

Another feature of Fordism that Fiat tried to imitate, and succeeded in imitating, was the strategy of vertical integration. Since Fiat had become a stock company in 1906, it absorbed several other shops. It owned plants for trucks, buses, tractors, aeroplanes, and railway engines and carriages; it had its electricity generating station. Mining properties, along with steel and iron foundries and mills, allowed the company to control its working materials from mine to finished product. As two American engineers put it (J. A. Lucas and F. E. Bardrof, in a series of articles published in *American Machinist* in 1924), 'the Ford plan has a counterpart in Italy'.[19]

Fiat had little recourse to subcontracting. But the foundation of the Italian Bedaux Society provides evidence that it was the intention of the company to enter, with its consultants and technicians, into other minor shops, not only to sell a management consultancy service, but also to establish new ties and study the opportunities for productive collaboration, in relation to subcontracting. In-house production, however, remained predominant until the 1970s, when Fiat's recourse to subcontractors increased because of the lower labour cost in small non-unionized firms.

The Bedaux system measured the performance of each worker (or group of workers) in points (corresponding to a minute of basic work); this way, according to the experts of the Bedaux Society, comparisons of productivity of the various sub-cycles of the production process were made easier. The basic work rate was 60 Bedaux points (Bx), and this corresponded to the contractual hourly pay. If workers produced more points, they received a proportional bonus (reduced by a 25% quota that was paid to the foremen; for example, 75 Bx represented a 20% increase in production, with 15% of hourly pay going to the worker and 5% to the foreman; from 1934 onwards the quota for the foreman was abolished and paid to the worker).

In the machine shops, the Bedaux system worked as genuine piece-work for workers, who in some areas were skilled turners adjusting the machines themselves, and in other areas were low-skilled machine oper-ators, for whom the set-up of machines was carried out by skilled workers paid on an hourly basis. In the latter areas, the Bedaux system aimed at having each worker operate more than one machine (two or three), be-cause of the distinction between the operating time of a machine and the working time of its operator.

In the case of the low-skilled or unskilled assembly-line workers, the Bedaux system worked as a sort of fixed bonus system, very similar to the Fordist practice. Its role in the assembly line was to indicate a standard performance (a sort of target time) for individual operations or segments of the line. After a period of experimenting and training, the standard performance was made to correspond to 80 Bx, that is, to a 33 per cent increase beyond the basic work rate. There were buffer stocks between segments of the line. If a group of workers on a given segment kept up with the standard pace, they received a 33 per cent bonus. If they did not, the foreman had to analyse the causes. If the causes lay in technical difficulties (faulty pieces, delivery of materials and tools, breakdowns), the workers received the bonus anyway (for idle times due to longer breakdowns, workers received only the basic hourly pay). In contrast, if it was the workers' fault, they only received the bonus corresponding to the Bx points reached. In the case of individual incompetent workers, they were dismissed or transferred to other tasks.

The pace of work flow at Lingotto was quite slow in comparison to American standards, not only because of smaller production volumes and more frequent set-up of machines, but also because the subdivision of worker tasks continued to be less. The shortest cycle time (for the smallest models) on the tact moving mixed assembly line was 3.5 minutes (com-pared to 26 seconds on the specialized line for the Fiat 500 transmission by the late 1960s).[20]

In the 1930s at Fiat the assembly line was in its early stages, yet it was not as primitive as at Lancia, the only other Italian car manufacturer to introduce the assembly line; at Lancia, by the end of that decade, there was a 30-minute tact line (this was a very slow pace, maybe also because of the quality requirements of a high-class product).

By the mid-1930s Fiat decided to build a new factory in the south of Turin, in Mirafiori. The new plant was opened in 1939, but its assembly lines only really came into full production at the end of the 1940s. The reference model was the Ford River Rouge plant. The aim was to over-come the obstacles to flow of materials due to the layout on several floors used at Lingotto. Mirafiori was designed for mass production. Fiat management intended to broaden the large-scale output of small cars, extending the experience of the Topolino model. The war and postwar troubles delayed the project until the mid-1950s.

The Fordist mass production methods, however, were implemented only in the manufacturing of the smallest models (2,695,197 Fiat 600 were produced from 1955 to 1970, in all their various versions, and with periodical technical improvements; 3,612,928 Fiat 500 were produced from 1957 to 1975). These models had their own assembly lines, and their component parts were increasingly produced by low-skilled workers operating special-purpose machines and transfer machines (first introduced in 1955),[21] in dedicated machine shops that, unlike Lingotto's more flexible layout, produced only the component for one model (specialized linear layout). On the other lines, several models of medium- and large-size cars continued to be produced in small lots, using semi-special machines. Between 1949 and 1961 Fiat's full-line policy had 65 models of car (19 of which were basic models). The mass production of medium-size cars began in the late 1960s, and throughout the 1970s some assembly lines (at Mirafiori as well as at Lingotto and in the new Rivalta plant) still continued to manufacture different models, for niche production of vans, large cars, and sports cars. In the big Mirafiori plant, which had more than 50,000 employees in 1970, mass production and niche production coexisted, as well as large-volume and small-lot production.

The confrontational model of industrial relations rooted in the Italian labour movement deeply affected the reorganization of production management techniques, starting from the 1950s. Taylorist and Fordist methods were intended to ensure the highest level of control by management over work and workers. Workers' tasks were subdivided as far as possible, in order to simplify actions and to make the production flow independent of workers' manual skills or practical technical knowledge. The aim of management was to reduce to the lowest possible level the important role skilled workers had played in the 1930s and 1940s, when the productivity and efficiency of the shops largely depended upon them. Most skilled workers had always been, as it were, militant socialists and communists, as well as trade union members and activists. The company aimed to get rid of their influence within the production process. In the early 1950s, over 2,000 skilled Fiat workers were fired on political grounds. Unskilled workers, mainly new immigrants coming from rural areas, were less influenced by the labour movement's urban tradition, and were therefore more responsive to Fiat's social policy (company benefits, etc.).

A strictly hierarchical management system had already been set up in the 1920s at Lingotto. Discipline throughout the plant was similar to that in a military organization. Employees of one department were not allowed to enter other departments. All persons had to wear insignia, indicating their rank and working place. There were four different insignia, for higher executives, department superintendents, foremen, and workers. Orders were short but definite, and lesser executives and workmen were expected to implement them without comment.

In the 1950s and 1960s, Fiat top managers dreamt of a factory running like clockwork, and acted with a sort of mechanical rationality that stressed the synchronization of simple, standardized tasks performed by unskilled and interchangeable workers. They did not even think of a devolution of responsibilities to middle management, or of any sort of autonomous worker participation. Fordist technologies and a rigid discipline were seen as responses to traditional difficulties in labour relations.

Pay levels were linked to productivity growth. A piece-work system was reintroduced in 1946; although it was no longer called Bedaux, it was substantially similar. Worker performance was still measured in points, the basic level being 100 instead of 60. The standard performance for assembly lines and for work at special automatic machines was fixed at 133 points (again, a 33% increase). From 1949 onwards, a collective bonus system, linked to the whole plant output, was introduced in addition to the piece-work system.[22] As a new wave of time and motion studies was started by the early 1950s, bones of contention arose about fixing basic levels of work rates, and, whenever the factory did not run like clockwork, it was the job of the foremen to raise line speeds to recover standard output after stoppages due to technical problems.

For a long period after the defeat of industrial unionism by the mid-1950s, Fiat workers were unable to protest against this system and against the poor quality of working conditions. But worker discontent was widespread, and little by little it spread to the new unskilled workers. When the political situation and the labour market conditions changed in the late 1960s, the company did not succeed in stopping the new wave of conflicts, not even through the tardy implementation of a human relations system (in the wake of Elton Mayo), and new personnel management techniques.

10.6. Developments in the 1970s and 1980s

Radical rationalization processes in the fifties and sixties gave rise to a continuing deterioration in working conditions. The percentage of skilled workers decreased dramatically. Apprenticeships disappeared, and vocational training played a weak role. Career opportunities for blue-collar workers within the company were poor. Since the degree of mobility from shop floor to office was very low, in the second half of the sixties upgrading had in itself little significance for workers.[23] Company benefits were aimed at encouraging the workers to identify with the company and with job stability (but without any promise of lifelong employment). In the management's eyes, this was to replace the call for a higher quality of work in terms of skill required and job satisfaction.

But this attempt to obtain a sort of passive consent from the workers failed, as became clear by the end of the 1960s, when the company welfare system could no longer cope with the dramatic increase in the number of employees (in 1967, with the opening of the new Rivalta plant, 20,000 worker were hired). The social tension behind a first strike in 1962 had been underestimated, so when a new wave of strikes began in 1968, management was surprised by the anti-company attitudes the strikers expressed. The harshness of working conditions, along with the unsolved social problems affecting the newly immigrant workers, triggered off a sort of anti-industrial attitude. Both new unskilled and senior workers joined in protesting against the organization of work and in refusing to do work with a low professional content; this was one of the main character-istics of the 1968–80 labour struggles.

In June 1969 an agreement with the unions marked the first beginnings of a new system of representation based on shop stewards and works councils. Fiat, which had let the CI be strangled to death because it did not want an increase in the number of its members,[24] thus ended up in the space of a couple of years having to conduct negotiations with a small group of 12 to 15 representatives on the workers' side, but with the presence in the same building of between 250 and 300 shop stewards who closely followed and controlled the progress of the discussions.

The union strategy was to pose the problem of work organization in terms of improvement in the quality of jobs and in professional skill, which should then lead to promotions in the classification scale.[25]

The catch phrase 'The new way to make an automobile' was launched; it tried to achieve the introduction of new forms of job enlargement, enrichment, and rotation. But as the assembly lines worked at very fast rates (with cycle times of about one minute), it was difficult to modify the distribution of jobs. Some experiments were introduced on new lines, where islands were set up on a linear line, in order to allow job rotation, but the results were not very good from the point of view of productivity; generally speaking, the technological conditions did not allow significant change.

The impending crisis of the Ford system in the 1970s was linked to the inability of its inflexible production methods to respond to the new fea-tures of markets; but it was aggravated by the workers' struggles.

Starting from the late 1960s, union demands aimed at obtaining de-tailed contractual regulations in order to limit management power over work and workers. A rigid Taylorist organization was offset by equally rigid rules to safeguard workers and prevent a free-wheeling use of the

TABLE 10.4. *Number of cars and trucks produced by Fiat,*
1962–73

	Cars	Trucks
1962	768,540	48,813
1963	941,890	69,843
1964	913,690	42,478
1965	994,300	35,163
1966	1,151,900	40,990
1967	1,279,380	50,070
1968	1,346,000	67,328
1969	1,249,137	60,186
1970	1,391,674	60,337
1971	1,438,000	61,109
1972	1,489,631	64,057
1973	1,572,964	68,500

Source: Fiat in cifre (Turin, 1974).

workforce: restrictions on overtime, on internal transfers, on shifts, and so on.

Since the production system was not flexible, the only possibility of responding rapidly to the demands of the market was by transferring manpower. The need for internal mobility was opposed by all the rules for the protection of the worker. As time passed, this question became one of the bitterest bones of contention, because the inflexibility of the labour force persisted, at a time when the market was becoming more volatile and segmented and required ability to respond more promptly. Since worker expectations of an improvement in working conditions were not being realized, worker apprehensions contributed to keeping the attachment to the protective rules alive. When the 1980 crisis on European markets clearly illustrated the company's serious loss of competitiveness, which was primarily ascribed to workers' indiscipline, Fiat decided to take decisive action in order to restore the authority of middle-ranking management.[26]

After the trade unions suffered a major defeat in 1980, which marked the end of a twelve-year long wave of struggles, Fiat tried to recover productivity and competitiveness through a capital-intensive form of modernization.

The logic that underlay the technological innovation in the 1970s was a growth of output linked to an improvement in working conditions to ease social tensions; in the 1980s this logic changed to one of trying to obtain the greatest possible reduction in manpower. In 1979 the number of cars produced per worker was 14; in 1986 it had doubled, rising to 29; in the same years the number of workers was reduced from 115,000 to 60,000. The reorganization took place in two directions. First, a different policy of make or buy was introduced, with greater use of subcontracting, along with a process of reshaping relationships with the suppliers. Secondly, the company invested in product innovation, and the renewal of plants. The factories at Termoli, for engines, and Cassino, for car bodies, were the most technologically advanced, using avant-garde automation.

The most important innovation was perhaps the simplification of the product. The Fiat car in the 1980s was no longer a linear assembly of single pieces; a range of subsystems of complex components (engine, brakes, gears, drives, and dampers, every one of which constituted a sort of finished product) was now assembled with interchangeable pieces and special pieces for the different models. The result was a modular fan of production that simultaneously allowed both economies of scale and flexibility to meet the changing demands of the market, and a more rapid innovation of the product. In 1982 Fiat had 12 basic models with 42 common components and 52 specific to the models; in 1986 the basic models were reduced to 9 and were assembled with 49 common and 13 specific parts. The total number of variations added up to 80.[27]

Some observers defined this system as neo-Fordism, in that it pursued economies of scale and still coped with the need for flexibility by large stocks; from the point of view of the production management system, it was also considered a computer-aided neo-Taylorism, since workers' tasks and the division of labour did not substantially change. Other scholars called it flexible integration (a third approach between neo-Fordism and flexible specialization)[28] in that it fostered flexibility at the general level of the company, where a central coordination integrated the various plants and shops, some of which had a flexible fan of production, others a standardized one.[29]

By the late 1980s, however, the double challenge of flexibility and quality made Fiat aware of the necessity to shift from a technocentric to an anthropocentric production system. Difficulties and shortcomings experienced in the highly automated and yet traditionally managed production systems, along with the influence of the Japanese model, led the company, by the end of the 1980s, to plan replacing the hierarchical-functional organization by the integration of functions at all levels, and especially between staff and line. The hierarchical chain has been made shorter (at present it has gone from seven to five levels) in order to smooth the decision-making process and to solve the problems where they arise. The layout of machinery has been made more flexible, according to the principles of cellular manufacturing (or group technology).[30] Fiat is now asking its workers for cooperation and shifting responsibilities from higher up to lower down.[31]

This process offers the opportunity for changing the confrontational model of industrial relations, as long as the trade unions agree with the new lean production or just-in-time techniques.

Indecision and uncertainty, however, are still alive. They are reinforced by the contrasting signals that come from Fiat top managers, as well as from the trade unions. The uncertainty on both sides is more understandable if it is remembered that a strongly confrontational system of industrial relations has lasted for nearly a century; during this time there have been alternating periods of victory and defeat, by one side or the other, with little or no development of a culture of participation on either side. The trade unions have only recently, and with difficulty, overcome the antagonistic position they held in the 1960s and 1970s, and it was only after 1985 that they accepted negotiations on the flexibility of the workforce.

The suspicious attitudes of the worker organizations towards technological and organizational changes are reinforced by previous experiences of the short-run effects the changes had on working conditions. The major waves of change in the history of Fiat (during World War I, in the 1930s, in the mid-1950s, and in the early 1980s) have all occurred when the worker movement has been in a period of defeat, and with legal or *de facto* suppression of workers' rights. Therefore, there has been no experience of cooperation and bargaining between company and unions over the direction and effects of technological change.

On the other hand, the new production methods are now only in their early stages, and there is no shortage of obstacles to their implementation. Therefore it is very difficult to foresee what the consequences of the new anthropocentric production systems will be in shaping new patterns of labour relations.

10.7. Conclusions

In Italy, as in Japan, annual production of cars and trucks reached one-million units in the early 1960s, some six or eight years later than in the United Kingdom, West Germany, and France. Yet in Italy more than 80 per cent of the total national production was Fiat-made. Starting from 1955, Fiat's average annual production of the smallest models (Fiat 600 and Fiat 500) was more than 150,000 units (of each model). Therefore, the volume appropriate to the efficient use of mass production techniques was reached by Fiat no later than its major European counterparts. It can be said that at Fiat the Ford system was fully applied by the late 1950s, in terms of specialized linear production, subdivision of tasks, unskilled labour, and direct shop-floor control, even if the scale of production remained decidedly lower in comparison to American standards, and even though Fiat's product policy was more similar to a softened Sloanism (a wide range of models, without frequent model changes but with small progressive improvements), Moreover, to produce different models in different volumes for different market segments, different methods were adopted in different shops or plants, as would be expected.

In the interwar period, Fiat's adoption of Fordism was selective. Given the characteristics of its domestic and foreign markets, production volumes could not be high enough to fully exploit the benefits of mass production. The aspects of the Ford system that were adopted included the layout of machinery according to flow in the production process, a moving assembly line, and interchangeable parts. Yet the standardization of products, parts, worker tasks, and work rates still remained limited. In particular, most machines were multi-purpose and semi-special, thus allowing flexibility in production.

Given the low degree of machinery specialization, the rationalization process was carried out by implementing Taylorist methods aimed at organizing men. Through the methods and motions study, the Bedaux system (which was also introduced by Morris, Rover, and Citroen)[32] established standard times for the various jobs; these times gave rise to a stricter piece-work system and were used to balance the assembly lines, to make comparisons of productivity in the various shops, and to estimate costs, thus forming the basis of a budgetary control system.[33]

Like the British and French manufacturers,[34] Fiat did not introduce a standard day rate. The decision to keep up piece rates stemmed from the kind of machine that was most commonly used. A day rate was found to be unsuitable when workers were operating semi-special and flexible machines, which required frequent set-up and adjustment, unless labour relations were fully cooperative. Although under fascism workers would have been unable to resist direct shop-floor control, production methods linked to market conditions held Fiat back from immediately implementing a complete control system. Little by little, as more standardized tasks were replacing those that required a greater degree of worker skill, the payment-by-results system was gradually transformed into a fixed bonus system that was not substantially different from a day rate; this process accelerated in the 1950s and was completed in the 1960s. Until piece-work was formally abolished in the early 1970s, both Fiat managers and workers continued to call the existing pay system 'piece-work', but this was no longer genuine piece-work.

Fiat tried to soften craft demarcations by introducing a very gradual classification scale, with many intermediate basic pay levels among categories of workers, and used individual upgrading to reward discipline and commitment rather than professional skill. The main unions themselves had shifted from a craft-based to an industry-based operation many years before. Nevertheless, in the interwar period, given the existing work organization, the distinction between skilled and semi-skilled and un-skilled labour remained quite clear-cut, in that skilled workers, unlike the others, still played a central role in production and maintained autonomy and decision-making authority.

By introducing mass production methods as soon as possible, Fiat aimed to reduce the importance of skilled workers, most of whom were union members and activists. In its post World War II troubles, the company decided to fight against, rather than compromise with, the workers, and eventually managed to defeat the leftist militant unionism and to reduce the number and role of skilled workers.

Thus the Italian experience after World War II was quite different from both British traditional methods and the West German co-determination system.[35] Since Fiat's Taylorist-style division of work did not gain worker consent and involvement, labour relations did not improve, and the confrontational industrial relations system in Italy did not undergo any process of modification. In the long run, neither the Italian precocious adoption of direct shop-floor control nor the British tardy attempt to introduce Fordist methods eased worker unrest. The wave of strikes in the 1970s provides evidence that Fordism in Italy, and perhaps in Europe, ultimately failed to conciliate workers with mass-production unskilled jobs by promising workers a share in mass consumption.

NOTES

1. Giovanni Agnelli, the founder of Fiat, visited the Ford Highland Park plant in 1912. See V. Castronovo, *Giovanni Agnelli: La Fiat dal 1899 al 1945* (Turin, 1977), 47. Taylor's *Shop Management* and *The Principles of Scientific Management* were translated into Italian in 1915.
2. 'Lavoro a cottimo', *Bollettino della Lega Industriale*, July 1914, 63.
3. D. Bigazzi, 'Gli operai della catena di montaggio: la Fiat 1922–43' [Assembly-line workers at Fiat], in *La classe operaia durante il fascismo* [Working class under fascism], ed, G. Sapelli, *Annali Fondazione Giangiacomo Feltrinelli* (1979–80); id., 'Management Strategies in the Italian Car Industry 1906–1945: Fiat and Alfa Romeo', in *The Automobile Industry and its Workers: Between Fordism and Flexibility*, eds. S. Tolliday and J. Zeitlin (Cambridge, 1986). A general survey on rationalization processes in Italy in the interwar years is available in G. Sapelli. *Organizzazione, lavoro e innovazione industriale nell' Italia tra le due guerre* [Work organization and industrial innovation in Italy in the interwar period] (Turin, 1978).
4. S. Musso, *La gestione della forza lavoro sotto il fascismo* [Workers and fascism] (Milan, 1987).

5. S. Musso, 'Il cottimo come razionalizzazione' [Piecework and rationalization], in *Torino fra liberalismo e fascismo* [Turin between liberalism and fascism], eds. U. Levra and N. Tranfaglia (Milan, 1987), 233; S. Schweitzer, *Des engrenages la chaine: Les usines Citroen 1916–35* (Paris, 1982).

6. On the juridical aspects see G. C. Jocteau, *La magistratura e i conflitti del lavoro durante il fascismo 1926–34* [Work conflicts and magistrature] (Milan, 1975).

7. On collective bargaining under fascism see S. Musso, 'Norme contrattuali e soggetti delle relazioni industriali dalla fine degli anni trenta alla caduta del fascismo' [Industrial relations in the fascist period], *Movimento operaio e socialista*, 1–2 (1990).

8. D. Bigazzi, 'Organizzazione del lavoro e razionalizzazione nella crisi del fascismo 1942–43' [Rationalization in the crisis of the fascist regime], *Studi Storici*, 2 (1978).

9. S. Musso, 'Produzione bellica e problemi di organizzazione del lavoro (1938–1948' [War production and rationalization], *Storia in Lombardia*, 1–2 (1993).

10. For a general history of Fiat see Castronovo, *Giovanni Agnelli*; P. Bairati, *Vittorio valletta* (Turin, 1983).

11. Cited in Progetto Archivio Storico Fiat, *1944–1956. Le relazioni industriali alla Fiat* [Industrial relations at Fiat] (3 vols.; Milan, 1992), I, 541.

12. On workers' struggles in the postwar years and in the 1950s see L. Lanzardo, *Classe operaia e partito comunista alla Fiat: La strategia della collaborazione 1945–9* [Workers and the communist party at Fiat] (Turin, 1971); A. Accornero, *Gli anni '50 in fabbrica: Con un diario di commissione interna* [Internal commissions in the 1950s] (Bari, 1973); G. Della Rocca, 'L'offensiva politica degli imprenditori nelle fabbriche' [Employers' offensive against militant workers], in *Annali Fondazione Giangiacomo Feltrinelli* (1974–5); E. Pugno and S. Garavini, *Gli anni duri alla Fiat. La resistenza sindacale e la ripresa* [The hard years at Fiat: Trade union resistance and the revival of struggles] (Turin, 1974); R. Gianotti, *Lotte e organizzazione di classe alla Fiat (1948–70)* [Workers' struggles, trade unions and politics at Fiat] (Bari, 1970); G. Petrillo, *La capitale del miracolo: Sviluppo lavoro potere a Milano 1953–62* [Economic growth, work and power in Milan] (Milan, 1992).

13. G. Sapelli, 'La cultura della produzione: "autorità tecnica" e "autonomia morale", in *Le culture del lavoro: L'esperienza di Torino nel quadro europeo* [Workers' and employers' attitudes towards work and productivity], eds. B. Bottiglieri and P. Ceri (Bologna, 1987); G. Berta, 'Le Commissioni interne nella storia delle relazioni industriali alla Fiat' [CI and the history of industrial relations at Fiat], in Progetto Archivio Storico Fiat, *1944–56: Le relazioni industriali alla Fiat*, III, 26–7; G. Bonazzi, *Il tubo di cristallo: Modello giapponese e Fabbrica Integrata alla Fiat Auto* [The Japanese model and the integrated factory at Fiat] (Bologna, 1993), 66–7.

14. Berta, 'Le Commissioni interne', 33.

15. Castronovo, *Giovanni Agnelli*, 339–45.

16. Bigazzi, 'Gli operai della catena di montaggio'.

17. *Fiat in cifre* (Turin, 1974).

18. Musso, 'Il cottimo come razionalizzazione'.

19. J. A. Lucas and F. E. Bardrof, 'The Impressive Story of Fiat', *American Machin-*

ist, 24 Apr., 1 May, 22 May, 19 June, 3 July, 10 July, 24 July, 7 Aug., 21 Aug. 1924.

20. F. Levi, 'Torino', in F. Levi, P. Rugafiori, and S. Vento, *Il triangolo industriale tra ricostruzione e lotta di classe* [Social unrest in Milan, Turin, and Genoa in the aftermath of World War II] (Milan, 1974), 244; I. Oddone, A. Re, and G. Briante (eds.), *Esperienza operaia, coscienza di classe e psicologia del lavoro* [Worker experiences and labour process] (Turin, 1977).

21. O. M. Sassi, 'Considerazioni sul progresso tecnologico alla Fiat nella produzione automobilistica' [Technological innovation at Fiat], in *Il progresso tecnologico e la società italiana* (3 vols.; Milan, 1960–2), vol. II, 1961; S. Leonardi, *Le macchine utensili e la loro industria* [The machine-tools industry] (Milan, 1961); F. Amatori, 'Impresa e mercato: Lancia 1906–69' [Lancia: The company and the market], in *Storia della Lancia: Impresa tecnologia mercati, 1906–69* (Milan, 1992).

22. Progetto Archivio Storico Fiat, *1944–6: Le relazioni industrali alla Fiat*.

23. M. Paci, *Mercato del lavoro e classi sociali in Italia* [The labour market and social classes in Italy] (Bologna, 1972).

24. In 1969 the Mirafiori plant reached its peak number of employees, with 47,500 workers and 5,000 clerical staff. There were only 18 workers and 3 clerical staff members of the CI to represent this huge body of employees. See Berta, 'Le Commissioni interne', 8.

25. On workers' struggles in the 1970s, see T. Dealessandri and M. Magnabosco, *Contrattare alla Fiat: Quindici anni di relazioni industriali* [Bargaining at Fiat], ed. C. Degiacomi (Rome, 1987); G. Della Rocca, 'Italy', in *Industrial Relations in Information Society: A European survey*, ed. G. Berta (Rome, 1986); I. Regalia, M. Regini, and E. Reyneri, 'Labour Conflicts and Industrial Relations in Italy', in *The Resurgence of Class Conflict in Western Europe since 1968*, eds. C. Crouch and A. Pizzorno (2 vols.; London, 1978–9), vol. I, 1978. Specifically on Fiat, G. Contini, 'The Rise and Fall of Shop-Floor Bargaining at Fiat 1945–80', in *The Automobile Industry and Its Workers: Between Fordism and Flexibility*; M. Revelli, *Lavorare in Fiat* [Work and workers at Fiat] (Milan, 1989). On how the Fiat foremen experienced the worker struggle see the memoirs of L. Arisio, *Vita da capi: L'altra faccia di una grande fabbrica* [The life of the foremen within the factory] (Milano, 1990). On the workers' defeat in 1980 see A. Baldissera, *La svolta dei quarantamila: Dai quadri Fiat ai Cobas* [The turning point of the march of the 40,000] (Milan, 1988). On the trade unions' policy see A. Accornero, *La parabola del sindacato: Ascesa e declino di una cultura* [The rise and fall of the trade unions] (Bologna, 1992).

26. The loss of competitiveness was also due to the delay in investment in new models by the mid-seventies. Furthermore, the Italian lira's entry into the European Monetary System in 1979 meant that it was no longer possible to regain competitiveness on foreign markets by devaluation.

27. R. Locke and S. Megrelli, 'Il caso Fiat auto', in *Strategie di riaggiustamento industriale* [Strategies of industrial reorganization], eds. M. Regini and C. F. Sabel (Bologna, 1989). See also V. Comito, *La Fiat tra crisi e ristrutturazione* [Fiat between crisis and reorganization] (Rome, 1982); G. Volpato, *L'industria automobilistica* [The car industry] (Padua, 1983); A. Becchi Collidà and S. Negrelli, *La transizione nell' industria e nelle relazioni industriali: l'auto e il caso*

Fiat [Changing industrial organization and industrial relations] (Milan, 1986); P. Bianchi, *Divisione del lavoro e ristrutturazione industriale* [Industrial reorganization and the division of work] (Bologna, 1984).

28. The obvious reference is M. J. Piore and C. F. Sabel, *The Second Industrial Divide: Possibilities of Prosperity* (New York, 1984).

29. Locke and Negrelli, 'Il caso Fiat auto'.

30. Bonazzi, *Il tubo di cristallo*, 81–2. Wemmerlöw, N. Hyer, 'Research Issues in Cellular Maufacturing', *International Journal of Production Research*, 3 (1987).

31. On the implementation of lean production at Fiat see Bonazzi, *Il tubo di cristallo*; C. Cerutti and V. Rieser, *Fiat: qualità totale e fabbrica integrata* [Total quality and integrated factory] (Rome, 1991), 70–3; V. Rieser, *La fabbrica oggi: Lo strano caso del dottor Weber e di mistre Marx* [Today's factory] (Siena, 1991); A. R. Calabrò and G. Della Rocca, 'Quale gerarchia? Capi ed esperti alla ricerca del proprio ruolo nei sistemi tecnologici integrati' [Middle management and the itegrated factory], *Studi organizzativi*, 2 (1992).

32. See A. Exell, 'Morris Motors in the 1930s', *History Workshop Journal*, 6 (1978); L. Lee Downs, 'Industrial Decline, Rationalization and Equal Pay: The Bedaux Strike at Rover Automobile Company', *Social History*, 1 (1990); Schweitzer, *Des engrenages à la chaine*; B. Mottez, *Systèmes de salaire et politiques patronales* (Paris, 1966).

33. P. Zotti, 'Il sistema del salario ai punti' [The Bedaux system], *L'Organizzazione*, 2 (1931); R. Malinverni, 'Per la razionalizzazione delle nostre aziende: Il sistema del "controllo budgetarie"' [Rationalization in management], *Rivista di politica economica*, 11, (1930).

34. See W. Lewchuk, *American Technology and the British Vehicle Industry* (Cambridge, 1987); P. Fridenson, 'Automobile Workers in France and their Work, 1914–83', in *Work in France*, eds. S. L. Kaplan and C. J. Koepp (Ithaca, 1986); T. Fujimoto and J. Tidd, 'The UK and Japanese Auto Industry: Adoption and Adaptation of Fordism', paper presented at Gotenba City Conference, 1993.

35. W. Lazonick, *Comparative Advantage on the Shop Floor* (Cambridge, Mass., 1990).

11

Two Kinds of Fordism: *On the Differing Roles of the Industry in the Development of the Two German States*

WERNER ABELSHAUSER

11.1. Introduction

Germany was one of the European countries that succumbed early to the fascination of Fordist production methods. German entrepreneurs made pilgrimages in the 1920s to the USA, seeking to discover the secrets of the economic future at the holy cities of capitalism, which of course included the Ford plants at Highland Park and River Rouge. When Henry Ford's autobiography was translated into German (*Mein Leben und Werk*) and published in November 1923, it became a best seller from the very beginning. The book triggered the first of several waves of fact-finding missions to Detroit by German engineers, trade unionists, managers, and journalists. When they returned to Germany almost all of them were totally convinced of the revolutionary character of this new method of production. As early as 1925 'Fordismus' was even a popular topic for lectures at the Berlin Humboldt University.

Creating the preconditions for a new age in automobile production was a strong motive behind numerous efforts to concentrate car production capacity in Germany. After the postwar stabilization of the economy, the large German banks—above all the Deutsche Bank—toyed with the idea of consolidating the most important German car producers into a large automobile trust, able to meet the challenge of Ford; Daimler, Benz, Opel, and BMW were still only medium-sized companies. Daimler and Benz merged in 1924–6. But all further efforts in this direction were hampered by unfavourable economic circumstances, and they ended abruptly in spring 1929 when Adam Opel AG was sold to General Motors at a large profit. This effectively ended serious discussion of a German Auto Trust and intensified what was known as the creeping crisis of the German automobile industry.

The gap between the high standards involved in the debate on the technical and theoretical aspects of 'Fordism' and the appalling realities of business life, which did not allow any successful implementation of Fordism at all, could no longer be bridged. When the German automobile industry finally caught up with Fordist methods of production after the depression of the early thirties, it introduced one of Fordism's most original creations: the Volkswagen plant, with its one-car programme, the Beetle. It was not until the 1950s however, that Volkswagen could begin its international success story.

In East Germany, at the same time, there developed a further curiosity in the history of Fordist mass production. Standardization and norming of industrial production was pushed forward relentlessly, also (though not principally) in the car industry. East Germany was, however, in the main directed towards its own small domestic market, and in any case limited the concept of Fordism exclusively to its microeconomic and technical aspect.

The German case is, therefore, an especially fruitful ground for insights into the implementation and adaptation of the Fordist system, in the context of the 'Wirtschaftswunder' in the west and of the planned economy in the east. The example of Volkswagen can in addition cast light on the particular characteristics of West German Fordism, in which the traditions of the German industrial economy mingle with the laws of a new globally applicable method of production.

11.2. 'Abortive Fordism': The German Car Industry in the Interwar Period

Before the end of the Great Depression in 1933 the German car industry did not play the leading role in industrial development that its American counterpart had during the 'Golden Twenties'. Certainly, Germany did not lack the general microeconomic preconditions for the successful implementation of Fordist mass production, such as qualified labour, technical and organizational know-how, and, last but not least, a strong interest in new modes of industrial relations in German industry. However, the macroeconomic framework did not favour Fordism. German domestic reconstruction after World War I was badly shaken by hyperinflation (1923) and slowed down by the Reichsbank's policy of using high interest rates to channel foreign credits into Germany. At the same time foreign trade did not provide the growth in demand necessary for the realization of Fordist economies of scale and scope. There were other difficulties. The triumphal progress of Fordism in the USA depended on the need to integrate a continually growing, industrially inexperienced, labour force into the production process. The German labour market in the 1920s and 1930s had completely different problems. The rationalization movements in Germany aimed at a better use of the relatively large core of qualified, skilled workers in production processes—that is, at fully exploiting its reserves of productivity. The labour market did not in itself provide any particular impulse towards Fordism. In addition, there virtually insurmountable obstacles to the expansion of the car industry on the demand side, given the conditions of the time. Running costs (maintenance, fuel, insurance, and tax) in particular meant that cars were out of the reach of broad segments of the population; also there were infrastructural shortcomings (Edelmann 1989: 234).

As a result, the introduction of assembly lines into the German car industry was slow. Although Adam Opel started assembly line work in 1923, many German car companies only started to make use of this organizational innovation as late as the beginning of the Great Depression in the early 1930s (Kugler 1987: 334 f.).

Against this background, the productivity gap between the USA and Germany was obvious. According to calculations by the German association of car dealers in 1925 the Ford Motor Company needed a labour input of 5.75 workers to produce one car; the German company Audi needed more than 350, and Benz as many as 450 workers.[1] The effect of Fordist rationalization was, nevertheless, striking wherever German car companies seriously tried to catch up. From 1924 to 1929 prices fell to half their basis level, output rose by about 300 per cent, and employment rose by about 50 per cent. However, even the most advanced German car maker, the Adam Opel AG at Rüsselsheim, had not yet reached a level that could be easily compared with Detroit's. Although Opel did introduce some kind of flow production in 1923, the result seems to be quite different from the conveyor belt system of the Fordist type. It is reported that in 1925 the conveyor belt at Opel was no longer than 45 metres (Kugler 1985: 53). Until 1929 production was still organized separately in different workshops, and only the finishing assembly process made use of the conveyor belt. Opel was also far from achieving the 'Fordist' one-minute cycle time. At Rüsselsheim a new car came off the production line only every 4.5 minutes, for a total of 105 cars a day, compared with the 7,000 Ford Model T's per day. At the end of the 1920s, however technical rationalization was far advanced. According to a statistical survey published by the German metal workers union, out of 29 companies in the automobile industry (Ledermann 1933: 27), 7 (24.2%, with 2,467 workers) had not modernized their equipment at all; 4 (13.8%, with 12,119 workers) had partially modernized their equipment; and 18 (62%, with 30,902 workers) had fundamentally modernized their equipment.

This technical progress did not coincide with economic success. On the contrary, the Great Depression of the early 1930s led the German motor car industry into a sharp contraction and concentration process. After numerous bankruptcies and mergers, only about a dozen companies (from the 86 existing in the mid-1920s) survived the slump. Joining Daimler-Benz, Opel, and the German Fordwerke, founded in 1929, was a new company, the Auto-Union AG. It was the result of a merger of four Saxonian automobile producers (the Zschopauer-Motoren-Werke J. W. Rasmussen, Horch, Wanderer, and Audi) that took place in 1932. This core group became a leading sector in Germany's recovery from crisis. Even more importantly, it benefited from the extremely favourable policy of the Third Reich towards the motorization of German society and of the Wehrmacht after 1933. The regime, and notably Hitler, put great pressure on the automobile industry to rise to the challenge of Fordism by using the 'Volkswagenwerk', founded in 1938 with a planned final capacity of 1.5 million cars a year, as an economic lever. However, the American-equipped and state-financed Volkswagenwerk could not start production before 1940 when, for obvious reasons, it concentrated on the production of war vehicles. Like most German industries, Volkswagen was not crucially hit by bombing attacks. All the air raids on the plant and the subsequent looting destroyed only 10 per cent of the machine tools (Reich 1990: 168). As a result, it was possible to restart production as early as August 1945 in order to produce 20,000 passenger cars for use by Allied military personnel. Though the German car industry was well prepared to catch up before 1945, there was as yet no proper breakthrough in mass production.

11.3. A Macroeconomic Framework to Fordism: The Two German States after 1945

11.3.1. West Germany

While stagnation characterized German economic development in the interwar period, economic growth became the leitmotif of postwar West German history. After the prewar level was regained at the beginning of the 1950s, GNP and national per capita income increased almost fourfold by the early 1980s. Economic growth increased the welfare of the individual and eased the social struggle over wealth distribution; it stabilized the Federal Republic's political system, and yet simultaneously altered the economy, society, and environment more radically than had the war and the preceding Great Depression put together. This experience was diametrically opposed to the reality of German economic development between the wars.

There is much to be said for seeing the essential basis for postwar economic development as a period of reconstruction (Abelshauser 1993; 1982: 34–53). According to this view, after heavy disruptions it became possible to return rapidly to a growth path whose course followed a growth that was economically possible and that had actually occurred during the 'golden age' of German capitalism before World War I. A particularly drastic disruption of economic growth undoubtedly occurred at the end of World War II. For a few years then the German economy was hindered from making full use of the productive potential which had been able to advance almost untouched by the wide-ranging economic disruption and breakdown associated with the war and its aftermath.

This applied in the first instance to the flow of technological advances, and also to the availability of a skilled labour force, essential for converting productive potential into economic growth. The state of resources at the starting-point after the war shows that it also applied to some extent to the material factors of production. The capital stock of West German industry in 1945 was about 20 per cent above the prewar level (1936), and the labour force potential had not worsened qualitatively; on the other hand, it had considerably increased in quantity as a result of the displacement of population from east to west (see Abelshauser 1982: 35–8). At the end of World War II, the discrepancy between actual performance and the potential performance level of the West German economy was therefore especially great. But even before the war the performance of German industry was below the level it had reached in 1914, when World War I began. The dynamics of industrial reconstruction in the interwar period was frequently and severely disturbed by both external and internal factors, such as inflation and world economic crises (see Figure 11.1). Even the 'economic miracle' of the Third Reich, which occurred after the cyclical downturn of the Great Depression had ended, did not succeed fully in overcoming the long-term lag in development (Abelshauser and Petzina 1980). This accumulated 'progress surplus' (measured against the actual result of the process of production) made it possible to achieve much higher growth rates in per capita GNP after World War II. A precondition was, however, that a start be made towards recovery, and the extent to which this development potential could be utilized depended, in turn, above all on the state of demand and the speed of accumulation of real capital.

These preconditions existed in West Germany at the beginning of the 1950s. From 1947 a recovery was in progress; this stabilized after the passing of the Korean crisis and received new and lasting stimulation from the changed conditions of the world market. There were practically no bounds to economic growth from the resource side. Production grew rapidly to absorb the excess capacity of both capital and labour supply. Until the mid-1950s, the growth of production was mainly due to better use of plant. Only subsequently was a large expansion of capacity needed

Werner Abelshauser

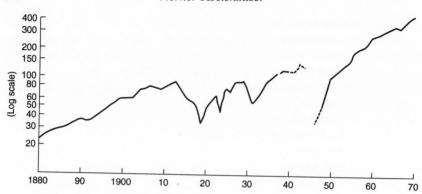

Fig. 11.1. *Growth of industrial production in the German Reich and the Federal German Republic 1880–1970 (1936=100)*

Source: Wagenführ, R., 'Die deutsche Industriewirtschaft' (Vierteljahrshefte zur Konjunkturforschung, Sonderheft 31), Berlin 1933, p. 64; Statistisches Bundesamt (ed.), 'Bevölkerung und Wirtschaft' (Statistik der Bundesrepublik Deutschland No. 199), Stuttgart 1958, p. 46; Statistisches Jahrbuch für die Bundesrepublik Deutschland 1972, passim.

to allow further production increases. A significant rise in marginal (gross) capital coefficients demonstrates this transition. In the first post-war cycle of the West German economy, from 1951 to 1955, an annual average of just 2.4 per cent of GDP was invested to achieve a 1 per cent increase in real GDP. In the third cycle (1959–1963), the rate of investment needed to achieve this averaged 4.5 per cent of GDP (*Sachverständigenrat zur Begutachtung der gesamtwirtschaftlichen Entwicklung* 1964/65: 52). This marked a transition from the extensive phase of 'reconstruction' in the narrow sense of the term to a phase of capital-intensive growth; but it did not mean that the conditions for growth that related particularly to the consequences of war were no longer relevant. On the contrary, their specific effects only now became apparent, even though most contemporaries did not appreciate them. The actual 'economic miracle' began, and it created the preconditions for the introduction of Fordism into West Germany.

The exuberant growth of the West German economy up to this point had been possible because resources could be used to the full to meet an almost unlimited demand. But now the reserves of intangible assets accumulated over time became a noticeable factor in economic growth. A readily available supply of high-quality labour facilitated a much quicker than normal adaptation of capital supply to technical progress and to the changing structure of demand. Economic growth up to the mid-1950s mainly brought about the restoration of the traditional sectoral structure of German industry; the strong structural tensions between industrial branches that had arisen after 1945 as a result of the partition of Germany were overcome. Now the structure of supply also changed, as productive capacity was further expanded. This structural change in industry towards new products and forward-looking technologies made high demands on the qualification structures of the labour force. In this context an intangible asset could be called upon, one that had accumulated despite all the disruptions in economic development: the potential embodied in the West German labour force.

These reserves were rebuilt in the postwar period. In 1946, 7.1 million people, who had lived elsewhere before the war began, lived in West Germany. By 1950 another 2.5 million refugees and people displaced by ethnic purges were added. A further 3.6 million Germans moved from the GDR to West Germany in the years from 1950 to 1962, and their mobility and qualifications far exceeded the average. A phase of development lays heavy demands on the capabilities of management and workforce in the handling of structural economic change; in this context human resources become crucial for realizing changes for growth. In this area, not least because of the division of Germany, the Federal Republic had greater reserves than its Western European neighbours. It was thus able to achieve relatively high increases in productivity even after the initial recovery phase. This brought about a productivity and price advantage on the world market and became the most important precondition for its strong world market position. This impulse for growth from the external economy contributed to the pace of domestic development.

West Germany now entered the 'mass consumption era' considerably later than had other industrialized countries (this had occurred in the United States during the 'roaring twenties'). The characteristic of this stage of industrial development was the distribution of consumer durables among broad sections of the population. The leading sector in this consumer boom was the automobile, alongside domestic appliances such as refrigerators and vacuum cleaners. The spread of the automobile in the USA had already had far-reaching economic consequences, from road building to new patterns of settlement. To this were added the first elements of the leisure society such as package tours, the marketing of mass sports, and show business. In the minds of most Germans of today, these benefits of the 'consumer society'—as well as the tendency to adopt an American cultural model—are closely associated with the 1950s and are seen as that decade's trademark.

Many of the characteristics of the consumer society of the Federal Republic, however, recalled the developments of the 1930s, which also featured many cultural imports from the USA (Schäfer 1981). The 1930s' cult of the automobile was steered by the NS regime—by means of massive pressure on the price of popular models (the Opel P4, the DKW Reichsklasse, the Opel Kadett, etc.) and the development of simpler and more economical models (the KdF car,[2] later called the Volkswagen)—in the direction of mass consumption (though the war meant that extensive plans had to be scrapped). Nevertheless, motorway construction continued, in competition with the armaments industry, until 1942. In 1939 there were 3,065 km (1,916 miles) of motorways, with 1,849 km (1,152 miles) under construction. Of this, 2,100 km (1,352 miles) were in West Germany. The number of car owners grew at a somewhat faster rate than in comparable neighbouring countries such as Great Britain. Despite the ever-lengthening shadow of the armaments industry, the success of Coca-Cola (1929–42), of Hollywood films, and of other products of the US entertainment industry (up to 1941), and massive advertising for home-ownership, cars, radios and televisions, cameras, and kitchen appliances created a climate of consumerism.

A novelty for the 1950s, by contrast, was that the possession of long-lasting consumer goods was no longer limited to middle and higher income groups. It spread to virtually all sections of the population with the growth in real incomes. In the second half of the 1950s, the possession of some consumer goods, such as televisions, gramophones, and refrigerators, was by and large no longer a function of social status. 'Democratization of consumption' reasserted itself. Since the Industrial Revolution, the phenomenon of the proliferation of consumption goods once mainly reserved for the privileged use of the ruling classes had been a repeated occurrence. The failure of cars conceived for a particular class, such as the 'cabin cruisers' of Gutbrod or the Maico of Messerschmidt, as well as the variety of motorization represented by the motorbike, is characteristic of this development, as is the triumphal advance of the Volkswagen Beetle, based on functional technology and an image that transcended social status. Neither the disappointed expectations of the 300,000 prewar Volkswagen savers, nor the associations of the former KdF car with Adolf Hitler's Third Reich, could halt its progress.

Between 1951 and 1961 the number of private cars increased sevenfold, from 700,000 to over 5 million. From 1954 onwards registrations of private cars exceeded those of motorbikes; in 1957 the total number of cars overtook the number of motorbikes for the first time. As a consequence, the level of motorization rose from 12.7 to 81.2 private cars per thousand population, although clearly it still lay below that of comparable countries (Siebke 1963: 79ff.). The scale of the need to catch up was illustrated by the fact that the Federal Republic achieved in 1960 the level of motorization that the United States had attained in 1920.

The proportion of employees among the new car-owning population rose at the same time, from 8.8 per cent in 1950 to 53 per cent in 1960; they thereby increased their proportion of total car ownership from 12 per cent to 54 per cent. 'Mini' cars such as Borgward's Lloyd, BMW's Isetta, or Glas's Gogomobil set the trend (see Table 1), and 'bowed low to the slowly

TABLE 11.1. *Passenger cars in the Federal Republic of Germany 1959, by size*

Class (by cc)	Car type	% of the cars included[a]	vH % of the cc class
Class I up to 999 cc Small cars	DKW Meisterklasse und F 93	3.0	14.9
	BMW Isetta 250 und 300		
	BMW 600	3.7	18.3
	Fiat/NSU 600/Jagst	4.3	21.2
	Lloyd LP 400	3.9	19.6
	LP 600	5.2	26.0
	total	20.1	100.0
Class II 1,000 to 1,499 ccm Medium-sized cars	Borgward Isabella	1.9	2.7
	Fiat/NSU 1.100/Neckar	1.3	1.9
	Ford 12 M 15 M	4.5 } 6.6 2.1	6.5 } 9.6 3.1
	Opel Olympia 51 LZ und P Olympia-Rekord	8.1 } 16.8 8.3	11.8 } 23.8 12.0
	VW Standard und Export	42.8	62.0
	total	69.0	100.0
Class III over 1,500 cc Large cars	Daimler Benz 170 V und 170 S	2.2	20.4
	180	1.6	14.2
	180 D	2.8	26.0
	Ford 17 M	1.8	16.8
	Opel Kapitän—51 LV und P	2.5	22.6
	total	10.9	100.0
	total	100.0	

Source: Lehbert 1962: 31.

[a] Cars included = 71% of total car registrations.

but powerfully increasing purchasing power of the masses' (Sachs 1984: 84f.). In the 1950s the private car became a key element of social mobility and prestige, of the bourgeois sense of freedom, and of job opportunities. The consequences for town planning, housing policies, leisure time, communication behaviour, economic structure, environment—indeed for virtually every area of human life—revolutionized everyday life. In the fifties the share of individual transport in total passenger transport rose from 33.1 per cent to 63 per cent, while that of rail fell from 37.5 per cent to 17.1 per cent and of local public transport from 28.8 per cent to 18 per cent (Bundesminister für Verkehr 1972: 28f.).

There was no doubt that the main pillars of the consumer society—motorization, tourism, and the mass media—first had a deep impact on public consciousness and in the collective sense of life during the 1950s and 1960s, and can therefore correctly be seen as results of the economic miracle after World War II. Their beginnings in the National Socialists' 'economic miracle' of the thirties are not without significance. These facilitated innovations in all three areas that strongly reflected German traditions and experience. The American car industry's model policy was less influential for Germany in the 1950s than in the 1920s and 1930s. In the 1960s, the direction of influence was even reversed, as is demonstrated by the export success of the Volkswagen, and later of prestige models such as Daimler Benz, BMW, or Porsche.

The internal dynamic of the West German economic miracle was homemade, in that the long-term development deficit was overcome and all the reserves of productivity of the West German economy were mobilized. But the unparalleled expansion of the world market in the 1950s and 1960s, and West Germany's integration into this development, belong to the framework that made this miracle possible. The annual rate of growth of exports by volume for the 16 OECD countries between 1913 and 1950 was a mere 1 per cent; from 1950 to 1973 it was an impressive 8.6 per cent. Germany profited from this development, to which it was itself a contributor. The rates of growth for its own exports by volume lay between −2.8 per cent and +12.4 per cent (Maddison 1989: 67). The greatest benefits from this extraordinary dynamism accrued to engineering products, such as special-purpose machines, office and telecommunications systems, and domestic appliances, as well as road vehicles. The advantages of this development for the last-mentioned were ambivalent. On the one hand manufacturers had ready access to the necessary strategic inputs, for instance raw materials, quality steels, or special-purpose machines; on the other hand, they were obliged to include the world market in their sales planning, in order thereby to achieve the economies of scale that were the core characteristic of the Fordist production method. The heading 'vehicles and aircraft' (and in the 1950s aircraft were of minimal significance) became the most important subcategory in German foreign-trade statistics. Its proportion of exports rose from 4.8 per cent in 1950 to 14.4 per cent in 1965. In 1936 it had represented just 2.6 per cent (*Statistische Jahrbücher für die Bundesrepublik Deutschland,* 1952 and 1966, passim). Undoubtedly, the macroeconomic framework after 1945 was well suited to the introduction of Ford mass production techniques.

Although the philosophy of West Germany's new postwar economic order ('Soziale Marktwirtschaft') gave absolute priority to the recovery of the consumer goods industries, it was not until 1951 that real consumption regained prewar (1936) levels. As far as the extent of motorization is concerned, the USA's 1920 level was reached by West Germany only in 1960. Hence the potential for the development of domestic markets seemed promising. At the same time macroeconomic conditions had completely changed. In 1951 West Germany started to stage a comeback in the world market and the automobile industry was to play an important part. Compared with an average national export ratio of 16 per cent, at the end of the 1950s the German automobile industry sold more than 50 per cent of its production abroad, the Volkswagenwerk as much as 58 per cent. At the same time Volkswagen was present in about 160 foreign countries, and approximately half of West Germany's balance of payments surplus was earned by Volkswagen alone.

11.3.2. *East Germany (GDR)*

The question is often posed: why was the GDR economy so much less successful than the West German, though they both sprang from the same bankrupt heap of the Third Reich and carried the same inheritance? The most common answer given is that the GDR had a worse starting position than West Germany (disparities of partition, higher reparations burdens); that it could not participate in the benefits of the Marshall Plan; that it continually lost qualified workers to West Germany; and that it introduced a model of economic planning in 1949/50 that even then no longer corresponded to the standard of modern 'socialist' economic policy, as practised in Great Britain or Scandinavia at the time and as sought for by West German social democrats. According to the most recent economic history research, the essential difference in the economic performance of the two German states is not so much the worse starting position or the lack of foreign aid in the recovery, but the adoption of an obsolete economic policy system unsuitable for a highly developed and differentiated industrial economy. Little attention has hitherto been given to the question of what consequences the different degrees of integration of the two German states into the world economy had on their capacity for innovation, their economic structure, and their techniques of industrial production.

It is easier to trace shifts in the goals adopted over time in the practice of the GDR foreign trade monopoly, than to trace corresponding changes in the west. One can establish a clearly defined succession of phases following from the primacy of politics in foreign economic matters. The foreign economic policy of the Soviet Zone/GDR was initially directed principally towards the solution of the reparations question and to the fulfilment of 'alliance obligations'. At the start of the 1950s a different task came to the fore, that of tackling and overcoming the consequences of the partition of Germany for the GDR's economic structure and the impact of the 'economic war' that followed in the wake of an increasingly frosty Cold War. Foreign trade developments in East and West Germany to a certain extent showed some similarities in the second half of the 1940s, but the two parts of Germany rapidly grew apart in the first half of the 1950s.

Rigidly applied controls over the GDR's foreign trade limited its options in regard to indigenous automobile construction. The indirect effects of the state foreign economic monopoly deserve particular attention. The GDR's economic policy in the 1950s was focused on the development of a domestic heavy industry, as in almost all of the peoples' democracies. The monopoly on foreign exchange usage additionally limited the opportunities of GDR car manufacturing in two ways. In the first place, the industry lacked the means to build up an automobile accessories industry at home, and, secondly, the GDR car industry was not able to use export proceeds for its own investments, for instance in the import of parts.

The issues of modernization and the encouragement of innovation only came to the fore in the GDR in the second half of the 1950s, as the awareness of shortcomings in many areas of East German industry became more acute in political decision-making bodies. Foreign economic matters were now—particularly in the first half of the 1960s—purposely oriented towards the promotion of innovation and of technology transfer. The example of the car industry shows the degree of success with which this industrial policy goal was achieved: it failed in the face of an insurmountable challenge and a dearth of available means. In this case the specifics of the international division of labour within the Eastern Bloc must be taken into account. In 1954 the Council for Mutual Economic Assistance (Comecon) made its first attempts to establish multilateral production specialization within the eastern economic community, beginning with vehicles, tractors, agricultural machinery, and railway carriages (Neumann 1980: 204). Vehicles were classified according to model size, in order to overcome the great diversity of vehicle types produced in the Comecon countries and to consolidate all of them into the vehicle classes necessary for the economy. For commercial vehicles, these categories were determined by carrying capacity. For passenger cars, cylinder capacity was the classification criterion for the respective model types. This classification was intended to make it possible to share out vehicle production among the Comecon partners. Contrary to expectations, the specialization of car production within Comecon proved to be extremely difficult. Most countries decided to maintain or to continue the production of different vehicle models. Only Hungary bucked this pattern with a deliberate attempt at limitation. Coordinated specialization only resulted for a few important car industry inputs, such as ball bearings, which were constructed in the Soviet Union. Yet product-related specialization was the key issue. There is much to suggest that Comecon's recommendations for specialization did not have a significant impact on the deepening of the international division of labour within the Eastern Bloc countries. In the machine-building sector, which includes vehicle manufacture, production grew faster than foreign trade. In addition, less than half of the recommendations made were aimed at the concentration of products within any one country. Thirty per cent of the specialization proposals permitted the continued manufacture of a product in three or more countries (Neumann 1980: 207). Many individual Comecon members, particularly those that were still developing, considered the production of their own vehicles and tractors an essential part of furthering industrialization and attaining the level of economic development that existed in the other participating countries.

The foregoing validates the hypothesis that the political orientation of foreign economic matters and the peculiarities of the eastern segment of the 'world' market in the 1950s contributed to the creation of a development deficit in the industry of the GDR, which was clearly identifiable at the latest by the end of that decade. The external economy was not available as a means of furthering the process of modernization in the GDR for many years after the war. The West German automobile industry was a pillar and a promoter of important organizational, technical, and economic innovations in the 1950s. A comparison with this unparalleled expansion shows even more starkly the shortcomings arising from East German industry's more limited integration into the world market. The comparison casts much light on the technical processes that find their most visible expression in the establishment of automatic avenues of transfer. More significantly, it offers insights into the macroeconomic relevance of a new method of production that has as a precondition a specific degree of integration into the world economy. It demonstrates clearly that the introduction of Ford mass production techniques was not only a consequence of important innovations of a technical, organizational, and economic nature, but also a contributor to further fruitful innovations in other branches of West German industry: in industrial relations, in state welfare policy, in transportation, and in other infrastructural areas.

11.4. 'West German Fordism': The Case of Volkswagen[3]

As the West German automobile industry resumed production after 1945, it was quite familiar with the organizational and technological foundations of Fordist continuous production, which was based on the mass production of standardized models with the help of specially built machines and equipment, which in each case were applied exclusively to one workpiece. This state of the art was certainly true for the offshoots of US multinationals, the Ford Werke AG and GM's Adam Opel AG. This notwithstanding, surprisingly enough as early as 1949 Volkswagen took the lead in the field of car making. VW produced 40,000 units in 1949, most being exported to Belgium, Switzerland, and Holland. Fordwerke, the largest producer in 1945/46 under the allied government, in three years slipped to being the fifth largest producer in Germany, with a thousand Taunus cars per month. Opel was also passed by Volkswagen, which obviously had the right product and leadership during the 1950s (Wilkins and Hill 1964: 391).

There may have been two main factors in favour of Volkswagen. First, VW could now systematically make use of the benefits it had accumulated during the Third Reich. These benefits were highly regarded even by British industrialists who inspected the plant in 1946 within the framework of the British Intelligence Objectives Sub-Committee (BIOS). One report suggested that 'compared with other automobile factories in Germany, and visualizing the originally intended factory layout, the Volkswagen effort is outstanding and is the nearest approach to production as we know it' (Reich 1990: 172). Although British industrialists expressed reservations about the quality and design of the Beetle, they were nevertheless convinced that the condition of both plant and equipment was good and operations could be maintained satisfactorily for many years. An American team of industrialists, on inspecting the plant within the framework of the US Field Intelligence Agency Technical (FIAT), suggested the company had the most modern installation in the world and, had the war not intervened, would already have had a major impact on the world market (Reich 1990: 172).

Secondly, VW had the status of a private firm, but there were either private nor public owners claiming their shares of the profits. As a result, VW was able to finance investment in the modernization of equipment out of its earnings. The problem was rather that the necessary special-purpose machines were not yet available in adequate numbers, particularly where they had to be imported from the dollar area.[4] During the Korean crisis of 1951, problems in the supply of strategic inputs (such as processed sheet metal of specific widths) led to interruptions in production and even the closure of whole plants. On the demand side, the Korean boom of the early 1950s first created the necessary preconditions for the high levels of turnover that had been impossible in the interwar period, mainly on account of excessive running costs and inadequate infrastructure. The slowness of adaptation to full-scale Fordist production methods can be demonstrated by a typical indicator, the cycle-time of car production. In 1950 it was 2.8 minutes, still far from the classical one-minute cycle that VW was not to achieve until 1953/54 (calculation on the basis of Wellhöner 1993: appendix, table 5.3).

Following the economic breakthrough of Fordist production methods, the West German automobile industry experienced a qualitative leap forward that clearly demonstrated the superiority of these production techniques. Prices for cars fell in real terms between 1950 and 1962, during the West German economy's reconstruction phase, while the general consumer price index rose by around 27 per cent (Deutsche Bundesbank: 7). This is all the more remarkable because key inputs for the automobile branch's production process became considerably more expensive in the 1950s. The price of steel rose by about 100 per cent, while wages rose by around 150 per cent (VDA 1962: 3). Increases in productivity were the decisive factor in the exceptional performance of the car industry. Productivity in motor vehicle production rose by an average of 9.37 per cent a year between 1953 and 1962. The corresponding figure for car production between 1952 and 1970 was 9.4 per cent, while overall industrial productivity rose by 'only' 5 per cent annually in this period (Dieckmann 1970: 101). The car industry's share of GDP rose at the same time from 1.7 per cent in 1952 to 5 per cent in 1960 and even further to 8.9 per cent in 1968. Around 220,000 passenger and commercial vehicles were produced in the Federal Republic in 1950; by 1962 the figure was 2.1 million. In 1956 West Germany assumed the rank of the world's second largest producer of vehicles, behind the USA and ahead of the UK. Oligopolistic market organization became increasingly prominent in car production; five companies shared about 79 per cent of the market among themselves. Volkswagen itself accounted for 30 per cent.

From 1950 to 1954, that is, during the period of reconsolidation of Ford production methods in the German car industry, Volkswagen moved from 82,399 units to 202,174 per year. Opel increased its volume from 59,990 to 148,242. Fordwerke in 1950 built 24,443 passenger cars, and 42,631 in 1954, struggling hard with Daimler-Benz (1954: 48,816 units) for third place in the German automobile industry. Volkswagen had outdistanced its American competitors in Germany almost from the beginning. Moreover, whereas Fordwerke was unable to better its 1939 achievements until 1952, Volkswagen made every effort to keep up with the American standard of car making, which at that time was characterized by the introduction of Detroit automation, as shown in Ford's new Cleveland Engine Plant, which started work in 1951. VW, the latecomer, could therefore profit from the negative experiences of automation, which in 1954 ended in a call for more flexible production methods. Against this background Volkswagen adapted in 1954/55 the most advanced version of Ford's production system in a way that would become typical of German Fordism in the 1950s:

1. VW went back to the origins of Fordism by restricting its production lines to no more than a single model.
2. VW developed a special relationship with its labour force, cooperating with the trade unions at the company level.
3. VW paid a great deal of attention to its service network at home and abroad.

The reorganization of the Volkswagenwerk followed meticulous observation of the North American market and the organization of American enterprises, the results of which were carefully evaluated in summer 1954. With the exception of Chrysler, each case indicated a higher work performance than at Volkswagen, to the extent that the automation of production along American lines would bring a saving of about 10 hours per car.[5] A continually flowing production line system was sought, linking the individual process stages together by multi-station transfer machines. The multi-purpose machines utilized till then were replaced by flexible special-purpose machines in all cases where large production runs could be planned without frequent constructional adjustments. This led, for example, to the complete automation of cylinder manufacture.[6] In body manufacture, the three principle parts—front of car, rear, and roof—were now completely prefinished separately, before the final assembly. The assembly then followed with the help of circular welding facilities consisting of hydraulic welding presses. Other subsections of production were automated, for instance the shaping of parts, the welding of steering bolts, and the planing of gearboxes. Technical renovation, however, was limited solely to the manufacture of Type I—the Beetle. Alongside this, the old oval production lines and assembly blocks for the preparation of special models in small runs persisted, in job-lot production incorporating craftsmanship. Automation was not planned for the production of Type II, the VW van, in view of the relatively small production runs.

Throughout the whole period the company's production management concentrated mainly on the manufacture of Type I, the classic VW Beetle, whose share of the concern's total production exceeded 75 per cent up to 1961. But Type II, the van, was closely related to the Beetle in terms of construction. VW only began something along the lines of diversification of its products at the start of the 1960s. The parallel to the central role of Ford's legendary Model T is obvious. Nevertheless, Volkswagen laid greater emphasis on the technical improvement of its 'flagship' than had Henry Ford in relation to his 'Tin Lizzie'.

This reorganization was initiated as early as 1954 and was practically complete by the end of 1956. A parallel reorganization of the enterprise's executive structure was also undertaken, again based on the American example. Lower levels of management gained broader competence in relation to short-term decisions regarding the implementation of general guidelines, technical processing, and product development. The concern's leadership kept responsibility for developing fundamental business strategy and for evaluating success. The subordinate management levels were responsible for the organizational and technical implementation of projected company goals. The highest possible degree of self-reliance at each level was sought. The 'Technical Development' division was formed and equipped in such a way that it could work as an independent enterprise. Heinrich Nordhoff, for many years Chairman of the Volkswagen Concern's Board of Management, described it as 'a car factory in miniature with its own atmosphere, a flexible apparatus which is to handle tasks for which the production division neither has nor should have the personnel or material resources'.[7]

Wolfsburg, where the Volkswagenwerk was located, differed from its original American role model, the Ford plant in River Rouge. There was no intention to copy the US concern's highly integrated production style, in which all stages of production, from raw material to finished product, were organized 'under one roof'. Volkswagen still relied on a multitude of suppliers, whose relationship to the factory was characterized by a strong asymmetry. In many cases the suppliers depended on the Volkswagenwerk for over 50 per cent of their annual output. Volkswagen therefore had the opportunity to influence their suppliers' pricing policies directly. In 1960, for instance, the suppliers experienced pay rises of 8.5 per cent on average in their own sector, but could only pass on a price increase of 0.5 per cent in the invoices they issued to the Volkswagenwerk.[8] Volkswagen's favourable bargaining position is illustrated by a typical passage in the 1962 annual report of the Purchasing and Materials Administration Management: 'We decided on principle not to approve price increases, and so the prices of 1961, with a few minor exceptions, were not only maintained but in part even reduced.'[9]

The configuration of industrial relations involved a mixture of American and German traditions. The Volkswagenwerk's pay policy was in essence based on the 'Ford pay compromise'—that coupling of pay and productivity growth that was institutionalized by the 1948 agreement between General Motors and the United Auto Workers, which subsequently assumed the character of a model for such deals (Piore and Sabel 1989: 92ff.). As at Ford, the apportioning of privileges and positions within the works hierarchy was also linked to seniority, that is, to length of service in the works. The criterion for payment was thereby related to job description and not to personal qualifications. In contrast to the

American example, however, at VW the introduction of Ford methods of mass production was not intended to ensure the complete control of management over the workplace above and beyond the flow of work. It had to take account of the fact that in Germany the enterprise has historically been regarded as a community, in which the cooperation of labour and capital also implied a division of power in the workplace between management and workforce.[10] Modern institutional forms of German industrial relations still incorporated central elements of the classical paradigm of craftsmanship production, with its emphasis on technical precision work (Lazonick 1990: 162). Volkswagen sought to uphold this tradition. This is evident in both the expressly communitarian rhetoric of management and the correspondingly strong sense of 'corporate identity' among the workforce. Cooperation with the works council and with the in-company trade union throughout the whole of the 1950s led, furthermore, to the leading role of the Volkswagenwerk in microeconomic social policy. The dimensions of supplementary payments in excess of standard wage rates is best illustrated by a comparison with the company's net profits, as shown in the balance sheets. Between 1950 and 1962 these totalled DM689 million. In the same period, voluntary supplementary payments of the Volkswagenwerk to its workforce amounted to DM630 million.[11] The yearly payment of a share in profits accounted for the lion's share of this supplementary budget. During the 1950s VW paid almost as much in extra wages and benefits, beyond what had been agreed through collective bargaining, as in profits declared in the company's balance sheet. This provided a sound basis upon which productivity reserves within the labour force could be mobilized.

Volkswagen was oriented from the beginning towards the world market, as were the other major German car manufacturers. This was in complete contrast to the great American role model, which was largely able to rely on the huge American domestic market in the 1920s. Its European equivalent, the EEC's large internal market, was in no position to fulfil the same function for Volkswagen in the first years after its establishment. France and Italy persisted with their protectionist tradition. They closed off their respective domestic markets from German competitors by institutional measures. Furthermore, when the Rome Treaties were signed in 1957, only 16.4 per cent of West German car exports went to what then became the EEC, while 35 per cent went to other European countries, including, for instance, the important markets of Sweden, Switzerland, and Austria, which joined the rival free-trade area of EFTA. As a proportion of car exports by value, the EFTA area was still more significant than the EEC in 1962 (27.2 per cent and 25.6 per cent respectively).

Heinrich Nordhoff, Chairman of the Board of Management at Volkswagenwerk, had little to say for the EEC in 1963, under the impact of Great Britain's failed attempt at accession:

A Europe of the nation states, the little Europe constellation, which only suits French claims to hegemony, would be a disaster for Europe, and is unsustainable even as an interim solution. This system deepens differences that must be overcome, and is founded on an inferiority complex that is unknown to us in the German automobile industry. . . . The EFTA system works well, it is organized by economic experts, while there is too much political romanticism in the EEC system. (Nordhoff 1992: 317)

Like the rest of the automobile industry, VW was not consulted before the foundation of the EEC. It perceived a tendency in the economic integration of the Six towards a deepening of differences inside the European economic area. Through the automobile industry's association (VDA) it complained in 1964 that 'the automobile is the object of a worldwide interventionism, which is exceeded only by that on agricultural products' (VDA 1962/3, 2 and 1963/4, 2). The Dillon Round of GATT in 1961 brought a lasting reduction of the EEC's common external tariff on the car sector. The USA reduced its passenger car tariff from 8.5 per cent to 6.5 per cent, while the EEC cut car tariffs from 29 per cent to 22 per cent. For West Germany this meant an *upward* adjustment to the new common EEC tariff, as it had hitherto only imposed rates from 13 per cent to 16 per cent (VDA 1961/2, 15 and 1959/60, 79).

Against this background, Volkswagenwerk's export strategy assumed particular importance. It relied on a network of general importers and dealers who conducted the marketing of the vehicles delivered from Wolfsburg to the respective national markets according to centrally determined and relatively narrowly defined rules, but on their own account. In some important markets the company felt obliged to organize sales through its own subsidiary, and in some cases to conduct the assembly of Volkswagens abroad. In the case of Volkswagen Canada Ltd. (1952), Volkswagen of America Inc. (1955), and Volkswagen France S. A. (1960), the sheer importance of the market argued for this procedure. In the case of Volkswagen do Brasil (1953), South African Motor Assemblers and Distributors Ltd. (SAMAD) (1956), and Volkswagen (Australasia) Pty Ltd. (1957), the establishment of subsidiaries was a consequence of the respective economic policies, which were based on import-substituting industrialization and which therefore made market access difficult for importers. Volkswagenwerk yielded only reluctantly to this pressure and frequently decided against the construction of its own assembly plants.

An example of such a decision is that of Japan. Heinrich Nordhoff justified the decision thus in 1954:

On the question of Japan I am of the opinion that this country will continue to be uninteresting and economically weak until it is able to reestablish relations with the Chinese mainland. As long as Japan is forcibly attached to the American economic sphere, the only path is for the Americans to incessantly hand over dollar surpluses in order to steer a basically untenable position away from a catastrophe. I have the impression that Japan is experiencing increasingly deepening difficulties, and it seems that the precariousness of this position is beginning to be recognized in the USA. That could lead to the following conclusion: Japan can be a useful starting position for the future. One should therefore not abandon Japan entirely. Up to this point in time Japan will remain a very sick child, and it would be wrong to do more than the given situation would bear.[12]

As the Japanese government increased its pressure on VW to assemble and eventually produce for itself in Japan, Nordhoff disagreed flatly: 'This does not change our opinion on Japan. If the frontiers there are closed off, Japan will simply disappear as a market, nothing more.'[13]

The American market, by contrast, was a fixed part of the Volkswagenwerk's Fordist growth strategy from the beginning—not just a residual after the domestic German market. While 8.2 per cent of the concern's total exports went to the USA at the start of its automation drive, the US market reached a share of 31.2 per cent as early as 1962, taking 22.2 per cent of Volkswagen's domestic production of passenger cars. At the start of the 1960s almost every fourth Beetle was exported to the United States.

The increase in productivity made possible by the successful adoption of automation after 1955 was fundamental to this success. Allied with the overvaluation of the Deutschmark, it permitted a competitive pricing policy without having to resort to dumping. However, the Beetle's exploitation of a niche in the US market was decisive. US concerns left this gap until the introduction of the 'compact car' in 1959, and left room for 'second cars' even after this date, as optimistic prognoses show: 'surveys show that two out of three Volkswagens are sold to families owning more than one car. Multi-car ownership is rapidly expanding and with it the market potential for the Volkswagen.'[14] As Volkswagen began its export offensive in 1954, there were no 'compact' cars, and the market share of imported cars was a mere 0.6 per cent. At the beginning of the 1960s, smaller cars had a market share of around 35 per cent, of which compact cars accounted for 23.8 per cent (1960), and the Beetle for 3 per cent (1961). Until American producers were convinced that sales volume justified the investment and would return reasonable profit, they would not undertake the heavy expenditures for styling, engineering, and special tools and machines to produce a new car line. Even after the smaller US producers, Studebaker and American Motors, introduced the first compacts, the Lark and the Rambler, in 1957, and the three giant firms, General Motors, Ford, and Chrysler, followed with their own compacts in 1959, Volkswagen was able to increase its market share from 1.7 per cent (1958) to 2.8 per cent (1962). Other importers, such as Renault, experienced heavy reverses at the same time. The source of VW competitiveness lay partly in the improved productivity mentioned above, but also owed much to the quality of the service network that had been developed. The number of VW dealers in the USA rose from 347 in 1957 to 687 in 1962. In view of the explosive growth in sales figures, the number of workshop employees per 100 registered vehicles fell from 1.7 (1958) to 0.75 (1961), but it was then consolidated at the relatively high level of 0.8. The price Volkswagen paid for this high level of services was a chronic inability to supply cars in the numbers demanded. The number of exported cars per individual American dealer indeed rose from 161 in 1957 to 292 in 1962, but this did not come near to satisfying the dealers' entire delivery requirements. The company thereby kept faithful to a policy statement made by Heinrich Nordhoff at the beginning of the company's export offensive to the United State: 'It will . . . be necessary not to force Volkswagen's turnover in the USA excessively, but to lay the greatest importance on the expansion of the still inadequate service organization. Every reverse and any kind of dissatifaction with this area is to be absolutely avoided.'[15]

Volkswagen also resisted the temptation to produce the Beetle on the spot. Plans to use the assembly factory in New Brunswick, New Jersey acquired from Studebaker for that purpose were definitively abandoned in January 1956 on profitability grounds. Heinrich Nordhoff thereby remained true to the Fordist concept of large-scale production: 'Our decision not to take the risk in New Brunswick is based on the fact that we are determined not only to maintain our place on the American market, but to expand it. We do not want to burden it with the danger of a factory of our own whose unprofitability might indeed lead to the loss of the American market.'[16] Production or assembly in the USA would have slowed the reduction in costs at Wolfsburg, in that a growing share of production would have been removed from German manufacture. At the same time the production runs sufficient for Fordist mass production would not have been achieved. Volkswagen instead opted for the expansion of its service network, which was run as a simple trading company by Volkswagen of America from its headquarters in Englewood Cliffs, New Jersey.

11.5. 'Fordism in One Country': The Automobile Industry of the GDR

In certain ways the macroeconomic framework that had impeded the introduction of Ford mass production techniques in the German automobile industry between the wars persisted in East Germany after 1945. The intention of the East German planned economy of using the advantages of large scale and of rational production came to grief primarily because of the inadequacies of the eastern segment of the world market, in which the GDR's external economy had to operate. In addition it had decisive domestic economic weaknesses, such as the extent of the transport infrastructure and the promotion of individual transportation.

As a result, car production did not play an important role in the recovery of the East German economy during the 1950s, although after 1945 the GDR inherited about 30 per cent of the former German automobile industry (compared with 27.5 per cent of German net production). It is not yet completely clear whether this was due to political decisions explicitly taken by the communist party Politburo or to unfavourable conditions stemming from weak world-market integration and lack of sufficient energy inputs. The latter is certainly true until 1963, when the newly built Russian oil pipeline 'Friendship' finally improved the availability of energy inputs.

The GDR did try to utilize the advantages of mass production, where the possibilities of standardization and the adoption of large scale production could be used extensively. This was the case, for instance, in shipbuilding and later in the serial style construction of apartment blocks. In principle the GDR's automobile industry was also organized on Fordist lines. The car production programme concentrated mainly on two models (see Table 11.2). In the 1960s the construction of passenger cars was gathered together into a single, centrally led combine with headquarters in Karl-Marx-Stadt (Chemnitz). Up to this time the GDR automobile industry had of course developed more slowly than in the west, but it had not completely lost its capacity for innovation. The 'Trabant', a 500 cc two-cylinder two-stroke car, began production at the end of 1957, and it achieved an annual production of 100,000 units from the 1970s. It was the first German vehicle with a synthetic body as standard. This was a response to the shortage of sheet, which had also been a problem for the Volkswagenwerk at the beginning of the 1950s. The Trabant's chassis was built in the former Horch plant, and its body, construction, and final assembly took place at Audi. Both plants were later unified as VEB Sachsenring Zwickau.

The Wartburg 311 was foreseen as a car for the 'lower middle class', and was produced at the former BMW plant in Eisenach. Its future as an internationally competitive automobile came to an end at last at the end of the 1950s, as the development of a four-stroke engine for it was refused. Nevertheless, its successor, the Wartburg 353, which went into production in mid-1966, corresponded to the state of the art at that time. In cooperation with French and West German firms the entire body construction was modernized, as were to some extent the mechanical sections of engine and transmission production. Modern welding facilities and multi-purpose

TABLE 11.2. *Passenger cars in East Germany (GDR) 1945–89*

I. 'Wartburg' (VEB Autoworks Eisenach)

Type	Period	Class (by cc)	Total Production
321	(1945–50)	1971	8,996
327	(1948–56)	1971	505
340	(1949–55)	1971	21,249
F 9	(1953–5)	900	38,783
311/321	(1955–66)	900/991	288,535
313	(1957–60)	900	469
353	(1966–88)	991	1,224,662
1.3	(1988–9)	1272	n.a.

II. 'Trabant' (VEB Autoworks Sachsenring Zwickau)

Type	Period	Class (by cc)	Total Production
F 8	(1949–55)	690	26,267
F 9	(1949–53)	900	1,880
P 240	(1955–9)	2407	1,382
P 70	(1955–9)	690	36,796
P 50	(1957–9)	500	131,495
P 60	(1962–4)	595	106,117
P 601	(1964–89)	595	3,000,000

Source: Dünnebier and Kittler 1990: 59, 64, 74.

equipment were installed, as were continuous production lines for cylinder heads and turrets.

With annual production runs of 100,000 and 50,000 cars, respectively, there was no shortage of incentive for the adoption of Ford methods of mass production. Nor can the relevance of another Ford paradigm, the limited availability of a qualified workforce, be denied for East Germany in the 1950s, given the continuing migration of skilled workers within Germany from east to west. Other preconditions for the economic success of Fordist production techniques were, however, completely lacking. In particular, the 'Ford pay compromise' was lacking in the GDR system (with the exception of the early 1970s). On the contrary, a 'negative class compromise' typified the GDR; the performance of the workers adjusted downward to the low level and slow development of incomes. Even with special bonuses, access to higher consumption possibilities was scarce. The system of 'reproductive autarky' within the combine, which was intended to solve the problem of supply shortages, also stood in the way

of a rapid improvement in productivity. The car combine ultimately made
not less than 80 per cent of its own supplies (Wirtschaftsforschung GmbH
Berlin 1992, 13). By contrast, the share of own-manufacture of supplies at
leading car firms in the West was only 40 per cent to 30 per cent, and
among Japanese car companies it was only 25 per cent to 15 per cent.

Given the combined effects of these factors, the car industry was not of
major importance in the economy of the GDR. In both German states, car
and machine building (a single category) and the metal-processing industry accounted for a similar proportion of total industrial production: 25
per cent and 35 per cent, respectively. This was not the case of car construction on its own. In West Germany this accounted for some 40 per cent
of car and machine building production and nearly 30 per cent of the
metal-processing industry's production. In the GDR the figures were a
mere 18 per cent and 12.5 per cent respectively. Precisely the effort to
practice 'Fordism in one country' (see Voßkamp and Wittke 1990) makes
clear that the innovative power of the new production methods, as in-
stanced above all in the car industry in West Germany, did not lie primarily in the utilization of 'economics of scale'. Rather, the deepening
integration into the world market in the 1950s offered the precondition for
a high degree of division of labour, which brought considerable cost
advantages, eased the transfer of technologies across borders, and has-
tened the adoption of innovations under the pressure of competition.[17] In
contrast, the GDR had to develop and produce special machines and other
high-tech inputs on her own, whereas the West German automobile industry was able to import such sophisticated devices from the USA. But it
was exactly in this field that the GDR was lagging far behind international
standards, as her Central Office for Research and Development pointed
out in 1960: 'Our lag in automation is mainly due to failures of the
development of industrial capacity in the field of electronics and measuring techniques. Our developing capacity in these fields as well as our
production capacity regarding electronic modules are far from meeting
our needs' (see Mühlfriedel 1993: 168). The absence of a Fordist automobile industry played a key role—albeit a negative one—in inhibiting the
development of a modern industrial structure in East Germany.

11.6. Prospects

Spilling over from the car industry, Fordism became the principal method
by which industrial development in West German industry was made
more dynamic. Although the 'Economic Miracle' was not a result of
introducing Ford production methods, Fordism gave this period of West
German economic development its specific character. Important indus-

trial branches such as chemicals or electro-technology were also opened to Ford production methods. But West German industry's leading sector, the automobile business itself, decisively influenced West German industrial society in terms of continually developing the preconditions for its functioning. This was clearest in the expansion of infrastructure, from the motorways to the 'car friendly' town. But the experience of the strongholds of Ford methods of production also set the pattern for the economy as a whole in relation to wages policy. Up to the end of the 1960s the development of real wages ran virtually parallel to that of productivity in the economy as a whole, and in the following years the two curves did not deviate widely. The Fordist experience also made its mark on workers' co-determination (*Mitbestimmung*) in West Germany. In its parity-based form this was legally binding only in the coal and steel sectors, but it came to be adopted much more broadly by large enterprises. Unions and management shared responsibilities in the control of the workplace; they thereby established confidence and stability, which were essential prerequisites for medium-term and long-term planning and for investments in human capital. The new production methods also created the need for a macroeconomic policy that could contribute to the stabilization of the development of turnover. All these practices and political-target orientations have long survived the crisis of Fordism, which began in the mid-seventies, even though they could not always be successfully implemented. Fordism thus itself introduced the instruments that have enabled West German industrial society to deal with the consequences of Fordism's own decline.

NOTES

1. Petition to the German Reichstag concerning the amendment of tariff laws, 8 July 1925, mimeographed.
2. KdF is the abbreviation of Kraft durch Freude (Strength through Joy), a mass organization under the German Workers Front, the NS surrogate for the dismantled free trade unions. The task of the KdF was to organize the worker's free time and thereby to propagandize for the regime.
3. I thank Dr Volker Wellhöner for allowing me to ransack his collection of data from Volkswagen Archives within the context of our joint project on the impacts of the world market on the two German economies.
4. The Marshall Plan did not bring any relief in this respect. The Volkswagenwerk's application to import five Bullard-Sixspindle automatics and a number of Gleason automatics to a combined value of $354,000 was approved in April 1949. As the machines were only delivered in October and the dollar parity had been changed from DM 3.34 to 4.20 in the interim, the concern lost in the region of DM 300,000. This all but made machine imports uneconomical.

VW Archives, Purchases, miscellaneous departments 1946–9, letter to the Administrative Council in Frankfurt of 17 Nov. 1949.

5. VW Archives, internal correspondence and memorandum of the production management (to 31/12/54), report from Höhne to Nordhoff of 11/8/54.
6. Ibid.
7. VW Archives, Nordhoff lecture at the Technical University of Brunswick on industrial economic leadership, 15 June 1957, 8.
8. Ibid.
9. VW Archives, Purchasing Annual Report.
10. On differing practices in Great Britain and the USA see Lazonick 1990.
11. VW Archives, Personnel, Production and Finances Division Annual Reports, passim.
12. VW Archives 9b/4: von Oertzen 1954, letter from Nordhoff to von Oertzen of 2 Sept. 1954.
13. VW Archives, Dealers' Organization Abroad, 31 Dec. 1955, Notes to a letter from Till of 29 Mar. 1955.
14. VW Archives, Annual Report of Volkswagen of America 1962, 14.
15. VW Archives, USA Office 1 May 1954 to 21 Dec. 1955, letter from Nordhoff to van de Kamp of 24 June 1955.
16. VW Archives, Volkswagen of America 1955/56, letter to van de Kamp of 26 Jan. 1956.
17. The GDR was not able to profit from technological innovations on the western markets to the same extent that the Federal Republic did, because of the COCOM embargo among other factors. The Volkswagenwerk gives a clear example of the overriding importance of the import of quality steels, special machines etc., for the improvement of the firm's competitiveness.

REFERENCES

Abelshauser, W. (1982), 'West German Economic Recovery: A Reassessment', *The Three Banks Review*, 135 (Sept.), 34–53.
—— (1993), *Wirtschaftsgeschichte der Bundesrepublik Deutschland (1945–80)* 7th edn., Frankfurt am Main.
—— and Petzina, D. (1980), 'Krise und Rekonstruktion. Zur Interpretation der gesamtwirtschaftlichen Entwicklung Deutschlands im 20. Jahrhundert', in W. H. Schröder and R. Spree (eds.), *Historische Konjunkturforschung*, Stuttgart, 56–72.
Blaich, F. (1973), 'Die "Fehlrationalisierung" in der deutschen Automobilindustrie 1924 bis 1929', *Tradition*, 18, 18–34.
Bundesminister für Verkehr (ed.) (1972), *Verkehr in Zahlen*, Bonn.
Deutsche Bundesbank, *Geld und Bankwesen in Zahlen*, passim.
Dieckmann, A. (1970), 'Die Rolle der Automobilindustrie im wirtschaftlichen Wachstumsprozeß', in VDA (ed.), *Automobiltechnischer Forschritt und wirtschaftliches Wachstum*, Frankfurt am Main.

Dünnebier, M. and kittler, E. (1990), *Personenkraftwagen sozialistischer Länder*, Berlin.

Edelmann, H. (1989), *Vom Luxusgut zum Gebrauchsgegenstand. Die Geschichte der Verbreitung von Personenkraftwagen in Deutschland*, Frankfurt or Main.

Kugler, A. (1985), *Arbeitsorganisation und Produktionstechnologie der Adam Opel Werke (von 1900 bis 1929)*, Publication series of the International Institute for Comparative Social Research/Labour Policy, Wissenschaftszentrum Berlin, Berlin.

—— (1987), *Von der Werkstatt zum Fließband. Etappen der frühen Automobilproduktion in Deutschland*, *GG*, 13.

Lazonick, W. (1990), *Competitive Advantage on the Shop Floor*, Cambridge.

Ledermann, F. (1933), *Fehlrationalisierung—Der Irrweg der deutschen Automobilindustrie seit der Stabilisierung der Mark*, Stuttgart.

Lehbert, B. (1962), *Die Nachfrage nach Personenkraftwagen in der Bundesrepublik Deutschland*, Kieler Studien, 60, Tübingen.

Maddison, A. (1989), *The World Economy in the 20th Century*, Paris.

Mühlfriedel, W. (1993), 'Zur technischen Entwicklung in der Industrie der DDR in den 50er Jahren', in A. Schildt and A. Sywottek (eds.), *Modernisierung im Wiederaufbau. Die westdeutsche Gesellschaft der 50er Jahre*, Bonn.

Neumann, G. (1980), *Die ökonomischen Entwicklungsbedingungen des RGW*, I, 1945–58, Berlin.

Nishimuta, Y. (1991), 'German Capitalism and the Position of Automobile Industry between the Two World Wars', *The Kyoto University Economic Review*, 56/1, 15–24; 56/2, 15–28.

Nordhoff, H. (1992), *Reden und Aufsätze. Zeugnisse einer Ära*, Düsseldorf, Vienna, New York, Moscow.

Piore, M. J. and Sabel, C. F. (1989), *Das Ende der Massenproduktion*, Frankfurt or Main.

Reich, S. (1990), *The Fruits of Fascism. Postwar Prosperity in Historical Perspective*, Ithaca and London.

Sachs, W. (1984), *Die Liebe zum Automobil. Ein Rückblick in die Geschichte unserer Wünsche*, Reinbek.

Sachverständigenrat zur Begutachtung der gesamtwirtschaftlichen Entwicklung, *Jahresgutachten 1964/65*, Stuttgart/Mainz.

Schäfer, H. D. (1981), *Das gespaltene Bewußtsein. Deutsche Kultur und Lebenswirklichkeit 1933–45*, Munich, 146–240.

Siebke, J. (1963), *Die Automobilnachfrage*, Cologne.

Statistische Jahrbücher für die Bundesrepublik Deutschland (1952; 1966), Wiesbaden.

VDA (1960–4), *Annual Report for the Years 1959/60, 1961/62, 1962/3, and 1963/4*, Frankfurt or Main.

Voßkamp, U. and Wittke, V. (1990), 'Fordismus in einem Land—das Produktionsmodell der DDR', *Sozialwissenschaftliche Informationen*, 10/3, 170–80.

Wellhöner, V. (1993), 'Weltmarkt—"Wirtschaftswunder"—Westdeutscher Fordismus: Der Fall Volkswagen', unpublished manuscript, (Habilschrift), Bielefeld.

Wilkins, M. and Hill, F. E. (1964), *American Business Abroad. Ford in Six Continents*, Detroit.

Wirtschaftsforschung GmbH Berlin (1992), *Regionalstudie zur Entwicklung und Standortverteilung der Kraftfahrzeug—und ihrer Zulieferindustrie im östlichen Deutschland (Beiträge zur regionalen Wirtschaftsentwicklung, 4)*, Berlin.

12

Adoption of the Ford System and Evolution of the Production System in the Chinese Automobile Industry, 1953–93

CHUNLI LEE

12.1. Introduction

This paper aims to analyse the adoption of the Ford system in the Chinese automobile industry and its impact on the evolution of China's production system and division-of-labour structure. It attempts to show China's unique pattern of evolution, which started with the adoption of a full-fledged Ford production system in the First Automotive Works (FAW) in 1953. This production system dominated China's car industry for 30 years and was later forced to change in the intensified market competition.

FAW adopted a mass production system similar to that in Ford's River Rouge Plant, which represented the original Ford system.[1] This system, after being adopted from the Soviet Union, has exerted a profound impact on other Chinese car manufacturers up to the present. The evolution of this system in FAW epitomizes to a great extent the historical development of China's car industry.

Although some research has been conducted on certain aspects of the Chinese car industry, no one has examined the evolution of its production system from the perspective of the adoption of the Ford system in China.[2]

This paper will focus on the case of FAW, especially on the changes in its production and R&D systems. It will first give an overview of China's car industry by using the framework of a 'multi-layer division-of-labour structure'.[3] It will then analyse the development of FAW's production

I would like to express my sincere thanks to Prof. Takahiro Fujimoto, my academic adviser, and Prof. Kazuo Wada, both of the Faculty of Economics, University of Tokyo, for their valuable comments and insightful suggestions on this paper.

system—how it adopted the Ford production system from the Soviet Union, then responded to market competition first with the full model change and later a full-line strategy to increase flexibility in its production system. Finally it will compare the evolution any pattern in China's car industry to that in America's in order to highlight China's unique characteristics.

12.2. Overview: Development of the Chinese Automobile Industry

12.2.1. *Establishment of the Automobile Industry*

China had an opportunity to develop a direct relationship with Ford in the 1920s. At that time Ford planned to enter China, which was perceived as a promising market in the Far East, and build a KD assembly plant in Shanghai. However, the 1923 Kanto Great Earthquake in Japan wrought devastating damage on Tokyo's railway and tramway systems. Japan thus imported 800 units of the Ford Model T chassis to produce buses. The market in Japan developed rapidly, and as a result Ford established Ford Japan Co. and built a KD assembly plant in Yokohama in 1925. China therefore lost an early opportunity to introduce the Ford system to develop its car industry.[4]

In the 1940s the Nationalist government devised strategies to develop the Chinese car industry. The government Resources Committee entrusted the design of a car plant to America's Reo Auto Works, bought the design drawing of a finished car from America's Sterling Company, and planned to construct a car plant in Xiangtan City in Hunan Province. Meanwhile, the China Automotive Corporation was established in Chongqing. The government also planned to start KD production of the Benz car. But the outbreak of civil war prevented all these plans from materializing. China did not have any car industry before 1949, except for a small number of car repair shops.[5]

In 1949 the People's Republic of China was founded. In the following year Mao Zedong visited the Soviet Union and signed the Sino-Soviet Alliance Agreement of Friendship and Cooperation with Stalin. This led China to construct its first automobile plant, the First Automotive Works in Changchun in 1953. FAW was the largest of the 156 projects of technological and financial assistance that the Soviet Union undertook in China.

This symbolized the starting of the Chinese car industry. It also meant that China introduced the Ford production system through the Soviet Union. When the Ford system was introduced into the Soviet Union from the mid-1920s to the 1930s, two of the three car makers in the Soviet Union received direct technological assistance from Ford. They adopted the integrated production system that incorporated the whole process from parts manufacturing to final assembly, and imitated the production technology of the American car manufacturers.[6] With the assistance of the Soviet Union, FAW adopted the same highly vertically-integrated production system as the Soviet car makers.[7] Its mother plant was the old 'Stalin Auto Plant' in the suburbs of Moscow, now the ZIL Auto Company.

12.2.2. Formation of a Multi-Layer Division-of-Labour Structure

By 1958, several medium-sized car manufacturers appeared.[8] Of these, Nanjing Auto Co. and Jinan Auto Co. had the same integrated production system as FAW, and were often regarded as mini FAWs. Before the 1980s large and medium-sized car makers produced basic models in different segments for the national market. Instead of competing with each other, they coexisted because they produced different products (see Fig. 12.1).

During the Cultural Revolution (1966–76), the Chinese government advocated 'building one auto plant in each province'. As a result, many small local car manufacturers came into existence, and began to produce derivative models of Jiefang (FAW), Yuejin (Nanjing Auto), etc. for regional markets. They coexisted in different regional markets without competition (see Fig. 12.1). From this period, an extensive reproduction system established through the construction of new plants became the mainstream, determining the decentralized character of the Chinese car industry.[9]

As Sino-Soviet relations deteriorated and the Vietnam War intensified, China adopted the so-called 'three-line construction' policy, building industrial bases in mountain areas.[10] The largest project undertaken under this policy was the Second Automotive Works (SAW), now the Dongfeng Motor Corporation. From 1969 to 1975 SAW constructed another River Rouge-style plant with highly integrated production logistics in a mountain area in Hubei Province. After SAW started its operation in 1975, China's car industry moved toward a multi-layer division-of-labour structure. As indicated in Fig. 12.1, this structure consists of two large national manufacturers directly under central government control, a few medium-sized manufacturers, and many small local manufacturers in each province under the control of local governments. There was no direct competition between each layer. The large and medium-sized manufacturers coexisted by producing different products. Meanwhile, protected by local governments, small local makers were able to survive in each regional market. Lack of competition caused by partition of the domestic market and economic isolation between different regions made this struc-

Fig. 12.1. *The multi-layer division-of-labour structure in China's car industry (truck)*

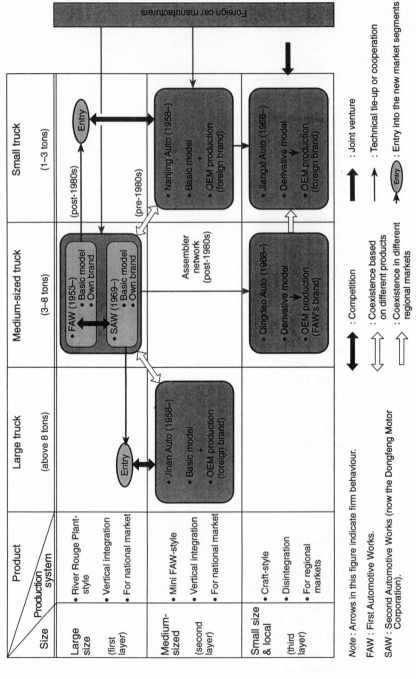

Note : Arrows in this figure indicate firm behaviour.

FAW : First Automotive Works.

SAW : Second Automotive Works (now the Dongfeng Motor Corporation).

Source: Yearly editions of *(Zhongguo qiche gongye nianjian)* (The Chinese Automobile Industry Yearbook).

ture come into existence in the first place. The relatively simple technology involved in truck manufacturing helped to maintain this structure. All these factors strengthened and stabilized the multi-layer division-of-labour structure in China's car industry.

12.2.3. Introduction of Foreign Capital and Restructuring of the Car Industry

The multi-layer division-of-labour structure underwent restructuring in the 1980s along with the formation of a national market, the government's 'business group policy', and especially the introduction of foreign capital and technology.[11]

In the early 1980s, the Chinese government pursued a business group policy under the influence of Japanese management practice. This policy aimed to systematize products through restructuring the whole car industry. Under the supervision of the China National Automotive Industry Corporation (CNAIC), seven large business groups were established, including the nationwide local car and parts manufacturers.[12] These manufacturers engaged in the OEM production of major models for the core companies like FAW and SAW, thus forming the domestic production network (e.g., the relationship between FAW and Qingdao Auto; see Fig. 12.1).[13]

Under the government's policy of 'opening to the outside world', foreign car manufacturers from Europe, America, and Japan began to enter China and started local production with foreign capital and technology in the early 1980s. From this time China switched the emphasis of car production from commercial vehicles to passenger cars. The Chinese government put forward the Big 3, Small 3, and Mini 2 policy, which aimed to consolidate passenger car production among eight car makers.[14] In passenger car production, Chinese car manufacturers adopted an incremental localization strategy emphasizing participation in the global network.[15]

In commercial vehicle production, two different patterns of technology transfer could be observed. Large car manufacturers like FAW and SAW that possessed developing capability mainly produced their own models, engaged in the independent development of products, and introduced foreign technology selectively. On the other hand, the medium-sized car manufacturers and some small local manufacturers introduced foreign technology through setting up joint ventures or tying up technically with foreign car companies. Backed up by technology transfer, they started to produce foreign vehicles in order to compete with large manufacturers. In response to this challenge, FAW and SAW adopted a full-line strategy and entered the markets of small and large trucks respectively, intensifying direct competition with the medium-sized and small makers (e.g., Nanjing and Jinan Auto, see Fig. 12.1).

The original multi-layer division-of-labour structure was thus re-structured into two groups: the large and small manufacturers group stressing production of domestic vehicles; and the medium-sized and small manufacturers group emphasizing OEM production of foreign brands. As a result, the entry of foreign makers accelerated further dispersion of car makers in local areas and an increase in the number of firms (see Note 9).

Through the overview of China's car industry, we can see FAW played an indispensable role in the development of the whole industry. The next Section will be devoted to analysing the case of FAW, the prototype of the Ford system in China.

12.3. Evolution of the Production System in the First Automotive Works

The evolution of FAW's production system can be divided into two periods—establishing and stabilizing the Ford system before 1986, and building up a full-line system after 1986, the year FAW carried out the full model change on Jiefang. Market competition was the most important factor in this evolution process.

12.3.1. Adoption of the Ford Production System from the Soviet Union[16]

FAW adopted a mass production system similar to that at the River Rouge Plant of Ford in 1953. The system was characterized by mass production of a single product and high vertical integration. The single model FAW produced was the Jiefang (Liberation) CA10 4-ton truck, which was based on the ZIS (later ZIL) 150 4-ton truck developed in the Stalin Auto Plant in the 1940s. Since this model was nicely suited for the bad road conditions and diverse natural conditions in China, it became the major means of road transportation. It was produced for 30 years from 1956 to 1986 without full model change, and 1.28 million units were produced in total. This model dominated the Chinese car market until the late 1970s. There were three improvements on this model (in 1958, 1981, and 1982), but all of them were minor changes of the CA10.

Chen Zutao, the first director of FAW's Engineering Department, said in 1983:[17]

FAW had only one product. For some 20 years FAW continued producing the single model Jiefang CA10, and did not realize the goal of having multiple models and more variations. Although we introduced the passenger car Red Flag and a 2.5-ton off-road vehicle later, they had little interchangeability with our previous models. Therefore we had to build new plants every time we introduced a new model. It was extremely inefficient.[18]

FAW built up its production logistics in a concentrated local area through highly vertical integration. As late as in 1985 the production logistics at FAW were not much different from those in its early stage of operation. It incorporated 26 plants, including casting, forging, machining, and final assembly plants. In 1985 FAW's in-house parts manufacturing rate was 60 per cent, other parts being supplied by 95 parts makers around the country. In the same year it employed 70,000 people and produced 85,000 units of vehicles, 99 per cent of which were Jiefang trucks (including chassis of the same model).[19]

Chen also commented on FAW's production logistics:

The plant engineering at FAW reflected the Soviet design policy in the 1950s. The major shortcoming is it was too general. Because we wanted to build our auto industry in a short time, it was right to construct a general auto plant at that time. But we did not take the chance to move forward to mass production through specialization. That hindered the further development of the auto industry. FAW's planned capacity was 30,000 units a year. Thanks to the efforts made by employees to tap production potentials, the annual output increased to 70,000 units. But it was difficult to go beyond that.[20]

FAW introduced Soviet-style business and work organization along with Soviet technology. In the business organization, which consisted of many functional departments, the president had highly centralized authority. FAW's first president supervised the translation of 'Organizational Design', a sixteen-volume work on management theory in the Soviet Union. The book covered the areas from organizational design to operation management, like job classification, wage system, operation manual, etc. On the basis of this management 'Bible', FAW built its work organization.

FAW roughly divided its blue-collar workers into two types: direct worker and quasi-direct worker. These workers generally performed one job throughout their lifetimes. All of them had to receive basic training at the FAW Technician Training School. Mentorism was the dominant means of skill training for entry-level workers who had finished the training school. FAW also divided jobs and wages into eight grades according to the difference of skill, and promoted workers on the basis of skill check.

The R & D system, which separated product development from manufacturing and market needs and was directly controlled by the central government, reflected the philosophy of the planned economy. In FAW's early period, product development was carried out solely by the

Changchun Automobile Laboratory, which was then responsible for developing products for the major car companies in China. The newly developed models were allocated to each car manufacturer on the basis of the government's centralized plan. FAW's engineering department only had the technological capability required for maintaining routine production; it did not possess the capability for product development. This was one of the main reasons for the technological stagnation at FAW.

Chen also talked about the centralized R & D system's influence on FAW:

We had very limited R & D capability to bring about technological innovation. FAW only had the capability to engage in 'simple reproduction', but little capacity to carry out 'extensive reproduction'. Because there were few product and process design engineers, and very limited equipment and facilities for R & D in FAW, it did not have the capability to develop new products or implement large-scale technological innovations. The Soviet auto makers then also lacked the means of technological progress. We should have paid more attention to this point.[21]

This rigid Ford production system was maintained in an environment in which FAW dominated the domestic market and large and medium-sized car manufacturers coexisted without competition into the early 1980s. But it was forced to change after market competition intensified in the mid-1980s.

12.3.2. *Response to Market Competition I: The Full Model Change*[22]

The direct competition with SAW brought a chain of reactions at FAW. Its first response was to make the full model change on Jiefang in 1986, which could be regarded as a historical turning point for FAW. The success in the model change stemmed from FAW's accumulation of product development capability through establishing a strong in-house R & D system. But the rigidity that existed in the Ford production system made FAW pay a high cost for the model change.

(a) *Facing the Direct Competition from SAW*[23]

The challenge from SAW triggered a chain of reactions at FAW. SAW was a purely domestic car maker boasting the best of China's machinery-industry technology in the 1970s. In 1978, under the government's centralized plan, SAW put the Dongfeng (Aeolus) EQ140 medium-sized truck (5-ton) on the market. Upon entering the 1980s, SAW rapidly expanded its market share with the technologically superior Dongfeng, and in 1986 it eventually outstripped FAW to become the top maker.[24] It is interesting to note that Donfeng was a new model developed by Changchun Automobile Laboratory for FAW's full model change. Thus the direct competition between FAW and SAW, rather than resulting from economic reasons, was caused by this unforeseen hitch in the government's centralized plan. In the same period, because of the government's openness policy, foreign vehicles poured into China, changing China from a seller's market to a buyer's market.

FAW, now competing with both domestic and foreign car makers, slumped into a management crisis. The monthly sales of its Jiefang decreased from 6,000–7,000 to 200–300 units. The inventory at one time peaked at 20,000 units.[25] Facing this situation, FAW had to make the first full model change on its Jiefang in thirty years.[26]

(b) Model Change under an In-House R & D System

FAW merged with the Changchun Automobile Laboratory, the largest R & D institution in China, in 1980 under the government's direction. The laboratory had 1,700 employees in 1985, 780 (46%) of whom were engineers.[27] The product development capability attained by establishing a strong in-house R & D system helped FAW to implement a series of technological innovations that began from the model change.

The new product developed independently by the laboratory was the Jiefang CA141 truck (5-ton). Seventy-eight per cent of the parts for the new Jiefang were newly designed. Because of funding constraints, FAW mainly invested in improving the efficiency of engine, body, and chassis. For the first time FAW used the CAD (computer-aided design) method for product design and part of the CAM (computer-aided manufacturing) method for manufacturing.

FAW changed its technological strategy from the so-called 'technology push' (complete adoption and imitation of the Soviet technology in the early period) to 'demand pull' (emphasis on independent product development to meet market needs). It only introduced the important technology and equipment that could not be produced domestically, e.g., imported transmission technology from Hino, and so on.

FAW sharply reduced the number of special-purpose machines that had been used in the early period and increased general-purpose machines during the model change. Of the 7,600 new machines only 1,800 were special-purpose machines. All of them were manufactured by FAW in-house. Only 4.7 per cent of the equipment was imported; 46.7 per cent of the machines were rebuilt.[28]

However, although FAW has obtained technological development capability, parts makers have not. In the process of development, FAW purposely increased the proportion of ordered parts, decreasing the in-house parts manufacturing rate from 60 to 50 per cent. But there is still little involvement by parts makers in the product development process.

(c) Limits of the Ford System and Cost of Technological Change

The full model change had a profound impact on FAW's production system. The last old Jiefang truck rolled off the assembly line on 26 September 1986, ending the 30-year history of the truck. In order to assemble and test new equipment, production stopped for 3 months until 1 January 1987, when FAW began to manufacture the new Jiefang. Because FAW had only one assembly plant producing the old Jiefang, there was little interchangeability in the existing manufacturing processes. As a result, FAW was forced to ramp up by shutting down.[29]

Moreover, its existing 12,000 machines, most of which were special-purpose machines for the old Jiefang, were obsolete. Except for the 4,600 units that were repaired and remodelled, over 60 per cent of the machines were abandoned. This is reminiscent of the high cost of the model change from Model T to Model A that took place at the River Rouge Plant of Ford in 1927. At that time only 30 per cent of the machines were usable, and production stopped for 6 months. If we compare these two events, can't we call it a historical irony?

Abernathy suggested the famous 'productivity dilemma' proposition. According to him, a trade-off relation exists between productivity and innovation.[30] This dilemma, along with the lack of investment decision authority, was the technological reason why FAW did not engage in the model change for 30 years. The lack of flexibility that existed in the Ford production system brought about its failure. The great cost incurred by FAW in the model change led it to abandon the Ford system.

12.3.3. *Response to Market Competition II: Establish a Full-Line System*

The intensified market competition and the costly model change on Jiefang also forced FAW to change its operation strategy from single-product mass production to the introduction of a full line to increase the number of models and their variations. This strategy was another response by FAW toward market competition in addition to its full model change. Its purpose was to hedge the risk of specializing in a single product by entering new markets. The present strategic goal of FAW is to transform itself from a single-product manufacturer to a general car maker.

(a) Introduce a Full-Line Strategy and Lean Production[31]

FAW attempted to inject some flexibility into its production system, only to end up with different effects in assembly and in parts production. In assembly, in order to respond to the diversification of products, FAW formed an in-company product-focused plant network by producing one model in one plant. The new production logistics at FAW consists of 28 parts manufacturing plants and 9 assembly plants.[32] In commercial vehicles FAW covers each segment of large, medium-sized, small, and light trucks with 4 basic models, and over 30 variations of the models. In passenger cars it has 3 basic models: the new Red Flag, Audi, and Golf (Jetta). All models, except VW's Audi, Golf, and Jetta, were developed by the Changchun Automobile Laboratory.[33] With the line-up of these 7 models FAW has established a full-line system.

In parts production, FAW has attempted to adopt the 'lean production' system that characterizes Toyota.[34] FAW introduced the philosophy of the Toyota production system as they invited Taiichi Ohno, one of the earliest proponents of the Just-in-Time method in Toyota, to conduct one-week seminars and on-the-spot technological instructions in 1978 and 1982 in FAW.[35] According to a report in 'People's Daily', FAW has started to run trials in 9 workshops located in 5 parts plants from 1992. The transmission plant, which received technological assistance from Hino Motor Co. of the Toyota Group, is regarded as a model plant of this system.[36]

FAW implemented the 'flow production' method, changing from the original method of upstream process pushing the downstream process to that of downstream process pulling the upstream process. This method largely reduced the inter-process inventory and use of floating funds . . .[37] FAW also changed from one worker handling one machine to one worker handling several different kinds of machines. It decreased direct workers on the line by making them work to their full capacity within the work time. As a result, it increased its productivity.[38]

This shows that in some of FAW's plants a single worker is in charge of multiple machines and workers engage in multi-task operations. For instance, in the transmission plant a single worker handles 3 to 5 different kinds of machines.

FAW has also changed its production and parts supply methods.

FAW changed from the big-lot production of a single product to small-lot production so as to produce multiple products, reduce inventory space, and decrease the workload of parts movement. It made an effort to stabilize the daily production volume, and changed its parts supply method from a storage to a 'kanban' delivery, which is set to match the 'tact' of production.[39]

All this demonstrates that FAW has made considerable efforts to increase flexibility in the production system in order to overcome the rigidity of its original mass production system.

(b) Expansion Strategy of the Business Organization:
Different Strategies for Different Products

FAW adopted two different strategies to expand its business organization: to expand the production of its home-based product, medium-sized trucks, to the regional markets; and to enter new market segments like small trucks and passenger cars by adopting a diversification strategy.

FAW's expansion strategies vary for different products—small and light trucks, medium-sized trucks, and passenger cars. The reason for these different strategies can be found in differences in the company's present expertise in product and production technology and its development strategy priorities.

(i) Medium-Sized Trucks: Formation of a Nationwide Assembler Network[40] In medium-sized trucks, in which it has a good mastery of product and production technology, FAW has established a nationwide assembler network for expansion into the new regional markets. But the production of chassis is still concentrated in FAW. This strategy is made possible by the fact that a truck's body can be separated from its chassis.

In the early 1980s the FAW Group was established under the influence of Toyota's operation strategy, in which the body assemblers of the Toyota Group engage in the consignment production of Toyota's models.[41] The primary objective of the FAW Group was to establish an assembler network nationwide by providing chassis made in FAW to the existing small local assemblers, which in turn engaged in OEM production of Jiefang variations (e.g., Qingdao Auto; see Fig. 12.1). This strategy can be seen as a reaction to SAW, which had taken the same strategy earlier.

(ii) Small and Light Trucks: M&A and Technological Transfer   In small and light trucks, which FAW aims at developing in the future but in which FAW has little expertise at present, it is establishing a hierarchical relationship in the business organization through M&A. FAW has changed its operation strategy from constructing new plants to merging existing makers and establishing a supplementary production system. In 1991 FAW merged with two car makers and two parts makers in Jilin Province and built a production base for manufacturing 60,000 units of small and light trucks annually.

The Jilin Light Truck Plant, one of the two car makers with which FAW merged, started to produce Carry, a Suzuki light truck (0.5t. and 0.75t.), under licence in 1984. After the merger, FAW conducted technological instruction in the plant, restructuring its integrated production system. This plant now only assembles vehicles and produces bodies (small Jiefang, 1 t., van-type). Other parts that were manufactured in-house are now supplied by different makers—engines and chassis by FAW, transmissions by Changchun Gear Plant, and axles by Changchun Light Truck Plant. In the same manner, FAW made the Changchun Light Truck Plant the base for manufacturing the 2-ton small truck (small Jiefang).[42] FAW's entry in the small and light truck markets caused direct competition with Nanjing Auto and other existing small truck manufacturers, leading to the restructuring of the relatively stabilized multi-layer division-of-labour structure (see Section 12.2).

(iii) Passenger Cars: Participation in the Global Network Because FAW almost has no technological capability to engage in large-scale production of passenger cars by itself, its strategy for passenger cars is to build a mass production system by participating in the multinational corporations' global network.

FAW-VW Co. Ltd., a joint venture with VW, has become an important part in Volkswagen's corporate strategy of establishing an in-company division-of-labour system on a global scale. It is especially important at a time when VW is looking for a solution to the problem of the high cost of domestic labour.[43] From the very beginning of its establishment, FAW-VW has planned to export motor vehicles to the Asian and Pacific region. The engine plant is scheduled to produce 270,000 engines in 1996. Apart from being used by FAW-VW, the extra engines will be supplied to subsidiaries of the VW group in Shanghai, Mexico, and Brazil.

FAW purchased the equipment from Volkswagen's Westmoreland Plant in Pennsylvania, which was producing Golfs (local name: Rabbits) after VW retreated from the US market, and put the newly acquired equipment into use in FAW-VW. Meanwhile, FAW constructed VW-style large passenger car plants (119 hectares) close to its headquarters plant.[44] Here the Chinese mass production system established in FAW merged with a VW-style mass production system that had been widely adopted in Europe in the 1950s.

In passenger car production, FAW has also made efforts to increase its flexibility by making the engines of its new Red Flag, the Audi, and the small and light trucks interchangeable. The engine for the 4 models is the CA488 engine (local name) of Chrysler's Dodge 600. All engine transfer lines and production equipment at FAW's Second Engine Plant were purchased second-hand from Chrysler.[45]

The changes in FAW's operation strategy after the 1980s show that the competition between FAW and SAW has caused a series of chain effects. This competition has not only accelerated the evolution of FAW's production system itself, like the model change and full-line strategy, but also promoted restructuring of the original multi-layer division-of-labour structure in the whole car industry.

12.4. Conclusion

By examining the changes in FAW's production system, R & D system, and business organization in different periods, we understand how the Ford production system has evolved in China. In order to illustrate the characteristics of China's production system, we will compare the evolution pattern in China's car industry to that in America's, which is regarded as the standard Ford system.

If we ignore differences in passenger car and truck production, we will find the major difference is China's reversal of the 'normal' development sequence. As is generally known, the American car industry started from the emergence of large numbers of small car makers engaged in craft production. The craft production system was eventually absorbed by the mass production system, established by the Big Three from the 1910s to the 1920s. Abernathy suggested the 'product and process life-cycle' model. He defined craft production as a 'fluid' state and mass production as a 'specific' state, and proposed that the production system would in the long run irreversibly evolve from the 'fluid' to 'specific' state. Meanwhile, the number of firms in the industry would decrease in this evolution process.[46]

In contrast, the craft production system was not absorbed by the mass production system in China, nor did the overall production system evolve from the 'fluid' to 'specific' state. Instead, it skipped the 'fluid' state and started from the 'specific' state with the establishment fo FAW, a proto-type of the traditional Ford system in China. Later on, medium-sized car makers and small local makers appeared and grew in parallel with FAW. The mass production system, represented by FAW and SAW, and the small-volume production and craft production system, represented by medium-sized and small car makers, coexisted by the late 1970s. The coexistence of the 'specific' and 'fluid' state is peculiar to China and cannot be found in America.

This phenomenon can be explained by two major factors: the stabilized multi-layer division-of-labour structure before the 1980s and the compli-cated competition pattern caused by the introduction of foreign capital and technology after the 1980s. These factors explain why the craft pro-duction system has not been absorbed by the mass production system, why the decrease in the number of firms posited by Abernathy cannot be observed in China's car industry.

The evolution in FAW's production system occurred in the wide con-text of the development of China's car industry. By analysing the case of the First Automotive Works, we can conclude that the dynamism of competition was and will be the most important factor contributing to the development of the production system and technological progress in the Chinese automobile industry.

NOTES

1. The typical Ford production system represented by Ford's River Rouge Plant in the early 1920s has the following main characteristics: standardized single

product, interchangeable parts, highly vertical integration, continuous assembly line, and extensive use of special-purpose machine tools. See D. A. Hounshell, *From the American System to Mass Production 1800–1932: The Development of Manufacturing Technology in the U.S.* (Baltimore, 1984), ch. 6 for details.

2. For the existing research on the Chinese car industry, see T. Tajima, 'Chugoku jidosha sangyo no tenkai to sangyo soshiki' [Development of China's auto industry and industrial organization], *Shakai Kagaku Kenkyu* [Social Science Study], 42/5 (1991). Also see C. Lee, 'Chugoku jidosha sangyo ni okeru chukan soshiki to bungyo kankei' [Intermediate organization and division-of-labour system in the Chinese automobile industry], *Kikan Chugoku Kenkyu* [China Research Quarterly], 22 (1992).

3. In this paper the 'multi-layer division-of-labour structure' refers to the coexistence or coordination relations between car manufacturers of different size in China (see Fig. 12.1). It is different from the concept of division of labour between assembler and supplier in Japan.

4. T. Fujimoto and J. Tidd, 'Ford system no donyu to genchi tekio: nichi-ei jidosha sangyo no hikaku kenkyu' [The UK and Japanese auto industry: adoption and adaptation of Fordism], *Kikan Keizaigaku Ronshu* [The Journal of Economics], 59/2 (1993), 42.

5. The First Automotive Works, *Zhongguo qiche gongye de yaolan—diyi qiche zhizaochang jianchang sanshi zhounian jinian wenji (1953–83)* [Cradle of the Chinese automobile industry—Commemorative articles on the 30th anniversary of FAW (1953–83)] (Privately published, 1983), 32, 53.

6. Nissan Motor Co., *Jidosha Sangyo Handbook* [Automobile Industry Handbook], 1992/3 (Tokyo, 1993), 36.

7. At the time, the Soviet Union did not have capabilities to provide China with large quantities of KD parts. This also necessitated building an integrated plant in China.

8. The leading medium-sized car manufacturers included Nanjing Auto Co. (the Yuejin 2.5-ton small truck), Jinan Auto Co. (the Yellow River 8-ton large truck), Shanghai Auto Co. (the Shanghai SH760 passenger car), and Beijing Auto Co. (the Beijing 212 off-road vehicle).

9. The number of Chinese car makers and total annual production volumes are: 1956: 1 company/1,600 units; 1960: 16/22,000; 1970: 45/87,000; 1980: 56/220,000; 1990: 117/510,000; 1992: 126/1,080,000; 1993: 127/1,300,000 units. From yearly editions of *Zhongguo qiche gongye nianjian* [The Chinese Automobile Industry Yearbook].

10. The 'three-line construction'. The north-eastern part of China that is closest to the Soviet Union is the first line; Beijing and the east coastal region is the second line; and the south-west and north-west interior area is the third line—an economic policy based on military and strategic considerations.

11. The 'business group policy' aimed to consolidate the existing companies into business groups to solve the problem of dispersion.

12. CNAIC is an administrative organization in charge of the overall Chinese car industry. It was first established by the government in 1964 and disbanded in 1966 when the Cultural Revolution began.

13. Other core companies were Nanjing, Jinan, Beijing, and Shanghai. Concerning the division-of-labour system in business groups, see Lee, 'Intermedi-

ate organization'.

14. The 'Big 3' were Shanghai VW Motor Co. (Santana), FAW-VW Auto Co. (Golf/Jetta), and Shenlong Motor Co. (joint venture between SAW and Citroen, ZX car); the 'Small 3' were Beijing Jeep Co. (Chrysler, Cherokee), Guangzhou Peugeot Auto Co. (Peugeot 505), and Tianjin Auto Co. (licensed production with Daihatsu, Charade); and the 'Mini 2' were Chang'an-Suzuki Auto Co. (Chongqing, Alto) and Guizhou Aircraft Industry Co. (Fuji Heavy Industries, licensed production, Rex). In addition, Toyota also announced its plan to set up a joint venture with Tianjin Auto to produce the Corona. From *Nihon Keizai Shimbun* [Japan Economic Journal], 17 June 1994.

15. For Chinese passenger car production, see C. Lee, 'Chugoku joyosha seisan ni okeru kokusanka senryaku to supplier network' [Localization strategy and formation of a supplier network in the Chinese automobile industry], *Sangyo Gakkai Kenkyu Nenbo* [Yearbook of the Society for Industrial Studies], 9 (1994).

16. The material in this section is based on The First Automotive Works, *Zhongguo qiche gongye de yaolan*, 22–59.

17. Chen was one of the major technological coordinators of FAW with the Soviet Union. He was a former president of CNAIC.

18. The First Automotive Works, *Zhongguo qiche gongye de yaolan*, 39.

19. The State Council Research Centre for Social, Economic, and Technological Development, *Yiqi zai gaige kaifang shiqi de jishu gaizao—laizi shengchan diyixian de diaocha baogao* [Technological innovation of FAW in the era of reform and openness—Report from the first-line production] (Beijing, 1988), 4–10.

20. The First Automotive Work, *Zhongguo qiche gongye de yaolan*, 39.

21. Ibid.

22. This material in this section is based on Part 3 of the State Council Research Centre, *Yiqi zai gaige kaifang shiqi de jishu gaizao*, 129–62 unless otherwise noted.

23. The data in this section is based on State Council Research Centre, *Yiqi zai gaige kaifang shiqi de jishu gaizao*, 4–5. As for the production system in SAW, see Lee, 'Intermediate Organization'.

24. Production volumes at FAW and SAW in selected years were: 1980—FAW: 66,000, SAW: 31,500; 1986—FAW: 61,600, SAW: 87,300; 1990—FAW: 69,400, SAW: 108,000; 1993—FAW: 170,000, SAW: 180,200. From yearly editions of *Zhongguo qiche gongye nianjian* [The Chinese Automobile Industry Yearbook].

25. The State Council Research Centre, *Yiqi zai gaige kaifang shiqi de jishu gaizao*, 180, 296.

26. According to Li Zhiguo, the first vice-president who was in charge of the model change, FAW put forward twelve proposals for model change on Jiefang in the years following its introduction in 1956. Because FAW had no authority to make the investment decisions, none of these proposals was approved. The majority of state investment was directed towards new car makers like SAW. From *Zhongguo qiche gongye nianjian 1986* [The Chinese Automobile Industry Yearbook 1986], 49–50.

27. The Changchun Automobile Laboratory, *Changchun qiche yanjiusozhi 1950–85* [A history of Changchun Automobile Laboratory 1950–85] (Privately published, 1985), 39–42.

28. The State Council Research Centre, *Yiqi zai gaige kaifang shiqi de jishu gaizao*, 40–1, 134–5.

29. Pattern of ramp-up: see K. B. Clark and T. Fujimoto, *Product Development Performance* (Boston, 1991), 188–204.

30. According to Abernathy, as productivity increases because of the standardization of product design, production processes become inflexible. At this time, any technological change will have a profound impact on the overall system. As the cost of technological change becomes extremely high, such change is suppressed. He termed this the 'productivity dilemma'. See W. J. Abernathy, *The Productivity Dilemma* (Baltimore, 1978), 3–4.

31. The material in this section is based on the State Council Research Centre, *Yiqi zai gaige kaifang shiqi de jishu gaizao*, 4–5, 68–70 unless otherwise noted.

32. Models produced in each plant are the following: (1) The First Passenger Car Plant (new Red Flag, 2.2 litres); (2) The Second Passenger Car Plant (Audi 100, 2.2 litres); (3) off-road vehicle plant (from 1963, CA30 2.5t.); (4) medium-size truck assembly plant (CA141 5t., CA151 6t.); (5) Changchun Large Truck Plant (CA155 8t.); (6) Changchun Small Truck Plant (1046 2t.); (7) Jilin Light Truck Plant (1026 1t.); (8) Sichuan Special-purpose Vehicle Plant; (9) Dalian Bus Plant (mainly producing chassis). In addition to these plants, FAW set up the FAW-VW Automotive Co. Ltd. in 1991 to begin the KD production of Golf and Jetta. Audi has been produced in FAW under a VW licence since 1988.

33. The new Red Flag is the only purely domestic passenger car in China today. FAW released this model in March 1994. The old Red Flag was the first Chinese-made passenger car introduced in 1958. Both models were developed independently by the Changchun Automobile Laboratory.

34. For the factors that make up the 'lean production' system, see J. Womack, D. T. Jones, and D. Roos, *The Machine That Changed the World* (New York, 1990). This paper emphasizes the Just-in-Time method, the core of this system, as compared with the traditional Ford production system.

35. From author's interview with Li Zhiguo, the first vice-president of operations management at FAW, 6 Sept. 1994.

36. From 'FAW implements the "lean production" system: challenging the traditional mass production system', *People's Daily* (China), 29 May 1994.

37. Taiichi Ohno posited the method of downstream process pulling the upstream process as a basic rule of Toyota's 'kanban' method, which was designed for a smooth 'flow production'. See T. Ohno, *Toyota Seisan Hoshiki* [Toyota production method], (Tokyo, 1978), 58–9. For the 'flow of production' in Toyota, see Ch. 1 in this volume for details.

38. *People's Daily*, 29 May 1994.

39. *People's Daily*, 29 May 1994.

40. In this paper, the product-focused plants network refers to FAW's in-company plants. The assembler network is the group of small local assemblers engaged in OEM production of FAW's brands. These assemblers are companies independent of FAW.

41. For the formation of assembler network and consignment production in Toyota, see Ch. 2 in this volume. Also see Y. Shioji, 'Toyota jiko ni okeru itakuseisan no tenkai' [The evolution of consignment production at Toyota Motor Company], *Keizai Ronso* [Economics Review], 138/5, 6 (1986).

42. The State Council Research Centre, *Yiqi zai gaige kaifang shiqi de jishu gaizao*, 43–4, 93–4.

43. Carl H. Hahn, former chairman of VW, was the top planner of VW's China project. He predicted China would become the world's largest vehicle market by the year 2015. He said companies that could obtain a competitive advantage in the Chinese market would increase their relative strategic importance in the global automotive sector. See C. H. Hahn, 'Facing the Future Challenges of the Global Automotive Market', a paper presented at the 1993 Global Automotive Conference—'Megatrends and the Auto Industry' (Brussels, 6–7 Dec. 1993).

44. The annual planned capacity of the new plant is 300,000 units (Golf/Jetta). Production is scheduled to start in 1994 and reach 150,000 units in 1996.

45. The State Council Research Centre, *Yiqi zai gaige kaifang shiqi de jishu gaizao*, 64–7.

46. According to Abernathy, the 'fluid' state is characterized by high flexibility and low productivity (craft production), and the 'specific' state is characterized by low flexibility and high productivity (mass production). He cited Ford's River Rouge Plant, which specialized in a single product, as an extreme example of the rigid mass production system. See Abernathy, *The Productivity Dilemma*, ch. 4.

INDEX